NanoScience and Technology

NanoScience and Technology

Series Editors: P. Avouris   K. von Klitzing   H. Sakaki   R. Wiesendanger

The series NanoScience and Technology is focused on the fascinating nano-world, mesoscopic physics, analysis with atomic resolution, nano and quantum-effect devices, nanomechanics and atomic-scale processes. All the basic aspects and technology-oriented developments in this emerging discipline are covered by comprehensive and timely books. The series constitutes a survey of the relevant special topics, which are presented by leading experts in the field. These books will appeal to researchers, engineers, and advanced students.

Sliding Friction
Physical Principles and Applications
By B.N.J. Persson
2nd Edition

Scanning Probe Microscopy
Analytical Methods
Editor: R. Wiesendanger

Mesoscopic Physics and Electronics
Editors: T. Ando, Y. Arakawa, K. Furuya,
S. Komiyama,
and H. Nakashima

Biological Micro- and Nanotribology
Nature's Solutions
By M. Scherge and S.N. Gorb

**Semiconductor Spintronics
and Quantum Computation**
Editors: D.D. Awschalom, N. Samarth,
D. Loss

Semiconductor Quantum Dots
Physics, Spectroscopy and Applications
Editors: Y. Masumoto and T. Takagahara

Nano-Optoelectronics
Concepts, Physics and Devices
Editor: M. Grundmann

Noncontact Atomic Force Microscopy
Editors: S. Morita, R. Wiesendanger,
E. Meyer

Nanoelectrodynamics
Electrons and Electromagnetic Fields
in Nanometer-Scale Structures
Editor: H. Nejo

Single Organic Nanoparticles
Editors: H. Masuhara, H. Nakanishi,
K. Sasaki

Epitaxy of Nanostructures
By V.A. Shchukin, N.N. Ledentsov and D. Bimberg

Nanostructures
Theory and Modeling
By C. Delerue and M. Lannoo

**Nanoscale Characterisation
of Ferroelectric Materials**
Scanning Probe Microscopy Approach
Editors: M. Alexe and A. Gruverman

**Magnetic Microscopy
of Nanostructures**
Editors: H. Hopster and H.P. Oepen

H. Hopster H.P. Oepen (Eds.)

Magnetic Microscopy of Nanostructures

With 179 Figures

Springer

Professor Herbert Hopster
University of California
Department of Physics and Astronomy
Irvine, CA 92697 USA
E-mail: hhopster@uci.edu

Professor Hans Peter Oepen
Universität Hamburg
Institut für Angewandte Physik
und Zentrum für Mikrostrukturforschung
Jungiusstr. 11
20355 Hamburg, Germany
E-mail: hoepen@physnet.uni-hamburg.de

Series Editors:
Professor Dr. Phaedon Avouris
IBM Research Division, Nanometer Scale Science & Technology
Thomas J. Watson Research Center, P.O. Box 218
Yorktown Heights, NY 10598, USA

Professor Dr., Dres. h. c. Klaus von Klitzing
Max-Planck-Institut für Festkörperforschung, Heisenbergstrasse 1
70569 Stuttgart, Germany

Professor Hiroyuki Sakaki
University of Tokyo, Institute of Industrial Science, 4-6-1 Komaba, Meguro-ku
Tokyo 153-8505, Japan

Professor Dr. Roland Wiesendanger
Institut für Angewandte Physik, Universität Hamburg, Jungiusstrasse 11
20355 Hamburg, Germany

ISSN 1434-4904
ISBN 3-540-40186-5 Springer Berlin Heidelberg New York

Library of Congress Control Number: 2004104060

This work is subject to copyright. All rights are reserved, whether the whole or part of the material is concerned, specifically the rights of translation, reprinting, reuse of illustrations, recitation, broadcasting, reproduction on microfilm or in any other way, and storage in data banks. Duplication of this publication or parts thereof is permitted only under the provisions of the German Copyright Law of September 9, 1965, in its current version, and permission for use must always be obtained from Springer. Violations are liable for prosecution under the German Copyright Law.

Springer is a part of Springer Science+Business Media

springeronline.com

© Springer-Verlag Berlin Heidelberg 2005
Printed in Germany

The use of general descriptive names, registered names, trademarks, etc. in this publication does not imply, even in the absence of a specific statement, that such names are exempt from the relevant protective laws and regulations and therefore free for general use.

Typesetting by the authors
Final layout: Le-TeX, Leipzig
Cover design: *design& production*, Heidelberg

Printed on acid-free paper   SPIN: 10926963   57/3141/ba - 5 4 3 2 1 0

Preface

In recent years, a new field in science has been growing tremendously, i.e., the research on nanostructures. In the early beginning, impetus came from different disciplines, like physics, chemistry, and biology, that proposed the possibility of producing structures in the sub-micron range. The worldwide operating electronic companies realized that this would open up new fields of application, and they proposed very challenging projects for the near future. Particularly, nanomagnetism became the focus of new concepts and funding programs, like spintronics or magnetoelectronics. These new concepts created a strong impact on the research field of fabricating nanoscaled magnetic structures. Simultaneously, a demand for appropriate analyzing tools with high spatial resolution arose. Since then, the development of new techniques and the improvement of existing techniques that have the potential of analyzing magnetic properties with high spatial resolution have undergone a renaissance. Aiming at systems in the range of some 10 nm means that the analyzing techniques have to go beyond that scale in their resolving power. In parallel to the efforts in the commercial sector, a new branch has been established in basic research, i.e., nanomagnetism, that is concerned with the underlying physics of the fabrication, analyzing techniques, and nano-scaled structures. The progress in one of these fields is inherently coupled with better knowledge or understanding and, hence, success in the other fields. The imaging technique – as a synonym for spatial resolution – plays a key role in this triangle.

In this book, we bring together the state-of-the-art techniques of magnetic imaging. We do not claim to present a complete survey of all the techniques that are around nowadays. The evolution is too fast to keep track of the development during the time it takes to edit a book. Nevertheless, we have put the emphasis on giving a comprehensive survey of the magnetic imaging techniques that have already demonstrated the high spatial resolution or, at least, have the potential to obtain it. Some techniques are well established nowadays and are already utilized in technologically oriented laboratories for commercial purposes. The majority of techniques presented are at the status of prototype basic research experiments.

It is the scope of the book to give a deeper insight into the technology and the understanding of the related effects. For the latter purpose, a more elaborate theory of operation seems unavoidable in some cases. Each of the techniques presented has its strength and drawback. It is not the intention of the book to give a ranking of the techniques with respect to certain properties, e.g., spatial resolution. It is, however, the aspect of complementariness that is the focus of our intention. The different techniques address different aspects of the physics of nanomagnetism. This is demonstrated with examples from recent investigations. The combination of different techniques will give the most complete information about the magnetism on the nanometer scale. In this sense, the book is meant as a state-of-the-art reference book for the magnetic imaging techniques available.

Chapters 1 and 2 deal with the application of synchrotron radiation. The tunability of synchrotron radiation allows elemental resolution by tuning to absorption edges. This, combined with the use of circularly polarized radiation (circular dichroism) and the spatial imaging of the emitted electrons leads to the possibility of magnetic imaging with elemental resolution (Chap. 1). Linear dichroism can be used to determine along which axis magnetic moments are aligned, e.g., in antiferromagnetic structures (Chap. 2).

The application of short laser pulses allows imaging by the magneto-optical Kerr effect with high temporal resolution, thus allowing magnetization reversal processes to be studied in detail. This is described in Chap. 3.

The current state of electron microscopies for magnetic imaging is described in Chap. 4 (Lorentz microscopy) and Chap. 5 (electron holography). Spin-polarized electron techniques have led to two new magnetic microscopies. In SPLEEM (Chap. 6), the spin dependence of low-energy electron diffraction off magnetic surfaces is used to image the surface magnetization. On the other hand, in SEMPA (or spin-SEM) a spin polarization analysis of the secondary electrons is performed as the primary electron beam in an SEM is scanned across the surface. Chapter 7 deals with the basics of SEMPA and its applications to fundamental research, while Chap. 8 discusses applications to magnetic storage media.

The invention of scanning tunneling microscopy (STM) has led to a large number of scanning probe microscopies. Chapter 9 discusses magnetic effects on the electron current due to local magnetoresistance. Spin polarized STM using magnetic tips is only in its infancy. It is the only technique capable of truly atomic resolution. Chapter 10 describes the present situation. The state and future prospects of magnetic force microscopy (MFM) is discussed in Chaps. 11 and 12. Chapter 13 discusses other scanning probe techniques for magnetic imaging.

Irvine, U.S.A. *H. Hopster*
Hamburg, Germany *H. P. Oepen*
May 2003

Contents

1 Imaging Magnetic Microspectroscopy
W. Kuch 1

1.1 Microspectroscopy and Spectromicroscopy – An Overview 2
    1.1.1 Scanning Techniques 2
    1.1.2 Imaging Techniques 3
1.2 Basics 6
    1.2.1 X-Ray Magnetic Circular Dichroism 6
    1.2.2 Photoelectron Emission Microscopy 9
1.3 About Doing XMCD-PEEM Microspectroscopy 11
    1.3.1 Experiment 11
    1.3.2 Data Analysis 13
1.4 Specific Examples 15
    1.4.1 Ultrathin fcc Fe Films 15
    1.4.2 Spin Reorientation Transition in Co/Ni Bilayers 19
1.5 Summary and Outlook 23
References 24

2 Study of Ferromagnet-Antiferromagnet Interfaces Using X-Ray PEEM
A. Scholl, H. Ohldag, F. Nolting, S. Anders, and J. Stöhr 29

2.1 Introduction 29
2.2 Photoemission Electron Microscopy 31
    2.2.1 X-Ray Absorption Spectroscopy 32
    2.2.2 X-Ray Magnetic Linear Dichroism (XMLD) 33
    2.2.3 X-Ray Magnetic Circular Dichroism (XMCD) 34
    2.2.4 Temperature Dependence of X-Ray Magnetic Dichroism 34
    2.2.5 Experiment 35
2.3 Antiferromagnetic Structure of LaFeO$_3$ Thin Films 37
2.4 Exchange Coupling at the Co/NiO(001) Interface 41
    2.4.1 Angular Dependence of Domain Contrast in NiO(001) 42
    2.4.2 Polarization Dependence of Domain Contrast 43

	2.4.3	Coupling Between Co and NiO–AFM Reorientation	44
	2.4.4	Interfacial Spin Polarization in Co/NiO(001)	46
2.5	Summary		47
References			48

3 Time Domain Optical Imaging of Ferromagnetodynamics
B.C. Choi and M.R. Freeman 51

3.1	Introduction		51
	3.1.1	Historical Background of Time-Resolved Techniques	52
3.2	Instrumentation		54
	3.2.1	Physical Principle of Magneto-Optic Effect	54
	3.2.2	Time-Resolved Experiments	56
	3.2.3	Experimental Apparatus	57
3.3	Representative Results in Thin Film Microstructures		61
	3.3.1	Picosecond Time-Resolved Magnetization Reversal Dynamics	61
	3.3.2	Precessional Magnetization Reversal and Domain Wall Oscillation	63
3.4	Conclusion and Outlook		64
References			65

4 Lorentz Microscopy
A.K. Petford-Long and J.N. Chapman 67

4.1	Introduction		67
4.2	Experimental Requirements		68
	4.2.1	Basic Instrumental Requirements	68
	4.2.2	Specimen Requirements	69
4.3	Basic Theory		70
4.4	Imaging Modes in Lorentz Microscopy		71
	4.4.1	Fresnel Mode	72
	4.4.2	Foucault Mode	73
	4.4.3	Low-Angle Electron Diffraction	74
	4.4.4	Differential Phase Contrast (DPC) Imaging	76
	4.4.5	Electron Holography	80
	4.4.6	In-situ Magnetizing Experiments – Use of the TEM as a Laboratory	80
4.5	Application to Spin-Valves and Spin Tunnel Junctions		82
4.6	Summary and Conclusions		85
References			85

5 Electron Holography of Magnetic Nanostructures
M.R. McCartney, R.E. Dunin-Borkowski, and D.J. Smith 87

5.1	Introduction		87
5.2	Basis of Electron Holography		89
	5.2.1	Theoretical Background	89

		5.2.2	Experimental Setup	92
		5.2.3	Practical Considerations	93
		5.2.4	Applications	94
	5.3	Applications to Magnetic Materials		95
		5.3.1	FePt Thin Films	95
		5.3.2	NdFeB Hard Magnets	96
	5.4	Magnetic Nanostructures		97
		5.4.1	Co Spheres	97
		5.4.2	Magnetotactic Bacteria	99
		5.4.3	Patterned Nanostructures	100
	5.5	Outlook		106
	References			107

6 SPLEEM
E. Bauer 111

6.1 Introduction 111
6.2 Physical Basis of Beam-Specimen Interactions 112
6.3 Experimental Setup and Procedure 119
6.4 Applications 123
    6.4.1 Single Layers 123
    6.4.2 Nonmagnetic Overlayers 128
    6.4.3 Sandwiches 129
    6.4.4 Other Topics 132
6.5 Summary 133
References 134

7 SEMPA Studies of Thin Films, Structures, and Exchange Coupled Layers
H.P. Oepen and H. Hopster 137

7.1 Introduction 137
7.2 Instrumentation 139
    7.2.1 Basics: Secondary Electron Emission 139
    7.2.2 Spin-Polarization Analyzer 143
    7.2.3 Electron Column 145
    7.2.4 Polarization Vector Analysis 147
7.3 Case Studies 148
    7.3.1 Ultrathin Films 148
    7.3.2 Films with Perpendicular Magnetization 149
    7.3.3 Films with In-Plane Magnetization 155
    7.3.4 Exchange Coupled Films 157
    7.3.5 Decoration Technique 160
    7.3.6 Imaging in Magnetic Fields 162
7.4 Conclusions 164
References 164

8 Spin-SEM of Storage Media
K. Koike 169

8.1 Introduction 169
8.2 HDD Recording Media 170
8.3 Obliquely Evaporated Recording Media 172
8.4 Magneto-Optical Recording Media 176
8.5 Concluding Remarks 178
References 179

9 High Resolution Magnetic Imaging by Local Tunneling Magnetoresistance
W. Wulfhekel 181

9.1 Introduction 181
9.2 Experimental Setup 185
9.3 Magnetic Switching and Magnetostriction of the Tip .. 187
9.4 Magnetic Imaging of Ferromagnets 189
9.5 Magnetic Susceptibility 195
9.6 The Contrast Mechanism 196
9.7 Conclusions and Outlook 200
References 201

10 Spin-Polarized Scanning Tunneling Spectroscopy
M. Bode and R. Wiesendanger 203

10.1 Introduction 203
10.2 Experimental Setup 205
10.3 Experiments on Gd(0001) 206
10.4 Domain and Domain-Wall Studies on Ferromagnets .. 210
10.5 Surface Spin-Structure Studies of Antiferromagnets .. 215
Conclusions 222
References 222

11 Magnetic Force Microscopy: Images of Nanostructures and Contrast Modeling
A. Thiaville, J. Miltat, and J.M. García 225

11.1 Introduction: The Magnetic Force Microscope 225
11.2 Principle of MFM 226
    11.2.1 MFM Layout 226
    11.2.2 Modes of Operation 228
11.3 Gallery of Nanostructures MFM Images 230
    11.3.1 Ultrathin Films 230
    11.3.2 Nanoparticles 231
    11.3.3 Nanowires 232
    11.3.4 Patterned Elements 232
11.4 MFM Contrast in Absence of Perturbations 234

		11.4.1 Two-Dimensional Case 235

        11.4.1 Two-Dimensional Case 235
        11.4.2 One-Dimensional Case 236
        11.4.3 MFM as a Charge Microscopy 240
  11.5 MFM Contrast in the Presence of Perturbations 242
        11.5.1 Tip Stray Field Values 242
        11.5.2 Forces in the Case of Perturbation 243
        11.5.3 Perturbations in Patterned Permalloy Elements 245
  11.6 Conclusion and Perspectives 248
References 249

12 Magnetic Force Microscopy – Towards Higher Resolution
L. Abelmann, A. van den Bos, C. Lodder 253

  12.1 Principle of MFM 253
        12.1.1 Mode of Operation 254
        12.1.2 Instrumentation 261
  12.2 MFM in Magnetic Data Storage Research 263
  12.3 Limits of Resolution in MFM 265
        12.3.1 Critical Wavelength 265
        12.3.2 Thermal Noise Limited Resolution 267
  12.4 Tip-Sample Distance Control 273
  12.5 Tips. 276
        12.5.1 Ideal Tip Shape 276
        12.5.2 Handmade Tips 277
        12.5.3 Coating of AFM Tips 278
        12.5.4 Tip Planes: The CantiClever Concept 279
References 282

13 Scanning Probe Methods for Magnetic Imaging
U. Hartmann 285

  13.1 General Strategies in Scanning Probe Microscopy 286
  13.2 Probe-Sample Interactions Suitable for Magnetic Imaging 287
  13.3 Scanning Near-Field Magneto-optic Microscopy 290
  13.4 Scanning SQUID Microscopy 292
        13.4.1 SQUID Basics 292
        13.4.2 Conventional Approaches 294
        13.4.3 Flux-Guided Scanning SQUID Microscopy 296
  13.5 Magnetic Resonance Force Microscopy 300
        13.5.1 Basic Concepts 300
        13.5.2 Global Experiments 302
        13.5.3 Mechanically Detected Magnetic Resonance
                 As an Imaging Technique 303
  13.6 Conclusions and Outlook 304
References 305

Index 309

Contributors

L. Abelmann
Systems and Materials for Information Storage,
University of Twente,
7500 AE Enschede, The Netherlands

S. Anders
IBM Research Division,
Almaden Research Center,
650 Harry Road, San Jose, CA 95120, U.S.A.

E. Bauer
Department of Physics and Astronomy,
Arizona State University,
Tempe, AZ 85287, U.S.A.

M. Bode
Institute of Applied Physics, Microstructure Advanced Research Center,
University of Hamburg,
Jungiusstr. 11, D-20355 Hamburg, Germany

J. N. Chapman
Department of Physics and Astronomy,
University of Glasgow,
Glasgow G12 8QQ, U.K.

B. C. Choi
Department of Physics,
University of Alberta,
Edmonton, Alberta, T6G 2J1, Canada

R. E. Dunin-Borkowski
Department of Physics and Astronomy,
Arizona State University,
Tempe, AZ 85287, U.S.A.

M. R. Freeman
Department of Physics,
University of Alberta,
Edmonton, Alberta, T6G 2J1, Canada

J. M. García
CNRS-Université Paris-sud,
Laboratoire de physique des solides, Bât. 510, Centre universitaire,
91405 Orsay cedex, France

U. Hartmann
Department of Experimental Physics,
University of Saarbrücken,
P.O. Box 151150, D-66041 Saarbrücken, Germany

H. Hopster
Department of Physics and Astronomy,
University of California,
Irvine, CA 92697, U.S.A.

M. R. McCartney
Department of Physics and Astronomy,
Arizona State University,
Tempe, AZ 85287, U.S.A.

K. Koike
Joint Research Center for Atom Technology, Tsukuba Central 4,
National Institute for Advanced Industrial Science and Technology,
1-1-1 Higashi, Tsukubaa, Ibaraki, 305-8562, Japan

W. Kuch
Max-Planck-Institut für Mikrostrukturphysik,
Weinberg 2, D-06120 Halle, Germany

C. Lodder
Systems and Materials for Information Storage,
University of Twente,
7500 AE Enschede, The Netherlands

J. Miltat
CNRS-Université Paris-sud,
Laboratoire de physique des solides, Bât. 510, Centre universitaire,
91405 Orsay cedex, France

F. Nolting
Swiss Light Source,
Paul-Scherrer-Institut,
CH-5232 Villigen-PSI, Switzerland

H. P. Oepen
Institute of Applied Physics, Microstructure Advanced Research Center,
University of Hamburg,
Jungiusstr. 11, D-20355 Hamburg, Germany

H. Ohldag
Advanced Light Source,
Lawrence Berkeley National Laboratory,
1 Cyclotron Road, Berkeley, CA 94720, U.S.A.

A. K. Petford-Long
Department of Materials,
University of Oxford,
Parks Road, Oxford OX1 3PH, U.K.

A. Scholl
Advanced Light Source,
Lawrence Berkeley National Laboratory,
1 Cyclotron Road, Berkeley, CA 94720, U.S.A.

D. J. Smith
Department of Physics and Astronomy,
Arizona State University,
Tempe, AZ 95287-1504, U.S.A.

J. Stöhr
Stanford Synchrotron Radiation Laboratory,
P.O. Box 20450, Stanford, CA 94309, U.S.A.

A. Thiaville
CNRS-Université Paris-sud,
Laboratoire de physique des solides, Bât. 510, Centre universitaire,
91405 Orsay cedex, France

A. van den Bos
Systems and Materials for Information Storage,
University of Twente,
7500 AE Enschede, The Netherlands

R. Wiesendanger
Institute of Applied Physics, Microstructure Advanced Research Center,
University of Hamburg,
Jungiusstr. 11, D-20355 Hamburg, Germany

W. Wulfhekel
Max-Planck-Institut für Mikrostrukturphysik,
Weinberg 2, D-06120 Halle, Germany

List of Acronyms and Abbreviations

AF	antiferromagnetic
AFM	atomic force microscopy
DOS	density of states
DPC	differential phase contrast
FM	ferromagnetic
GMR	giant magnetoresistance
LEEM	low energy electron microscopy
MBE	molecular beam epitaxy
MFM	magnetic force microscopy
ML	monolayer
MOKE	magneto-optic Kerr effect
MRAM	magnetic random access memory
MRFM	magnetic resonance force microscopy
PEEM	photoelectron emission microscopy
SEM	scanning electron microscopy
SEMPA	scanning electron microscopy with polarization analysis
SPLEED	spin polarized low-energy electron diffraction
SPLEEM	spin polarized low-energy electron microscopy
SPSTM	spin polarized scanning tunneling microscopy
SPM	scanning probe microscopy
Sp-STS	spin polarized scanning tunneling spectroscopy
SSP	scanning SQUID microscopy
STM	scanning tunneling microscopy
TEM	transmission electron microscopy
UHV	ultrahigh vacuum
XAS	X-ray absorption spectroscopy
XMCD	X-ray magnetic circular dichroism
XMLD	X-ray magnetic linear dichroism

1

Imaging Magnetic Microspectroscopy

W. Kuch

There are several well established techniques for spectroscopy of magnetic films and surfaces that are commonly employed when information about electronic states, binding properties, or element-resolved magnetic properties is required. The reduction in lateral size that goes along with the soaring extent to which magnetic elements and devices are used or planned to be used in technological applications in magnetic sensors, data storage, or magneto-electronics demands magnetic spectroscopic information on a microscopic lateral length scale. Thus, the combination of magnetic spectroscopy and microscopy into what is commonly termed microspectroscopy or spectromicroscopy would be ideal for the study of small magnetic structures.

    This chapter explains the combination of photoelectron emission microscopy (PEEM) and X-ray magnetic circular dichroism (XMCD) in absorption for imaging XMCD-PEEM microspectroscopy. In a PEEM, an electrostatic electron optics creates a magnified image of the secondary electron intensity distribution at the sample surface. When excited by soft X-rays, the image intensity can thus be regarded as a local electron yield probe of X-ray absorption. In XMCD, the measurement of the total electron yield of the sample is frequently used to determine the X-ray absorption as a function of photon energy and helicity of the circularly polarized radiation. Consequently, scanning the photon energy and recording PEEM images at each photon energy step for both helicities results in a microspectroscopic data set that allows one to extract the full information that is usually obtained from XMCD spectra for each single pixel of the images. Of particular interest is therefore the application of the so-called sum rules to extract the effective spin moment and the orbital moment, projected onto the direction of incoming light. This chapter starts with a short overview of magnetic microspectroscopy techniques in comparison to XMCD-PEEM microspectroscopy. The basics of the underlying spectroscopic and microscopic methods are briefly explained in Sect. 1.2. Important experimental aspects inherent to XMCD-PEEM microspectroscopy are discussed in Sect. 1.3. Finally, in Sect. 1.4, two recent examples of application of XMCD-PEEM microspectroscopy are presented, in which the method has proven beneficial for the study of interesting issues in the field of ultrathin magnetic films.

1.1 Microspectroscopy and Spectromicroscopy – An Overview

The terms "microspectroscopy" and "spectromicroscopy" both refer to techniques that combine spectroscopy and microscopy. "Spectromicroscopy" is commonly used to describe microscopic imaging techniques in which the image contrast is due to spectroscopic details. The acquired images are then related to a certain energy of either electrons or photons. "Microspectroscopy," on the other hand, is primarily used to describe techniques in which spectroscopic information is obtained from a small area on a sample. In terms of the dependence of information gained, spectromicroscopy is thus a technique that yields data as a function of the two space coordinates for a certain value of the energy coordinate, whereas microspectroscopy delivers data as a function of energy for a fixed pair of values of the space coordinates. The consequent extension of both spectromicroscopy and microspectroscopy would be to get the full spectroscopic *and* spatial information in the same measurement. That is, data are obtained as a function of all three variables, namely, the two space coordinates and energy. In that limit, "spectromicroscopy" and "microspectroscopy" become identical. The topic of this contribution is the combination of X-ray magnetic circular dichroism (XMCD) and photoelectron emission microscopy (PEEM) for the measurement of such a three-dimensional data set. It may be considered as either full-image microspectroscopy or full-energy spectromicroscopy, where we (arbitrarily) have chosen the former name, and thus will refer to it as "microspectroscopy". In all cases, the extension to full-image microspectroscopy, or imaging microspectroscopy, represents a considerably higher experimental effort, and it will be only practical if the gain in information makes it worthwhile.

In this section, a short overview of some microspectroscopic and spectromicroscopic techniques used for the investigation of magnetic samples is given, and the use of XMCD-PEEM as an ideal imaging microspectroscopic technique is motivated. A more comprehensive overview of spectromicroscopic techniques for non-magnetic applications can be found in [1].

1.1.1 Scanning Techniques

In microscopy, one can generally distinguish between scanning techniques and techniques that use parallel imaging. We will start with the scanning techniques. A scanning technique that is commonly employed at most synchrotron light sources and can be used for magnetic microspectroscopy is scanning X-ray microscopy (SXM). The incident X-ray radiation from the synchrotron is focused with appropriate X-ray optics, for example, by Fresnel zone plates, into a small spot on the sample. Depending on the photon energy range, spot sizes smaller than 200 nm have been achieved [2, 3]. In plain microscopy applications, the sample is scanned and the transmitted X-rays generate the microscopic image. With only minor modifications such setups can easily be used for spectromicroscopy or microspectroscopy. For magnetic microspectroscopy in the simplest case, the dependence of the transmitted, absorbed, or reflected X-ray intensity on photon energy is recorded. Magnetic contrast is obtained from the dependence on magnetization direction of the X-ray absorption cross section at elemental absorption edges when circular polarization is used (see Sect. 1.2.1). Another variant includes electron spectroscopy, where emitted electrons of a certain kinetic energy are detected [4]. Here, magnetic contrast can be obtained from magnetic dichroism in photoelectron spectroscopy, which is the difference in photoelectron intensity upon variation of magnetization direction or X-ray polarization [5]. Since the magnetic contrast is higher in absorption, only this has been used for magnetic imaging [6,7]. The advantage of scanning X-ray microscopy is that the microscopy component of the technique is completely in the excitation path, so that on the detection path standard spectrometers can be used to provide the spectroscopy component. The energy resolution for electron detection can thus be chosen to be the same as in plain photoelectron or Auger electron spectroscopy. For imaging microspectroscopy, however, the disadvantage, as in all scanning techniques, is that the time needed for a complete scan of both the sample position and the energy can be quite long.

Another scanning spectromicroscopic technique for imaging magnetic properties is spin-polarized scanning tunneling microscopy, in which the spin-dependence in electron tunneling between ferromagnets is used as a contrast mechanism (see Chaps. 9,10). In one approach, the tip magnetization is periodically reversed (Chap. 9), while another approach relies on differential electron tunneling spectroscopy (Chap. 10). In the latter, the bias voltage between a magnetic tip and the sample is set to an energy at which the dependence of the tunnel current on the direction of sample magnetization is maximized [8,9]. Without changing the experimental setup microspectroscopy can also be performed. For this, the tip position is kept fixed and the bias voltage is varied. In principle imaging scanning tunneling microspectroscopy is also possible. Due to restrictions in acquisition times, however, in most cases this is used only to find the best energy for obtaining magnetic domain images.

For the imaging of magnetic domains, laser scanning Kerr microscopy has also been used [10]. The magneto-optical Kerr effect using visible light is a commonly employed method to measure magnetization curves. The spectroscopic variant, Kerr spectroscopy, where the wavelength of the exciting laser light is scanned, is used for the characterization of electronic properties [11]. No reports exist, however, of imaging scanning Kerr microspectroscopy measurements.

As mentioned before, in general, the disadvantage of all scanning techniques for imaging microspectroscopy is that three parameters, namely, two space coordinates *and* the energy, need to be scanned step by step, which can make it a rather lengthy undertaking. Consequently, in most cases, the relation between effort and benefit does not favor imaging scanning microspectroscopy.

1.1.2 Imaging Techniques

Parallel imaging techniques have the advantage over scanning techniques in that for imaging microspectroscopy only the energy needs to be scanned, while at each energy step a complete image is acquired. Parallel imaging techniques may, therefore, be accelerated to achieve feasible measuring times even for full image microspectroscopy. The individual images are equivalent to two-dimensional sets of data points, which are acquired in parallel. Parallel imaging of magnetic spectroscopic information is based either on magneto-optical effects or on magneto-dichroic effects in electron spectroscopy after optical excitation.

An example of magneto-optical effects [12] is the magneto-optical Kerr effect using visible light, as already mentioned in the previous section. In the microscopic variant, optical microscopy is used to convert the magneto-optical information into a domain image of the sample [10]. The gain in information that would result from the combination of Kerr microscopy with wavelength scanning Kerr spectroscopy, however, does not seem worth the effort, since no imaging Kerr microspectroscopy has been reported in literature up until now.

This differs in the range of soft X-rays, where elemental core level absorption edges are accessed. The optical constants vary strongly in the vicinity of these edges and depend on the magnetization of the sample. The absorption of circularly polarized X-rays at the absorption edges depends on the relative orientation between light helicity and magnetization direction. This mechanism, which will be presented in the following section, is named X-ray magnetic circular dichroism (XMCD) in absorption, and is the first choice for obtaining magnetic contrast in imaging. It has been demonstrated, though, that in principle magnetic circular dichroism in angle-resolved photoemission can also be used to obtain magnetic contrast in photoelectron spectromicroscopy. Images obtained in an imaging hemispherical electron analyzer from Fe $3p$ photoelectrons after off-resonant excitation with circular polarization exhibited a weak magnetic contrast [13]. In a more recent paper, magnetic contrast was claimed even using unpolarized light from an X-ray tube and magnetic dichroism in Fe $2p$ photoemission [14]. The signal-to noise ratio, however, is significantly worse in the photoemission case compared to images obtained in the same instrument using XMCD in absorption to generate magnetic contrast [15, 16].

Two ways can generally be used to image the local X-ray absorption: either by imaging photons, or by imaging emitted electrons. Imaging the transmitted photons has been successfully performed for magnetic spectromicroscopic domain imaging in a transmission X-ray microscope [17]. The sample is thereby prepared such that its total thickness allows the transmission of soft X-rays, and a zone plate–based X-ray optics is used to create the image of the transmitted beam. In general, the drawback for microspectroscopy with photon imaging techniques stems from problems due to the energy dependence of the focal length of X-ray optics, in particular, of zone plates. The magnification and focusing of the resulting image, therefore, varies during a photon energy scan, which leads to a significant blurring of the image if no correction, for example, by a sophisticated image processing software, is performed.

In that respect, the imaging of the distribution of emitted electron intensity for microspectroscopic purposes is clearly easier. Since just the X-ray absorption needs to be detected as a function of photon energy, no explicit energy filtering of the electrons is necessary and the high intensity secondary electrons may be used. X-ray optics, if any, are used only for the illumination of the imaged area of the sample. Different types of electron optics have been employed successfully for XMCD-based spectromicroscopic imaging of magnetic domains, all of which are classified under the name "electron emission microscopy". While in most of the more recent work fully

Fig. 1.1. Magnetic domain image of a triangular microstructure of 30-nm-thick polycrystalline Co on Si using a PEEM and XMCD. Field of view is $40 \times 40 \mu m^2$

electrostatic photoelectron emission microscopes (PEEMs) (see Sect. 1.2.2) were used [18–23], an imaging hemispherical electron analyzer [15, 16], and in [24] a low energy electron microscope (LEEM, [25, 26]) have also been employed. Note that the latter is different from magnetic imaging by spin-polarized LEEM, which is presented in Chap. 6. Figure 1.1 shows an example of a domain image taken with a PEEM. It shows a lithographic triangular microstructure of 30-nm-thick polycrystalline Co. Different grayscale contrast represents different directions of magnetization, where bright means pointing up, dark pointing down, and intermediate gray indicates a horizontal magnetization direction [27].

The advantage of extending XMCD-based spectromicroscopy with electron detection to imaging microspectroscopy is obvious: Experimentally, it is quite straightforward if some aspects, as outlined in Sect. 1.3, are considered. XMCD is a widely used and comparably well understood spectroscopic technique, so there is a significant gain in quantitative information from full-image microspectroscopy compared with the acquisition of spectromicroscopic images; this will be discussed in Sect. 1.2.1. Finally, due to the availability of high-brilliance insertion-device beamlines at third-generation synchrotron radiation light sources, the time required for recording three-dimensional data sets for imaging microspectroscopy is approaching feasibility while still maintaining reasonable spatial resolution. This will be demonstrated by selected examples in Sect. 1.4.

A further advantage of X-ray absorption–based spectromicroscopy is that by using linearly polarized X-rays, a magnetic-dichroic signal can also be obtained from oxidic antiferromagnets [28, 29]. The use of this X-ray linear magnetic dichroism for the imaging of antiferromagnetic domains is outlined in Chap. 2.

In the remainder of this chapter, PEEM is assumed as the electron emission microscopy technique for magnetic X-ray absorption spectroscopy. This is due to the existing work in this field, although, most of what is described is also valid for any other type of electron emission microscopy.

1.2 Basics

1.2.1 X-Ray Magnetic Circular Dichroism

Since its experimental discovery [30], magnetic circular dichroism in soft X-ray absorption has developed into a widely used technique for the element-specific characterization of magnetic films and multilayers. This is in part due to the so-called sum rules that have been proposed to deduce quantitative magnetic information from XMCD spectra [31, 32]. Other reasons for the widespread use of XMCD are: The magnetism-related changes in the absorption cross section are quite large; there are several synchrotron radiation light sources around the world providing X-rays of tunable wavelength, and it is comparatively easy to measure X-ray absorption from the total photoelectron yield, where only the sample current has to be detected.

This section is aimed at providing the reader who is not familiar with XMCD spectroscopy with the basic ideas in order to follow the remainder of the chapter. More comprehensive introductions can be found elsewhere [12, 22, 33–36].

We will restrict ourselves to the $L_{2,3}$ absorption edges of $3d$ transition metals, i.e., the onset of excitation of transitions of $2p$ core electrons to empty states above the Fermi level. Let us for the moment consider absorption in a paramagnet. An explanation of X-ray absorption spectroscopy in a one-electron description is shown in Fig. 1.2. The left upper panel shows a schematic representation of the occupied density of states of the $2p$ core levels. The important point here is that because of spin-orbit interaction, the $2p$ states are energetically split into the clearly separated $2p_{1/2}$ and $2p_{3/2}$ levels. Any further splitting into sublevels is not important here. Absorption of X-rays by the excitation of electronic transitions from the $2p$ states is determined by the occupied density of states of the $2p$ core electrons and the unoccupied density of states available for these transitions above the Fermi energy (E_F). The latter is schematically shown in the upper right panel, where the shaded area represents the unoccupied states. The contribution from states of predominantly s, p character is represented by flat energy dependence, whereas d states are shown as sharp peaks around E_F. The resulting absorption spectrum is obtained from the convolution of the occupied density of states of the left upper panel and the unoccupied density of states of the right upper panel. Finite experimental photon energy resolution has to be taken into account by an additional convolution with a Gaussian. A typical $L_{2,3}$ absorption spectrum is shown in the bottom panel of Fig. 1.2. It is seen that the absorption signal related to transitions into empty $3d$ states shows up as two peaks at the energetic positions of the $2p_{1/2}$ and $2p_{3/2}$ states, whereas transitions into unoccupied s, p states give rise to a step-like background. Since the magnetic moment of the $3d$ transition metals is mainly governed by $3d$ valence electrons, the latter is usually subtracted as a step function with relative step heights of 2:1 [37], according to the occupation of the $2p_{3/2}$ and $2p_{1/2}$ core states, as shown in the bottom panel of Fig. 1.2.

If $2p \to 3d$ transitions are excited by circularly polarized radiation, these transitions exhibit a spin polarization because of selection rules [38]. In other words,

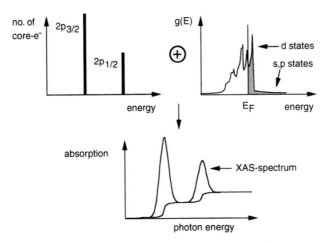

Fig. 1.2. Schematic explanation of X-ray absorption spectroscopy (XAS). The absorption spectrum shown in the bottom panel results from the convolution of the occupied density of states of the core levels (*upper left*) and the unoccupied density of states $g(E)$ of the valence states (*upper right, shaded area*). The contribution from s, p states is usually approximated and subtracted in the form of a step function (*bottom panel*)

for a certain light helicity, more electrons of one spin direction with respect to the direction of the incoming light are excited into the unoccupied $3d$ states than of the other spin direction. In a paramagnet, this does not lead to a change in absorption intensity, since the number of unoccupied states is equal for both spin directions. In a ferromagnet, however, the density of unoccupied states is different for electrons of spin parallel or antiparallel to the magnetization direction, leading to a spin magnetic moment defined by the difference in occupation. This is explained in Fig. 1.3. It shows a schematic representation of the spin resolved density of states, separated into density of states of majority spin electrons at the top and density of states of minority spin electrons at the bottom. If magnetization and light incidence are aligned with each other to some degree, there are consequently more possible transitions for one direction of light helicity than for the other. This leads to a difference in absorption for opposite light helicity. In a dichroism spectrum, calculated as the difference between absorption spectra for opposite helicity, a non-zero difference will show up at the energy positions of the peaks related to transitions from the $2p_{3/2}$ and $2p_{1/2}$ levels into the empty $3d$-like states. Since the spin polarization of $2p_{3/2} \rightarrow 3d$ transitions has an opposite sign than that of the spin polarization of $2p_{1/2} \rightarrow 3d$ transitions [12, 35], the dichroism at the L_3 and L_2 edge will have an opposite sign, i.e., the difference curve will show peaks of opposite sign at the energy positions of the L_3 and the L_2 edge. This is shown schematically for the (hypothetical) case of a material with only a spin moment μ_S in the top panel of Fig. 1.4. There, the difference curve between absorption spectra taken with opposite helicity of the circularly polarized light is depicted, which exhibits a positive peak at the L_3 edge, a negative peak at the L_2 edge, and zero elsewhere. The spin polarization of $2p_{1/2} \rightarrow 3d$ transitions is twice

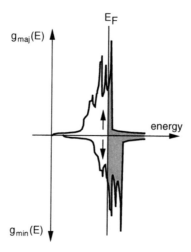

Fig. 1.3. Schematic representation of density of states of a ferromagnetic metal. Shown is the spin resolved density of states for majority electrons $g_{maj}(E)$ in the positive y direction, and the spin resolved density of states for minority electrons $g_{min}(E)$ in the negative y direction. The shaded areas are unoccupied density of states above the Fermi energy E_F available for $2p \rightarrow 3d$ transitions

as large as the spin polarization of $2p_{3/2} \rightarrow 3d$ transitions. On the other hand, the absorption at the L_3 edge is twice as high as at the L_2 edge because of core hole occupation (cf. Fig. 1.2). Together both lead to an equal size of the dichroism at the two edges, as schematically plotted in the topmost panel of Fig. 1.4.

$2p \rightarrow 3d$ transitions excited by circularly polarized radiation are not only spin polarized, but also show an orbital "polarization." This is a direct consequence of the absorption of a circularly polarized photon with angular momentum $\Delta m = \pm 1$ [12, 35]. Both the $2p_{1/2} \rightarrow 3d$ and $2p_{3/2} \rightarrow 3d$ transitions show the same sign and same magnitude of orbital polarization. If a sample possesses a non-zero orbital magnetic moment, this means that the unoccupied states (and also the occupied states) have a non-zero net angular momentum. Let us consider the (hypothetical) case of a metal with only an orbital moment and no spin moment. In this case, there will again be a non-zero dichroism at the L_3 and L_2 edges, but this time with an equal sign at the two edges. Because of the different number of $2p_{1/2}$ and $2p_{3/2}$ electrons, the resulting dichroism at the L_3 edge is twice as large as at the L_2 edge (middle panel of Fig. 1.4).

A real sample will have both spin and orbital magnetic moments. The two extreme cases shown in the top and center panels of Fig. 1.4 define an orthonormal basis for the measured XMCD spectrum from a real sample (bottom panel of Fig. 1.4), which will be a superposition of both. The experimental spectrum can thus be decomposed unambiguously into its spin and orbital basis functions. This is what is done by the so-called sum rules [31, 32].

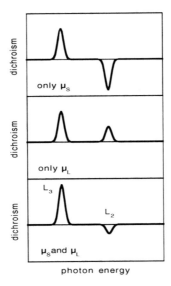

Fig. 1.4. Schematic explanation of sum rule analysis of XMCD spectra to obtain spin and orbital moments. Shown is the decomposition of an XMCD spectrum (*bottom*) into its components resulting from spin moment μ_S (*top*) and orbital moment μ_L (*center*)

It has to be mentioned that what is extracted as the "spin moment" from the sum rules is an effective spin moment $\mu_{S,\mathrm{eff}}$ [32], which includes the actual spin magnetic moment μ_S plus a contribution from the magnetic dipole term. The latter is zero in the bulk of cubic crystals, but can be of the same order as the orbital moment in ultrathin films [39].

Although the derivation of the sum rules was done under simplifying assumptions, and there has been some dispute about their applicability [34,37,40–43], they seem to yield reasonable results for the 3d transition metals [34,37,44–46]. Together with the element-selectivity of X-ray absorption spectroscopy at core level absorption edges they provide a quite powerful tool for the quantitative investigation of magnetic materials.

1.2.2 Photoelectron Emission Microscopy

Photoelectron emission microscopy (PEEM) belongs to the parallel imaging electron microscopies. The name "photoelectron" is due to its use in metallurgy in early years, when threshold excitation of photoelectrons at the vacuum level by illumination with Hg discharge lamps was used for the acquisition of work function contrast images [47]. For excitation with higher photon energies, which will be considered in this chapter, low energy secondary electrons are dominant in the imaging process. We nevertheless stick to the name PEEM, although in that case "secondary electron emission microscope" would be more correct.

After the introduction of ultrahigh-vacuum compatible instruments [48, 49], PEEM has been used for the study of surfaces and surface reactions [50–53]. Incorporating a magnetic electron beam splitter allowed the excitation by low energy electrons (LEEM, low energy electron microscopy [25, 54]), which yields additional information about the surface structure and morphology. LEEM can also be employed for magnetic imaging if spin-polarized electrons are used; this is described in Chap. 6. The availability of synchrotron light sources for excitation with X-rays of tunable energy opened a new field of application for PEEMs [55], in which resonant X-ray absorption at elemental core levels is used to image the distribution of different elements at the sample surface.

In a PEEM, in contrast to transmission electron microscopy, the electrons that are used for the imaging do not have a well-defined energy and momentum. To get a sharp image it is therefore necessary to limit the range of electron energies and emission angles. In a PEEM, this is achieved by passing the accelerated electrons through a pinhole aperture, the so-called contrast aperture. A higher lateral resolution is thereby achieved at the expense of intensity, and vice versa.

Figure 1.5 shows the schematic setup of an electrostatic PEEM [21]. The principle of other electrostatic PEEMs is similar, so the main points can be explained using

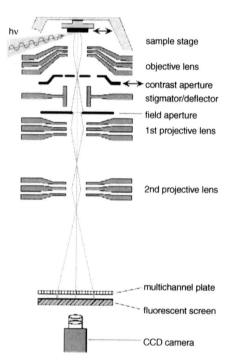

Fig. 1.5. Schematic set up of a photoelectron emission microscope (PEEM). Electrostatic electron lenses create an image of the electrons emitted at the sample surface at a fluorescent screen. (Reproduced from [21] with permission, Copyright (1998) by the World Scientific Publishing Company)

this type. In Fig. 1.5, the sample is shown at the top. It is illuminated by synchrotron radiation under a grazing angle to the sample surface, which is 30 °C in the present example. The sample is kept on ground potential, and electrons are accelerated toward the objective lens, an electrostatic tetrode lens. Typical acceleration voltages are 10–20 kV. The contrast aperture is located in the back focal plane of the objective lens. It selects only those electrons for imaging that originate from a certain range of emission angles. The size and lateral position of this aperture can be changed by moving a slider assembly carrying several apertures of different diameters. In an alternative design, the contrast aperture is located at the back focal plane of the first projective lens [23]. Astigmatism and small misalignments of the optical axis, caused, for example, by a misalignment of the sample, can be corrected by an electrostatic octupole stigmator and deflector. A variable field aperture in the image plane of the objective lens allows one to limit the field of view and to suppress stray electrons. Two electrostatic projective lenses transfer the image onto an imaging unit – which in our example consists of an electron multichannel multiplier and a fluorescent screen on the vacuum side – and charge coupled device (CCD) camera with a conventional lens optics outside the vacuum chamber. Alternative approaches use a glass fiber coupling between the screen and CCD camera [23].

Resolutions down to 20 nm in PEEM imaging using topographic or elemental contrast in threshold photoemission [56] and for excitation with synchrotron radiation [23, 57] have been reported. Even better resolution can be achieved in the LEEM mode, using magnetic objective lenses [25, 26, 58]. Presently, attempts are underway to push the resolution to below 5 nm by aberration correction [59]. In XMCD-PEEM magnetic microspectroscopy, however, intensity is a critical issue. To achieve reasonable acquisition times for the measurement of a complete microscopic and spectroscopic data set, lateral resolution will typically be selected to be a few hundred nanometers in practical microspectroscopy applications.

1.3 About Doing XMCD-PEEM Microspectroscopy

When employing magnetic dichroism effects for the spectroscopy of magnetic materials, the magnetic information is obtained from changes in the spectra that occur either upon changing the magnetization state of the sample or the polarization properties of the exciting radiation. This usually involves the measurement of relatively small differences between large signals and imposes high experimental requirements with respect to signal reproducibility, stability, and flux normalization. The additional imaging step in microspectroscopy is certainly not facilitating the fulfillment of these requirements. In the following, some of the crucial obstacles and their solutions specific to XMCD-PEEM microspectroscopy will be discussed.

1.3.1 Experiment

Let us first consider effects related to the incident radiation. The normalization to the flux of the incoming beam is not straightforward in microspectroscopy, in contrast to conventional absorption spectroscopy. In the latter, usually the photo yield from a suitable optical element in the beamline or from a specially designed flux monitor is recorded simultaneously to the sample signal. Since the entire photon beam is contributing to both the monitor signal and the signal from the sample, normalization is achieved by simply dividing one by the other. In imaging microspectroscopy, however, the *local* photon flux density is important, not the *integral* flux. It cannot be directly measured and may locally deviate significantly from the integral monitor flux signal. The cause of such deviations may be the radiation characteristics of the insertion devices used in third-generation synchrotron light sources and beamline X-ray optics. An inhomogeneous distribution of the photon intensity within the imaged area on the sample invalidates the normalization to a conventional beam monitor. The fact that the intensity distribution of undulators depends also on the relative photon energy with respect to the maximum of the undulator harmonics complicates matters further. Although in principle these effects cannot be avoided, it is possible from the experimental side to reduce the discrepancy between the local flux density and the integral flux measured by the monitor as much as possible, as described in the following. If the remaining error is below a few percent, it can be approximately corrected out in the course of data analysis, as will be explained in more detail in Sect. 1.3.2.

To get a better correlation between the local and integral photon flux it is important that the illuminated area on the sample be not much bigger than what is imaged in microspectroscopy. All the flux outside the field of view adds an irrelevant contribution to the monitor signal. Attention should also be paid to the adjustment of the imaged area to the center of the undulator radiation. Finally, a significant reduction in photon energy-dependent effects can be achieved when the movement of the insertion device gap can be synchronized with the scanning of the grating of the monochromator.

Local energy resolution is an important issue for those beamline optics where the light spot is an image of the exit slit. In such beamlines, the monochromator energy dispersion at the position of the exit slit is imaged onto the sample surface. As a consequence, the local resolution does not change when the slit size is varied. The more serious implication for microspectroscopy is that there is also photon energy dispersion across the image in this case. If the energy dispersion across the image is of comparable size compared to the width of the spectral features of the sample, this shift in energy across the image has to be considered in data analysis of the microspectra.

A point that is normally not given closer attention in microscopy, but which becomes essential in quantitative microspectroscopy, is the linearity of the image detection system. A typical detection system may consist of several components, for example, a multichannel electron multiplier, fluorescence screen, and CCD camera. Although it is desirable to reach intensities as high as possible to get short exposure and scan times, it may sometimes be necessary to sacrifice some output signal in order to operate in the linear range of the image detection system. This range can be determined beforehand by manipulating the incident beam flux in a controlled way and comparing image intensity and flux monitor signal.

1.3.2 Data Analysis

In principle, data analysis in microspectroscopy could follow the same procedures as already established for data analysis in conventional spectroscopy. The main difference, however, is that in imaging microspectroscopy there are *many* spectra, for example, as many as there are pixels in the image. Data analysis for extracting the important parameters from this large number of spectra should therefore meet the following two requirements: It should be quick, and it should be automatic. In this section, a method will be described that allows the automatic analysis of XMCD microspectra to extract spin and orbital magnetic moments by means of sum rules.

To explain the procedure, the example of XMCD-PEEM microspectroscopy from a trilayer of 1.2 atomic monolayers (ML) Ni, an Fe layer of varying thickness, and 6 ML Co on a Cu(001) single crystal surface will be examined [60]. Circularly polarized X-rays of opposite helicity were provided by one of the two helical undulators of beamline BL25SU of SPring-8, Japan. Single pixel spectra corresponding to $3 \times 3\,\mu m^2$ areas of the sample have been acquired, which have to be analyzed automatically using the sum rules.

A problem for the automatic data analysis is that the data can sometimes be rather noisy, especially at positions where the dichroism is small. In most cases when the sum rules are applied the goal is only to extract two numbers, namely, the spin and orbital moments. In the following, a method is described that makes use of the fact that usually the line shape of the helicity-averaged absorption spectrum does not change across a microscopic image. In this case, the full spectral information of the single-pixel spectra can be reduced to two parameters of interest by a simple fit procedure, thereby improving statistics.

For this procedure, first a template XMCD spectrum with a sufficiently good statistics for sum-rule analysis is generated by averaging many single-pixel spectra over a larger area. This averaged spectrum can be analyzed as described in Sect. 1.2.1. The resulting difference curve between the two spectra for opposite helicity is then cut into two parts at an energy between the L_3 and L_2 edges, such that the two parts represent the dichroism of the L_3 and L_2 absorption edges, respectively. All the single pixel spectra are first corrected for local photon flux effects using a linear correction as a function of photon energy, as described in [60], which is less than 5% of the raw intensity in the present example, and then normalized to an edge jump of one. The two parts of the template XMCD curve are then fit to the single pixel difference curve by simply scaling. The only two fit parameters in that fit are the two scaling factors needed at the L_3 and L_2 edges to reproduce the single pixel difference curve by the template difference curve. An example is shown in Fig. 1.6. It shows corrected Fe $L_{2,3}$ absorption spectra of three different $3 \times 3\,\mu m^2$ pixels, obtained from 1.2 ML Ni/x ML Fe/6 ML Co films on Cu(001) for different Fe layer thicknesses x [60]. Depicted are the data for incident light of positive (solid line) and negative helicity (dotted line). The two peaks correspond to the Fe L_3 and L_2 absorption maxima. The results obtained from that sample will be discussed in Sect. 1.4.1. The three spectra presented in Fig. 1.6 have been selected to show very different Fe magnetic moments. In the top row (panels (a), (c), and (e)), the single pixel absorption spectra for positive

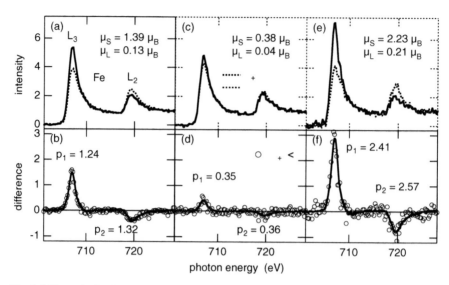

Fig. 1.6. Example for the sum-rule analysis of single pixel spectra. (a), (c), (e): single pixel absorption spectra for positive (*solid lines*) and negative helicity (*dotted lines*) from 1.2 ML Ni/Fe/6 ML Co/Cu(001) films for three different Fe thicknesses ((a) 10 ML, (b) 5 ML, (c) 2 ML). (b), (d), (f): difference between spectra for opposite helicity of panels (a), (c), and (e) (open symbols). The solid lines in (b), (d), and (f) are portions of a template difference curve, obtained from averaging over a larger area, scaled by the fit parameters p_1 and p_2 at the L_3 and L_2 edges, respectively, to match the single pixel data. The spin and orbital moments, calculated from p_1, p_2, and the spin and orbital moments of the template spectrum, are listed in panels (a), (c), and (e). (Reproduced from [60] with permission, Copyright (2001) by Elsevier Science)

(solid line) and negative helicity (dotted line) are shown. The bottom panels (b), (d), and (f) show the corresponding single pixel difference curves as open symbols. Please note the different vertical scale in the top and bottom panels. The solid lines in panels (b), (d), and (f) are the result of the template fits described above. The corresponding scaling factors for the dichroism at the L_3 and L_2 edges, p_1 and p_2, respectively, are also given. The solid line is, in other words, identical to the difference curve of the template spectrum, scaled by p_1 at the L_3 edge and by p_2 at the L_2 edge. The analysis of the template spectrum resulted in a spin moment of 1.1 μ_B and an orbital moment of 0.1 μ_B [60]. From these moments and the knowledge of p_1 and p_2, it is straightforward to calculate the spin and orbital moments for each of the single pixel fit results. For the three pixels presented in Figure 1.6, the resulting Fe moments are listed in the upper panels.

By this template fit procedure, stable fits are obtained even for noisy single pixel data and small dichroism. The scatter of individual data points in the difference spectrum is averaged out in an elegant way. Furthermore, the position of the zero line is maintained, and data points far outside the absorption edges are not considered in the fit. However, several positions of the image should be checked to make sure that the prerequisite for the fit, namely, the constant shape of the helicity-averaged absorption spectrum, is fulfilled.

1.4 Specific Examples

We will now discuss two specific examples in which the application of XMCD-PEEM microspectroscopy has proven beneficial to the study of the involved physical phenomena.

1.4.1 Ultrathin fcc Fe Films

Interesting systems for quantitative analysis of magnetic moments from microspectroscopy are ultrathin epitaxial Fe films. In these films, the interplay between structural and magnetic properties leads to a variety of different structural and magnetic phases. Whereas bulk Fe exists only in the bcc structure at temperatures up to 1,184 K, an fcc-like phase of Fe can be stabilized at room temperature when grown epitaxially on substrates with suitable surface lattice dimensions [61–71]. Room-temperature-grown films of Fe on Cu(001) single crystal surfaces are known to exhibit three structurally and magnetically different phases, depending on film thickness. For thicknesses below 4 atomic monolayers (ML), a fully ferromagnetic tetragonally expanded fcc-like structure is present (phase I) [62,63]. In the thickness range between 4 and 11 ML, a second phase (II), a relaxed fcc structure, is found, in which one observes a non-ferromagnetic behavior of the inner film layers [64–66]. For thicknesses above 11 ML, a third phase appears, a ferromagnetic (011)-oriented bcc phase (III) [67–69]. The magnetic behavior of these three phases is linked to the structure by the atomic volume. For different atomic volumes different magnetic ground states are theoretically predicted [72–74].

Especially the second phase of fcc Fe, where a large fraction of the film is non-ferromagnetic, has attracted a lot of interest. An antiferromagnetic coupling between ferromagnetic layers, separated by phase II fcc Fe, has been concluded from the measurements of magnetization loops of Co/Fe/Co/Cu(001) [75, 76] and Ni/Fe and $Ni_{81}Fe_{19}$/Fe multilayers [77, 78]. While total magnetization measurements of stacks of multiple magnetic layers can give only indirect evidence about the magnetic configuration, XMCD-PEEM microspectroscopy can provide element-resolved quantitative information. In this section, measurements of a Ni/Fe/Co/Cu(001) sample are presented [79], in which microspectroscopy at the Fe $L_{2,3}$ edges was used for determining the magnetic moments of the Fe layer. Element-resolved imaging of the Co underlayer and Ni overlayer revealed the presence of antiferromagnetic interlayer coupling.

To study the thickness dependence by microspectroscopy, the Fe and Ni layers were shaped into 255-µm-wide crossed wedges. They were prepared by placing a slit aperture of 0.5-mm width 1 mm in front of the sample and rocking the sample-slit assembly by ±7.5° about the long axis of the aperture during deposition. The thickness of the continuous Co underlayer was 6 ML, the Fe thickness 0–14 ML, and the Ni thickness 0–6 ML. Low magnification settings of the PEEM enabled imaging of a complete wedge.

Fig. 1.7 shows magnetic images of the Co and Ni layers at the onset of the crossed Ni and Fe wedges. The complete Fe wedge of 0–14 ML Fe is within the imaged area. The Fe thickness increases from right to left, as indicated at the upper axis. The imaged part of the Ni wedge corresponds to 0–4 ML thickness, increasing from top to bottom, as indicated at the right axis. The Fe thickness is such that all three phases, as mentioned above, are present within this wedge. The upper panel of Fig. 1.7 shows the magnetic asymmetry of the Co layer. It shows a nearly uniform bright contrast. It corresponds to a magnetization direction along the direction of the external field, which was applied after the deposition of the Co layer and again after completion of the trilayer. The lower panel shows the contrast obtained at the Ni L_3 edge. Note that here the grayscale is defined to include positive as well as negative values of the asymmetry, in contrast to panel (a). In a stripe located at Fe thicknesses

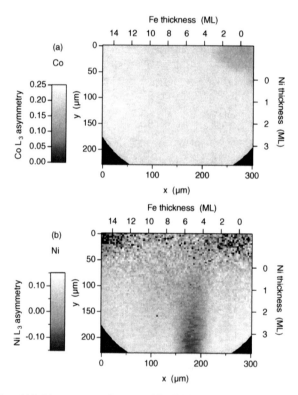

Fig. 1.7. Co (**a**) and Ni (**b**) asymmetry images of 0–4 ML Ni/0–14 ML Fe/6 ML Co/Cu(001). Ni and Fe thicknesses are indicated at the right and top axes, respectively. Different levels of gray correspond to different values of the magnetic asymmetry, as explained in the legend. Note that in panel (**b**) zero corresponds to an intermediate grayscale level, in contrast to panel (**a**), where zero is black. The Ni magnetization is antiparallel to the Co magnetization around 5 ML Fe thickness. (Reproduced from [79] with permission, Copyright (2000) by Elsevier Science)

between 5.0 and 6.5 ML, a negative (dark) Ni asymmetry is found, while the rest of the image exhibits an approximately uniform positive (bright) contrast. A negative asymmetry corresponds to an antiparallel orientation of the Ni layer magnetization at this position with respect to the Co layer magnetization. The Fe layer, consequently, mediates an antiferromagnetic interlayer coupling between the Co and Ni layers at this particular Fe thickness.

We will now turn our attention to the laterally resolved quantitative evaluation of the Fe magnetic moments at exactly the same region of the sample by XMCD-PEEM microspectroscopy. The result of a pixel-by-pixel sum-rule analysis is shown in the upper panel of Fig. 1.8. A total of 121 images for each helicity were recorded as a function of photon energy in the interval between 701 and 728 eV, similar to the spectra shown in Fig. 1.6. The size of a single pixel is $3 \times 3\,\mu m^2$, the exposure time was 10 s per image. Different levels of gray correspond to different values of the Fe effective spin moment, as explained in the legend. In the upper part of the image at zero or low Ni coverage the three different phases of Fe, as introduced above, are recognized by their distinctly different spin moments. They are labeled I, II, and III. Phase I extends up to a Fe thickness of approximately 3.5 ML and shows high spin moments of up to 2.5 μ_B. Phase II between approximately 3.5 and 11 ML is characterized by a low spin moment, whereas phase III again shows increasing spin moments up to 2.0 μ_B at Fe thicknesses above 11 ML. These numbers can be seen more easily from a line scan along the Fe wedge. The open symbols in Fig. 1.8 (b) represent a horizontal line scan obtained from a 7 pixel vertical average of the uncovered Fe/6 ML Co/Cu(001) sample in the region indicated by the upper white rectangle in panel (a). For increasing thickness of the Ni top layer, differences in phase II Fe moment are observed. The solid symbols of Fig. 1.8 (b) show a line scan

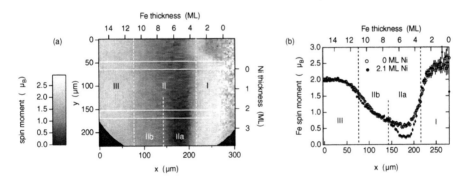

Fig. 1.8. (a) Result of a pixel-by-pixel sum-rule analysis for the Fe effective spin moment $\mu_{S,eff}$. Different values of $\mu_{S,eff}$ are represented by different levels of gray, as explained in the legend. The imaged region of the crossed Ni/Fe double wedge is exactly the same as in Fig. 1.7. Ni and Fe thicknesses are indicated at the right and top axes, respectively. Regions with different values of $\mu_{S,eff}$ are separated by dashed vertical lines and labeled I through III, IIa, and IIb. Horizontal line scans of $\mu_{S,eff}$ at the positions indicated by white rectangles in (a) are shown in (b). (Reproduced from [79] with permission, Copyright (2000) by Elsevier Science)

at around 2.1 ML Ni thickness (lower rectangle in panel (a)). From this line scan and also from the image plot of the Fe spin moments in panel (a) a further reduction of the Fe moment in phase II between about 4 and 7 ML compared to the uncovered Fe layer is observed. This region is labeled IIa, to distinguish it from the rest of phase II, where no change as a function of Ni overlayer thickness occurs (region IIb).

The observed Fe moments in phase I and III agree rather well with what is expected and known from literature [46,80]. The interesting behavior occurs in phase II. There is some dispute in the literature about the origin and location of the remaining moment of about 0.7–1.0 μ_B in phase II of the uncovered Fe/Co/Cu(001) sample: While from oxygen adsorption [75,76] and XMCD experiments [81,82] it had been concluded that the surface is not ferromagnetically ordered, and the ferromagnetism, consequently, had been attributed to the Co/Fe interface, other XMCD experiments [80], as well as measurements of photoelectron diffraction in magnetic dichroism [83] and spin-resolved valence band photoemission [84], provided evidence for the presence of a ferromagnetic layer on top of non-ferromagnetic underlayers, plus possibly ferromagnetism at the Fe/Co interface. The result from XMCD-PEEM microspectroscopy has to be interpreted as the exponentially depth averaged signal of all Fe monolayers in the film. A constant amount of ferromagnetic Fe at both interfaces plus non-ferromagnetic Fe in between would lead to a decreasing apparent magnetic moment with increasing Fe film thickness. This is not observed in the experiment (cf. Fig. 1.8). The situation is, therefore, probably more complicated. Layer-wise or bilayer-wise antiferromagnetic ordering of Fe moments [85–88], frustrations at interface steps or non-collinear moments [89] may have to be considered, possibly supported by further theoretical calculations.

By comparing Fig. 1.7 and Fig. 1.8, it becomes evident that region IIa with the extremely low Fe moment corresponds to the region of antiferromagnetic coupling between Ni and Co. Assuming ferromagnetic interface layers at both interfaces, this lowering of Fe moments could be explained by the strong direct coupling to the neighboring magnetic layers. In the case of antiferromagnetic alignment of Co and Ni, the Fe layers at both interfaces should be aligned opposite to each other, leading to lower apparent moments compared with the parallel alignment in region IIb. The question remains why without Ni overlayer no such decrease of the Fe moment by antiparallel alignment of the ferromagnetic Fe surface layer to the Co layer (and thus also to the Fe interface layer) is observed at the same Fe thickness. The reason could be some different influence of interface roughness and roughness-related magnetic frustrations in Fe/Co with and without Ni overlayer. There is also the possibility that without Ni, a 90 °C orientation of the Fe surface layer around 5.5 ML Fe thickness is present as the result of such competing frustrations, similar to the mechanisms leading to 90 °C interlayer exchange coupling [90,91].

In the present example of Ni/Fe/Co trilayers, XMCD-PEEM microspectroscopy was used for the thickness-dependent study of Fe moments. The effective spin moments of Fe in the three structural phases could be obtained from images showing the moments in a two-dimensional plot as a function of both Ni and Fe thicknesses. Characteristic changes connected with antiferromagnetic interlayer exchange coupling across the Fe layer could be observed. For this purpose, no special lateral resolution is required. In that case, microspectroscopy is just used as a very efficient way of parallel acquisition of a great number of XMCD spectra of crossed double wedges designed at length scales that are convenient for imaging with respect to intensity and lateral resolution. In the next section, an example of microspectroscopy with higher resolution is presented in which spin and orbital magnetic moments within magnetic domains are determined.

1.4.2 Spin Reorientation Transition in Co/Ni Bilayers

The control of the easy axis of magnetization is important for many applications in which magnetic ultrathin films are used. The direction of the easy axis is described by the angle-dependent part of the free energy, the so-called magnetic anisotropy energy (MAE). Minimization of the MAE with respect to the magnetization direction yields the easy axis of magnetization. The MAE is directly related to the anisotropy of the orbital magnetic moment [92]: The orbital moment should be higher for a direction of magnetization preferred by the MAE. This can be used to measure the angle dependence of the MAE in an element-selective and laterally resolved way by mapping the orbital magnetic moment by XMCD-PEEM microspectroscopy.

In the example presented in this section, Co/Ni epitaxial bilayers on Cu(001) have been investigated [93]. They were shaped into wedges of 255-μm widths, rotated by 90°C to each other, similar to the wedges described in the previous section. Co and Ni single films on Cu(001) exhibit a different behavior with respect to the easy axis of magnetization: Whereas Co/Cu(001) is always magnetized in the film plane ("in-plane") [94, 95], Ni/Cu(001) shows a perpendicular magnetization ("out-of-plane") over a wide range of thicknesses [96–99]. In Co/Ni bilayers, in-plane magnetization is, therefore, expected for large Co thicknesses and small Ni thicknesses, whereas out-of-plane magnetization should be present for small Co thicknesses and large Ni thicknesses.

Figure 1.9 shows the result of a pixel-by-pixel sum-rule analysis of the Ni spin moments of a region in a Ni/Co/Cu(001) crossed double wedge. The Ni thickness increases in the displayed area from 10.7 to 14 ML from left to right, as indicated at the top axis, and the Co thickness increases from 1.35 to 2.65 ML from bottom to top, as indicated at the right-hand axis. The image was obtained from 76,800 single pixel XMCD spectra of $370 \times 370\,\text{nm}^2$ size each. The resolution of the PEEM was adjusted to 500 nm, which resulted in a 30 s exposure time per photon energy step. A total of 105 images for each helicity between 845 and 890 eV photon energy were acquired using variable photon energy step spacings of 0.26 eV near the L_3 peak, 0.34 eV near the L_2 peak, 0.65 eV before the L_3 peak and in between the L_3 and L_2 peaks, and 1.4 eV in the post-L_2 region. Unlike the previous example, this sample was not magnetized by an external field before imaging. The resulting moments, therefore, include the cosine of the angle between the helicity of the incoming light and the local magnetization direction, which is different in the different domains of the as-grown domain structure. The microspectroscopic analysis thus does not yield the absolute moments, but a projection on the direction of incoming light, which, in the present example, was from bottom to top of Fig. 1.9, with an angle of 60°C to

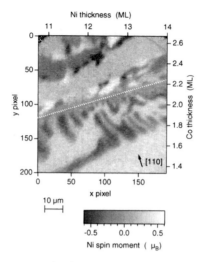

Fig. 1.9. Map of Ni spin moment projections onto the light incidence direction (from bottom to top, with an angle of 30 °C to the sample surface), resulting from the pixel-by-pixel sum-rule analysis of 76,800 single-pixel XMCD microspectra of a Co/Ni/Cu(001) crossed double-wedge sample. The Ni thickness is indicated at the top axis, the Co thickness at the right-hand axis. The grayscale to magnetic moment conversion is given in the legend at the bottom. At the white dotted line a spin reorientation transition between in-plane magnetization (top) and out-of-plane magnetization (bottom) occurs. Arrows in some domains indicate the magnetization directions. (From [93], Copyright (2000) by the American Physical Society)

the surface normal. Note that the grayscale used to represent the Ni spin moment projections is symmetric around zero.

Inspection of the domain pattern of Fig. 1.9 reveals two qualitatively different regions separated by the dotted line. In the upper part of the image, four different shades of gray are recognized, namely black, dark gray, light gray, and white. In the lower part, only two different shades of gray are found and the domains are more rounded. Quantitative analysis of the values of $\mu_{S,eff}$ for Ni in the single domains leads to the result that in the upper part of the image the magnetization direction is in-plane, oriented along the four <110> crystallographic directions, as indicated by arrows. Since the light incidence azimuth was deliberately rotated out of the crystal symmetry axes, each of these four directions results in a different grayscale representation. In the lower part of the image, the magnetization is out-of-plane, either parallel or antiparallel to the surface normal. The resulting absolute value of the Ni effective spin moment is 0.65 μ_B, constant over the entire image [93].

At the dotted line of Fig. 1.9, a spin reorientation transition between in-plane and out-of-plane magnetization occurs, either by varying the Ni thickness, or by varying the Co thickness. From the position and slope of that line, conclusions about the relative MAEs of the Co and Ni layers can be drawn [93]. The domain pattern on the out-of-plane side of the spin reorientation line is related to the spin reorientation transition: As the spin reorientation line is approached, more and more oppositely magnetized out-of-plane domains are formed. This can be explained by the competition between the magneto static energy, on the one hand, and the energy cost for creating domain walls, on the other hand. Closely spaced, alternating up and down magnetized out-of-plane domains have a lower magneto static energy than a single out-of-plane domain due to partial flux closure [100]. The formation of domains, on the other hand, requires the creation of additional domain walls of certain domain wall energy. Since in domain walls between two out-of-plane domains an in-plane component of the magnetization is present, this domain wall energy is directly related to the MAE of the system. Close to the spin reorientation transition the domain wall energy is consequently low, so that the formation of many small domains can be energetically favorable [101]. A more detailed discussion of the domain structure at spin reorientation transitions between in-plane and out-of-plane magnetization is given in Chap. 7.

Figure 1.10 shows the result of the pixel-by-pixel sum-rule analysis for the Ni orbital moments of the same region of the sample as Fig. 1.9. The same domain pattern is recognized, although the noise in the image is higher. The higher noise is a consequence of the application of the sum rules, where for the evaluation of the spin moment the peak areas of the L_2 and L_3 dichroism are added, whereas for the orbital moment they are subtracted [31]. The interesting quantity for the interpretation of the orbital moment, independent of magnetization direction, is the ratio of orbital to spin moment, $\mu_L/\mu_{S,\text{eff}}$. From the statistics of Fig. 1.10 it is clear that for the present 370×370 nm^2 pixel resolution a pixel wise interpretation of orbital moments would yield too big an error. To improve statistics in $\mu_L/\mu_{S,\text{eff}}$, the information from

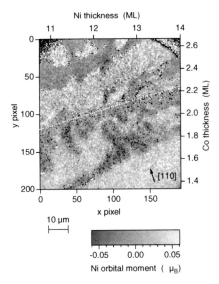

Fig. 1.10. As in Fig. 1.9, but for the Ni orbital moment μ_L. Different projections onto the direction of the incoming light are represented by different grayscales, as defined in the legend. (From [93], Copyright (2000) by the American Physical Society)

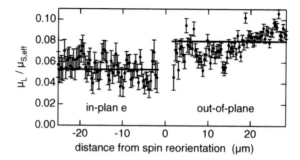

Fig. 1.11. Orbital to spin moment ratio $\mu_L/\mu_{S,\text{eff}}$ as a function of the distance from the spin reorientation line. Each data point is an average of 192 points along a line parallel to the spin reorientation transition in Fig. 1.9. Solid lines mark the average $\mu_L/\mu_{S,\text{eff}}$ ratio in the in-plane region (left) and in the out-of-plane region (right). (From [93], Copyright (2000) by the American Physical Society)

several pixels has to be averaged. The result of an averaging of pixels with a common distance from the spin reorientation line is shown in Fig. 1.11. Here, the orbital to spin moment ratio $\mu_L/\mu_{S,\text{eff}}$ is plotted as a function of the distance from the spin reorientation line. The left-hand side of Fig. 1.11 corresponds to the in-plane region of the image, the right-hand side to the out-of-plane region. Each data point contains information of 192 pixels along a line parallel to the spin reorientation transition. The horizontal solid lines in Fig. 1.11 mark the average on both sides.

Although there is still considerable scatter, the orbital moment in the out-of-plane region is seen to be distinctly higher than in the in-plane region by nearly 0.03 $\mu_{S,\text{eff}}$. This is interpreted in terms of the above-mentioned connection between the anisotropy of the orbital moment and the magnetic anisotropy energy. In the present example of Co/Ni/Cu(001), the MAE of the Ni layer alone is favoring an out-of-plane easy axis in the whole range of thicknesses considered here [93]. The observed in-plane magnetization is thus exclusively due to a stronger in-plane anisotropy of the Co layer, which overcompensates for the out-of-plane anisotropy of the Ni layer. The rigid magnetic exchange coupling between the two magnetic layers leads to a common easy axis of the bilayer, which results from the energy minimization of the summed MAE contributions of both layers. That means that in the upper part of the images the magnetization direction of the Ni layer is along the Ni hard in-plane direction, and in the lower part along the Ni easy out-of-plane direction. The Ni orbital moment is thus higher for a magnetization direction along the easy axis and lower for a magnetization direction along the hard axis.

Estimates for the size of the MAE can be made from the difference in orbital moment. A simplified theoretical description of the relation between the difference in orbital moment and the difference in MAE for two different magnetization directions resulted in a proportionality between both [92]. The proportionality factor, however, depends on some integrals over density of states in the valence bands [92] and may vary strongly between different samples, or even as a function of film thickness. In the present example, the change in orbital moment is 0.027 $\mu_{S,\text{eff}}$, or 0.018 μ_B. The proportionality constant between $\Delta\mu_L$ and MAE has been determined experimentally for Ni in Ni/Pt multilayers by independent measurements of both the orbital moments and the magnetic anisotropy [102]. Assuming that the same proportionality constant is valid for the present Co/Ni bilayers, an MAE of $+(47\pm 10)$ µeV/atom is obtained. In spite of the uncertainties connected to the use of the proportionality constant of a different sample, this value very well makes sense if we remember that XMCD measures the exponentially depth weighted average of the orbital magnetic moment within the probing depth and, thus, of the MAE. A literature value for the MAE of the inner layers of Ni/Cu(001) is $+34$ µeV/atom [99]. Considering the fact that in the present sample the interface between Ni and Co has been found to contribute a high positive MAE [93], the value of $+47$ µeV/atom is amazingly close to the expected anisotropy energy of Co/Ni/Cu(001).

In this example of Co/Ni/Cu(001), XMCD-PEEM has been used for a quantitative analysis of the Ni spin moment projections, the characterization of local magnetization directions, and the identification of a spin reorientation transition between in-plane and out-of-plane easy axes in the sample. The observed local moments' projections are consistent with an absolute value of the effective spin moment, which is constant across the imaged area of the sample, and domains with either $\langle 110\rangle$ in-plane magnetization or $\pm[001]$ out-of-plane magnetization. The connection of orbital moments and magnetic anisotropy adds an important feature to XMCD-PEEM microspectroscopy, which makes it an ideal tool for the study of local magnetic anisotropies in small magnetic structures.

1.5 Summary and Outlook

XMCD in X-ray absorption and PEEM can be combined into imaging XMCD microspectroscopy measurements at a reasonable experimental effort. The PEEM is thereby used as a parallel detector for local electron yield with microscopic spatial resolution. Scanning the photon energy of the incident X-ray beam and recording PEEM images at each energy step allows one to extract the full spectroscopic information inherent to XMCD from every position in the images, if some experimental requirements concerning evenness of illumination, flux normalization, as well as linearity and stability of the image detection system can be fulfilled. Quantitative information about the element-resolved magnetic moments, separated into spin and orbital contributions, can be extracted from a full image XMCD microspectrum through the use of sum rules. In cases where no special microscopic resolution is needed, this can be used for a quick but detailed map of the thickness dependence of thin film samples by microspectroscopy of wedge-shaped samples. This was illustrated by the example of Ni/Fe/Co ultrathin films on Cu(001), where the Ni and Fe layers were grown as crossed double-wedges.

The method can also be used to investigate micromagnetic phenomena by XMCD. The example of XMCD-PEEM microspectroscopy at the spin reorientation transition between out-of-plane and in-plane easy magnetic axes in Co/Ni/Cu(001) showed how the correlation between the anisotropy of the orbital magnetic moment and the magnetocrystalline anisotropy can be used to obtain element-resolved information about local magnetic anisotropy from XMCD microspectra.

Future improvements concerning the brightness of the illumination and the efficiency of the detection system will reduce the noise level of the microspectra and allow one to probe magnetic anisotropy at the microscopic level, for example, in domain walls or in technologically relevant magnetic microstructures. It will also enable one to choose a higher lateral resolution for microspectroscopy, maintaining reasonable acquisition times. The development of new PEEM instruments with aberration-correcting optics [59] will further push the limit for the attainable maximum resolution. Beamlines providing high-brilliance circularly polarized radiation with the possibility of quick helicity reversal [103] will help improve the accuracy of circular dichroism-based techniques.

The use of imaging energy filters [104–107] to select only electrons of a certain kinetic energy for the imaging process has interesting implications for magnetic microspectroscopy. On the one hand, it allows magnetic dichroism effects that occur only in photoemission [108] to be included in a microspectroscopic study. If the projective electron optics is tuned in order not to project the real space image plane onto the screen but the diffraction plane, on the other hand, photoelectron diffraction measurements [109–111] can be combined with XMCD. Positioning of the contrast aperture to certain features in the photoelectron diffraction pattern can then be used to obtain structure-sensitive XMCD microspectroscopy data and, thus, combine structural and magneto-spectroscopic information.

Finally, the inclusion of time-resolved stroboscopic pump-probe experiments that exploit the pulsed time structure of synchrotron radiation [112] into XMCD-PEEM microspectroscopy opens the way for the quantitative investigation of magnetization dynamics on microscopic length scales. Snapshots of the spatial spin and orbital magnetic moments' distribution in the course of reversible dynamic magnetization processes may be obtained in that way.

Acknowledgement. The author would like to thank J. Kirschner, F. Offi, S.-S. Kang, J. Gilles, S. Imada, and S. Suga for discussions and experimental work. Support from BMBF (No. 05 SL8EF1 9), DFG (Nos. Ki 358/3-1 and 446 JAP-113/179/0), JSPS, and JASRI is gratefully acknowledged.

References

1. B.P. Tonner, D. Dunham, T. Droubay, J. Kikuma, J. Denlinger, E. Rotenberg, and A. Warwick, J. Electron Spectrosc. Relat. Phenom. **75**, 309 (1995).
2. L. Casalis, W. Jark, M. Kiskinova, D. Lonza, P. Melpignano, D. Morris, R. Rosei, A. Savoi, A. Abrami, C. Fava, P. Furlan, R. Pugliese, D. Vivoda, G. Sandrin, F.-Q. Wei, S. Contarini, L. DeAngelis, C. Gariazzo, P. Natelli, and G.R. Morrison, Rev. Sci. Instrum. **66**, 4870 (1995).
3. J. Voss, J. Electron Spectrosc. Relat. Phenom. **84**, 29 (1997).
4. M. Kiskinova, Surf. Rev. Lett. **7**, 447 (2000).

5. L. Baumgarten, C.M. Schneider, H. Petersen, F. Schäfers, and J. Kirschner, Phys. Rev. Lett. **65**, 492 (1990).
6. Y. Kagoshima, T. Miyahara, M. Ando, J. Wang, and S. Aoki, J. Appl. Phys. **80**, 3124 (1996).
7. T. Warwick, K. Franck, J.B. Kortright, G. Meigs, M. Morenne, S. Myneni, E. Rotenberg, S. Seal, W.F. Steele, H. Ade, A. Garcia, S. Cerasari, J. Denlinger, S. Hayakawa, A.P. Hitchcock, T. Tyliszczak, J. Kikuma, E.G. Rightor, H.-J. Shin, and B.P. Tonner, Rev. Sci. Instrum. **69**, 2964 (1998).
8. O. Pietzsch, A. Kubetzka, M. Bode, and R. Wiesendanger, Phys. Rev. Lett. **84**, 5212 (2000).
9. M. Kleiber, M. Bode, R. Ravlic, and R. Wiesendanger, Phys. Rev. Lett. **85**, 4606 (2000).
10. A. Hubert and R. Schäfer: Magnetic Domains, Springer, Berlin (1998), and references therein.
11. D. Weller: Magneto-optical Kerr spectroscopy of transition metal alloy and compound films. In: H. Ebert and G. Schütz (ed.), Spin-Orbit-Influenced Spectroscopies of Magnetic Solids, Springer, Berlin (1996).
12. H. Ebert, Rep. Prog. Phys. **59**, 1665 (1996).
13. C.M. Schneider, Z. Celinski, M. Neuber, C. Wilde, M. Grunze, K. Meinel, and J. Kirschner, J. Phys.: Cond. Matt. **6**, 1177 (1994).
14. T. Kinoshita, K.G. Nath, Y. Haruyama, M. Watanabe, M. Yagi, S.-I. Kimura, and A. Fanelsa, J. Electron Spectrosc. Relat. Phenom. **92**, 165 (1999).
15. C.M. Schneider, K. Meinel, K. Holldack, H.P. Oepen, M. Grunze, and J. Kirschner: Magnetic spectro-microscopy using magneto-dichroic effects in photon-induced auger electron emission. In: B.T. Jonker et al. (ed.), Magnetic Ultrathin Films, Materials Research Society, Pittsburgh (1993).
16. C.M. Schneider, K. Holldack, M. Kinzler, M. Grunze, H.P. Oepen, F. Schäfers, H. Petersen, K. Meinel, and J. Kirschner, Appl. Phys. Lett. **63**, 2432 (1993).
17. P. Fischer, T. Eimüller, G. Schütz, P. Guttmann, G. Schmahl, K. Prueg, and G. Bayreuther, J. Phys. D: Appl. Phys. **31**, 649 (1998).
18. J. Stöhr, Y. Wu, M.G. Samant, B.B. Hermsmeier, G. Harp, S. Koranda, D. Dunham, and B.P. Tonner, Science **259**, 658 (1993).
19. W. Swiech, G.H. Fecher, Ch. Ziethen, O. Schmidt, G. Schönhense, K. Grzelakowski, C.M. Schneider, R. Frömter, H.P. Oepen, and J. Kirschner, J. Electron Spectrosc. Relat. Phenom. **84**, 171 (1997).
20. F.U. Hillebrecht, D. Spanke, J. Dresselhaus, and V. Solinus, J. Electron Spectrosc. Relat. Phenom. **84**, 189 (1997).
21. W. Kuch, R. Frömter, J. Gilles, D. Hartmann, Ch. Ziethen, C.M. Schneider, G. Schönhense, W. Swiech, and J. Kirschner, Surf. Rev. Lett. **5**, 1241 (1998).
22. J. Stöhr, H.A. Padmore, S. Anders, T. Stammler, and M.R. Scheinfein, Surf. Rev. Lett. **5**, 1297 (1998).
23. S. Anders, H.A. Padmore, R.M. Duarte, T. Renner, T. Stammler, A. Scholl, M.R. Scheinfein, J. Stöhr, L. Séve, and B. Sinkovic, Rev. Sci. Instrum. **70**, 3973 (1999).
24. T. Kachel, W. Gudat, C. Koziol, T. Schmidt, G. Lilienkamp, E. Bauer, and M. Altman, J. Appl. Phys. **81**, 5025 (1997).
25. E. Bauer, Rep. Prog. Phys. **57**, 895 (1994).
26. E. Bauer, Surf. Rev. Lett. **5**, 1275 (1998).
27. S. Imada, S. Suga, W. Kuch, and J. Kirschner, Surf. Rev. Lett. **9**, 877 (2002).
28. A. Scholl, J. Stöhr, J. Lüning, J.W. Seo, J. Fompeyrine, H. Siegwart, J.-P. Locquet, F. Nolting, S. Anders, E.E. Fullerton, M.R. Scheinfein, and H.A. Padmore, Science **287**, 1014 (2000).

29. F. Nolting, A. Scholl, J. Stöhr, J.W. Seo, J. Fompeyrine, H. Siegwart, J.-P. Locquet, S. Anders, J. Lüning, E.E. Fullerton, M.F. Toney, M.R. Scheinfein, and H.A. Padmore, Nature **405**, 767 (2000).
30. G. Schütz, W. Wagner, W. Wilhelm, P. Kienle, R. Zeller, R. Frahm, and G. Materlik, Phys. Rev. Lett. **58**, 737 (1987).
31. B.T. Thole, P. Carra, F. Sette, and G. van der Laan, **68**, 1943 (1992).
32. P. Carra, B.T. Thole, M. Altarelli, and X. Wang, Phys. Rev. Lett. **70**, 694 (1993).
33. J. Stöhr, J. Electron Spectrosc. Relat. Phenom. **75**, 253 (1995).
34. Y.U. Idzerda, C.T. Chen, H.-J. Lin, H. Tjeng, and G. Meigs, Physica B **208–209**, 746 (1995).
35. H. Ebert: Circular magnetic X-ray dichroism in transition metal systems. In: H. Ebert and G. Schütz (ed.), Spin-Orbit-Influenced Spectroscopies of Magnetic Solids, Springer, Berlin (1996).
36. J. Stöhr and R. Nakajima, IBM J. Res. Develop. **42**, 73 (1998).
37. C.T. Chen, Y.U. Idzerda, H.-J. Lin, N.V. Smith, G. Meigs, E. Chaban, G.H. Ho, E. Pellegrin, and F. Sette, Phys. Rev. Lett. **75**, 152 (1995).
38. U. Fano, Phys. Rev. A **178**, 131 (1969).
39. D. Weller, J. Stöhr, R. Nakajima, A. Carl, M.G. Samant, C. Chappert, R. Mégy, P. Beauvillain, P. Veillet, and G.A. Held, Phys. Rev. Lett. **75**, 3752 (1995).
40. R. Wuand and A.J. Freeman, Phys. Rev. Lett. **73**, 1994 (1994).
41. W. L. O'Brien, B.P. Tonner, G.R. Harp, and S.S.P. Parkin, J. Appl. Phys. **76**, 6462 (1994).
42. D. Rioux, B. Allen, H. Höchst, D. Zhao, and D.L. Huber, Phys. Rev. B **56**, 753 (1997).
43. J. Schwitalla and H. Ebert, Phys. Rev. Lett. **80**, 4586 (1998).
44. J. Vogel and M. Sacchi, Phys. Rev. B **49**, 3230 (1994).
45. X. Le Cann, C. Boeglin, B. Carrière, and K. Hricovini, Phys. Rev. B **54**, 373 (1996).
46. J. Hunter Dunn, D. Arvanitis, and N. Mårtensson, Phys. Rev. B **54**, R11157 (1996).
47. G. Möllenstedt and F. Lenz: Electron emission microscopy. In: L. Marton (ed.), Advances in Electronics and Electron Physics, Academic Press, London (1963).
48. H. Bethke and M. Klaua, Ultramicroscopy **11**, 207 (1983).
49. W. Engel, M.E. Kordesch, H.H. Rotermund, S. Kubala, and A. von Oertzen, Ultramicroscopy **36**, 148 (1991).
50. M.E. Kordesch, W. Engel, G.J. Lapeyre, E. Zeitler, and A.M. Bradshaw, Appl. Phys. A **49**, 399 (1989).
51. M. Mundschau, M.E. Kordesch, B. Rausenberger, W. Engel, A.M. Bradshaw, and E. Zeitler, Surf. Sci. **227**, 246 (1990).
52. H.H. Rotermund, S. Nettesheim, A. von Oertzen, and G. Ertl, Surf. Sci. **275**, L645 (1992).
53. S. Nettesheim, A. von Oertzen, H.H. Rotermund, and G. Ertl, J. Chem. Phys. **98**, 9977 (1993).
54. W. Telieps and E. Bauer, Ultramicroscopy **17**, 57 (1985).
55. B.P. Tonner and G.R. Harp, Rev. Sci. Instrum. **59**, 853 (1988).
56. Ch. Ziethen, O. Schmidt, G.H. Fecher, C.M. Schneider, G. Schönhense, R. Frömter, M. Seider, K. Grzelakowski, M. Merkel, D. Funnemann, W. Swiech, H. Gundlach, and J. Kirschner, J. Electron Spectrosc. Relat. Phenom. **88–91**, 983 (1998).
57. G. De Stasio, L. Perfetti, B. Gilbert, O. Fauchox, M. Capozi, P. Perfetti, G. Margaritondo, and B.P. Tonner, Rev. Sci. Instrum. **70**, 1740 (1999).
58. E. Bauer, Appl. Surf. Sci. **92**, 20 (1996).
59. R. Fink, M.R. Weiss, E. Umbach, D. Preikszas, H. Rose, R. Spehr, P. Hartel, W. Engel, R. Degenhardt, R. Wichtendahl, H. Kuhlenbeck, W. Erlebach, K. Ihmann, R. Schlögl, H.-J. Freund, A.M. Bradshaw, G. Lilienkamp, T. Schmidt, E. Bauer, and G. Benner, J. Electron Spectrosc. Relat. Phenom. **84**, 231 (1997).

60. W. Kuch, J. Gilles, F. Offi, S.S. Kang, S. Imada, S. Suga, and J. Kirschner, Surf. Sci. **480**, 153 (2001).
61. J. Thomassen, F. May, B. Feldmann, M. Wuttig, and H. Ibach, Phys. Rev. Lett. **69**, 3831 (1992).
62. M.T. Kief and W.F. Egelhoff, Jr., Phys. Rev. B **47**, 10785 (1993).
63. K. Heinz, S. Müller, and P. Bayer, Surf. Sci. **337**, 215 (1995).
64. P. Bayer, S. Müller, P. Schmailzl, and K. Heinz, Phys. Rev. B **48**, 17611 (1993).
65. D. Li, M. Freitag, J. Pearson, Z.Q. Qiu, and S.D. Bader, Phys. Rev. Lett. **72**, 3112 (1994).
66. M. Straub, R. Vollmer, and J. Kirschner, Phys. Rev. Lett. **77**, 743 (1996).
67. M. Wuttig, B. Feldmann, J. Thomassen, F. May, H. Zillgen, A. Brodde, H. Hannemann, and H. Neddermayer, Surf. Sci. **291**, 14 (1993).
68. J. Giergiel, J. Kirschner, J. Landgraf, J. Shen, and J. Woltersdorf, Surf. Sci. **310**, 1 (1994).
69. J. Giergiel, J. Shen, J. Woltersdorf, A. Kirilyuk, and J. Kirschner, Phys. Rev. B **52**, 8528 (1995).
70. M.-T. Lin, J. Shen, W. Kuch, H. Jenniches, M. Klaua, C.M. Schneider, and J. Kirschner, Surf. Sci. **410**, 290 (1998).
71. S.S. Kang, W. Kuch, and J. Kirschner, Phys. Rev. B **63**, 024401 (2001).
72. V.L. Moruzzi, P.M. Marcus, K. Schwarz, and P. Mohn, Phys. Rev. B **34**, 1784 (1986).
73. V.L. Moruzzi, P.M. Marcus, and J. Kübler, Phys. Rev. B **39**, 6957 (1989).
74. P.M. Marcus, S.L. Qiu, and V.L. Moruzzi, J. Phys.: Cond. Matt. **11**, 5709 (1999).
75. E.J. Escorcia-Aparicio, R.K. Kawakami, and Z.Q. Qiu, Phys. Rev. B **54**, 4155 (1996).
76. R.K. Kawakami, E.J. Escorcia-Aparicio, Z.Q. Qiu, J. Appl. Phys. **79**, 4532 (1996).
77. W. Kuch and S.S.P. Parkin, Europhys. Lett. **37**, 465 (1997).
78. W. Kuch and S.S.P. Parkin, J. Magn. Magn. Mater. **184**, 127 (1998).
79. W. Kuch, J. Gilles, F. Offi, S.S. Kang, S. Imada, S. Suga, and J. Kirschner, J. Electron Spectrosc. Relat. Phenom. **109**, 249 (2000).
80. D. Schmitz, C. Charton, A. Scholl, C. Carbone, and W. Eberhardt, Phys. Rev. B **59**, 4327 (1999).
81. W. L. O'Brien and B.P. Tonner, Surf. Sci. **334**, 10 (1995).
82. W. L. O'Brien and B.P. Tonner, Phys. Rev. B **52**, 15332 (1995).
83. X. Gao, M. Salvietti, W. Kuch, C.M. Schneider, and J. Kirschner, Phys. Rev. B **58**, 15426 (1998).
84. R. Kläsges, D. Schmitz, C. Carbone, W. Eberhardt, and T. Kachel, Solid State Commun. **107**, 13 (1998).
85. R. Lorenz and J. Hafner, Phys. Rev. B **54**, 15937 (1996).
86. T. Asada and S. Blügel, Phys. Rev. Lett. **79**, 507 (1997).
87. E.G. Moroni, G. Kresse, and J. Hafner, J. Phys.: Cond. Matt. **11**, L35 (1999).
88. D. Spisak and J. Hafner, Phys. Rev. B **62**, 9575 (2000).
89. R. Lorenz and J. Hafner, Phys. Rev. B **58**, 5197 (1998).
90. J.C. Slonczewski, Phys. Rev. Lett. **67**, 3172 (1991).
91. J.C. Slonczewski, J. Magn. Magn. Mater. **150**, 13 (1995).
92. P. Bruno, Phys. Rev. B **39**, 865 (1989).
93. W. Kuch, J. Gilles, S.S. Kang, S. Imada, S. Suga, and J. Kirschner, Phys. Rev. B **62**, 3824 (2000).
94. P. Krams, F. Lauks, R.L. Stamps, B. Hillebrands, and G. Güntherodt, Phys. Rev. Lett. **69**, 3674 (1992).
95. M. Kowalewski, C.M. Schneider, and B. Heinrich, Phys. Rev. B **47**, 8748 (1993).
96. F. Huang, M.T. Kief, G.J. Mankey, and R.F. Willis, Phys. Rev. B **49**, 3962 (1994).
97. W. L. O'Brien and B.P. Tonner, Phys. Rev. B **49**, 15370 (1994).

98. B. Schulz and K. Baberschke, Phys. Rev. B **50**, 13467 (1994).
99. M. Farle, B. Mirwald-Schulz, A.N. Anisimov, W. Platow, and K. Baberschke, Phys. Rev. B **55**, 3708 (1997).
100. Y. Yafet and E.M. Gyorgy, Phys. Rev. B **38**, 9145 (1988).
101. M. Speckmann, H.P. Oepen, and H. Ibach, Phys. Rev. Lett. **75**, 2035 (1995).
102. F. Wilhelm, P. Poulopoulos, P. Srivastava, H. Wende, M. Farle, K. Baberschke, M. Angelakeris, N.K. Flevaris, W. Grange, J.-P. Kappler, G. Ghiringhelli, and N.B. Brookes, Phys. Rev. B **61**, 8647 (2000).
103. M.R. Weiss, R. Follath, K.J.S. Sawhney, F. Senf, J. Bahrdt, W. Frentrup, A. Gaupp, S. Sasaki, M. Scheer, H.-C. Mertins, D. Abramsohn, F. Schäfers, W. Kuch, and W. Mahler, Nucl. Instr. and Meth. A, **467–468**, 449 (2001).
104. D.W. Turner, I.R. Plummer, and H.Q. Porter, Rev. Sci. Instrum. **59**, 45 (1988).
105. G.K.L. Marx, V. Gerheim, and G. Schönhense, J. Electron Spectrosc. Relat. Phenom. **84**, 251 (1997).
106. Y. Sakai, M. Kato, S. Masuda, Y. Harada, and T. Ichinokawa, Surf. Rev. Lett. **5**, 1199 (1998).
107. T. Schmidt, S. Heun, J. Slezak, J. Diaz, K.C. Prince, G. Lilienkamp, and E. Bauer, Surf. Rev. Lett. **5**, 1287 (1998).
108. G. Rossi, G. Panaccione, F. Sirotti, and N.A. Cherepkov, Phys. Rev. B **55**, 11483 (1997).
109. C.S. Fadley: Recent developments in photoelectron diffraction. In: J. Kanamori and A. Kotani (ed.), Core-Level Spectroscopy in Condensed Systems, Springer, Berlin (1988).
110. C.S. Fadley, Surf. Sci. Rep. **19**, 231 (1993).
111. H. Daimon, T. Nakatani, S. Imada, S. Suga, Y. Kagoshima, and T. Miyahara, Rev. Sci. Instrum. **66**, 1510 (1995).
112. M. Bonfim, G. Ghiringhelli, F. Montaigne, S. Pizzini, N.B. Brookes, F. Petroff, J. Vogel, J. Camarero, and A. Fontaine, Phys. Rev. Lett. **86**, 3646 (2001).

2
Study of Ferromagnet-Antiferromagnet Interfaces Using X-Ray PEEM

A. Scholl, H. Ohldag, F. Nolting, S. Anders, and J. Stöhr

This chapter discusses polarization dependent X-ray photoemission electron microscopy (X-PEEM) and its application to coupled magnetic layers, in particular ferromagnet-antiferromagnet structures.

2.1 Introduction

Over the last decades, magnetism has evolved into one of the cornerstones of information storage technology, providing the foundation of a $50 billion dollar per year worldwide storage business. Today's high-tech magnetic devices are based on thin films, often patterned into sub-micron sized cells, and rely on the existence of magnetically well-defined states. Over the last ten years, our understanding of the structure and properties of magnetic thin films and multilayers has progressed remarkably, yet one key problem in our understanding has remained, namely, the characterization and understanding of interfaces, omnipresent in modern magnetic structures. Many of today's forefront areas in magnetism require a better understanding of the spin structure at interfaces [1, 2]. Examples are giant magnetoresistance structures and spin tunnel junctions [3], as well as "spintronics" devices based on spin injection into a semiconductor [4, 5]. In these structures, spin transport across metal-metal, metal-oxide, metal-semiconductor, and semiconductor-semiconductor interfaces is believed to strongly depend on the magnetic properties of the interfaces.

Today's magneto-electronic devices typically contain ferro-, ferri-, and antiferromagnetic layers, and the characterization of their complex structure requires new experimental tools that are sensitive to all three flavors of magnetic order. While there are numerous techniques sensitive to ferromagnetic order in thin-film systems ranging from magnetic force microscopy, over the magneto-optical Kerr effect, to photoemission techniques, there is clearly a need for methods that are also sensitive to antiferromagnetic order. The determination of the electronic, chemical, or magnetic structure of interfaces is especially challenging, since the weak signal originating from the interface has to be detected and isolated from the large background of the bulk. For lack of better capabilities scientists have tried to circumvent this problem by studying the early stages of interface formation with surface science techniques, by investigating ultrathin model systems, or by studying simply the "bulk" materials forming the interface. Unsubstantiated assumptions are often made in the interpretation of experimental data or for model calculations about the existence of bulk-like structures all the way to the interface or about the existence of atomically abrupt interfaces without chemical intermixing or roughness. In this article, we shall show that in reality interfaces are quite different.

Here, we present the investigation of a specific interface problem that is of considerable scientific interest and technological importance: the unsolved origin of exchange bias in antiferromagnet-ferromagnet sandwiches [6,7]. Discovered in 1956 by Meiklejohn and Bean [8], exchange bias refers to the unidirectional pinning of a ferromagnetic layer by an adjacent antiferromagnet. Ferromagnetic films typically have a preferred magnetization axis, "easy axis," and the spins prefer to align along this axis. There are two equally stable easy spin directions (rotated by 180°) along this axis, and it costs the same energy and requires the same external field to align the spins along either easy direction. When a ferromagnet (FM) is grown on an antiferromagnet (AFM), the exchange coupling between the two systems leads to an increased coercivity of the ferromagnet. This is usually attributed to an increased interface anisotropy, resulting from the coupling of "interface moments" to the antiferromagnet. The ferromagnetic hysteresis loop is still symmetric, indicating two equivalent easy directions. If, on the other hand, the AFM-FM system is grown in a magnetic field or after growth annealed in a magnetic field to temperatures above the AFM Néel temperature, the hysteresis loop becomes asymmetric and is shifted from zero by a field H_B. This unidirectional shift is called exchange bias and reflects the fact that there is now a preferred easy magnetization *direction* for the FM. The ferromagnet is pinned by the antiferromagnet into this easy direction, which is opposite to the bias field direction H_B. The fact that the origin of the exchange bias effect is still hotly debated after more than forty years of research is due to the difficulty associated with determining the magnetic interfacial structure mentioned above. While there is little doubt about the crucial importance of the interfacial spin structure, and more specifically about the existence of uncompensated "interfacial spins" that give rise to the exchange bias phenomenon [9], two fundamental problems have prevented a solution of the puzzle.

The first is the difficulty of imaging the microscopic antiferromagnetic structure in thin films with conventional techniques. While the AFM domain structure in bulk single crystals has been studied since the late 1950s [10–12], little is known about the domain structure in thin films. For example, in well-annealed bulk NiO, the typical domain size is in the 0.1–1 mm range [10, 13] whereas in epitaxial NiO films, the domain size has been estimated to be less than 50 nm [14]. This size is below the spatial resolution of neutron diffraction topography (about 70 µm), X-ray diffraction topography (1–2 µm) [10], and of conventional [10, 13] and nonlinear [15, 16] optical techniques, which are limited by diffraction (about 0.2 µm). On a microscopic level, the exchange coupling across the AFM-FM interface is expected to proceed domain by domain, leading to a correlation of the AFM with the FM domain structure. Therefore, measurement of the antiferromagnetic domain structure is of crucial importance.

The second problem is the determination of the interfacial spin structure and its relationship with the magnetic structure in the AFM and FM films. Still little is known about the origin of uncompensated interfacial spins and their role in the observed coercivity increase and the exchange bias effect. Previous models have invoked statistical arguments for the existence of uncompensated interfacial spins. These models take into account one or more of the following aspects: uncompensated termination of bulk antiferromagnetic domains [9], spin-flop canting of antiferromagnetic spins [17], and interface and bulk defects [18, 19]. It has also been proposed that coercivity and bias arise from different mechanisms like canting and defects, respectively [18].

The present chapter takes a new look at an old problem, utilizing X-ray spectroscopy and microscopy methods that have only recently been developed. The spectromicroscopy approach, utilized here, combines two well-established concepts, polarized X-ray absorption spectroscopy (XAS) and electron microscopy [20, 21]. It has many key strengths needed ultimately to solve the exchange bias puzzle. Because of the elemental and chemical specificity of XAS, one can tune into specific layers by tuning the photon energy. X-ray polarization control opens the door for magnetic studies by using X-ray magnetic linear dichroism (XMLD) [22–26] on antiferromagnets, and X-ray magnetic circular dichroism (XMCD) [27–29] on ferro- and ferrimagnets. Electron yield detection provides limited sampling depth [20, 30], and for thin film sandwiches, the signal contains a sizeable interface contribution. An electron microscope, in our case detecting the secondary photoelectrons that are emitted after X-ray absorption, offers high spatial resolution.

This chapter first reviews the concepts of XMLD and XMCD spectroscopies and photoemission electron microscopy (PEEM). It then gives experimental details on polarization control and the present PEEM-2 microscope at the Advanced Light Source (ALS) in Berkeley, U.S.A. Experimental results are discussed next for the structure of selected antiferromagnetic surfaces and for their coupling to ferromagnetic overlayers. Finally, a summary and outlook are given.

2.2 Photoemission Electron Microscopy

X-ray magnetic circular dichroism, which was first observed at the Fe K edge in 1987 [27], is today a standard method for the study of magnetic thin films and surfaces [1]. The availability of polarized and tunable X-rays from today's brilliant synchrotron sources has played an important role in the success of X-ray dichroism techniques. The last ten years have seen great progress following the first XMCD spectroscopy measurements at the important transition metal L edges [28] and the first XMCD microscopy experiments [31]. Although called "photoemission electron microscopy", PEEM is actually an X-ray absorption technique, since contrast is generated by lateral variations in the X-ray absorption crosssection. Since PEEM is based on the absorption of X-rays in matter, we will first discuss X-ray absorption spectroscopy in general. We will furthermore only discuss the near edge structure within a few eV of the absorption edge, because it contains the information about the chemical and magnetic properties of the sample in which we are interested. A general overview of near edge X-ray absorption techniques can be found in [32], and XMCD spectroscopy has been reviewed in [29, 33]. We will only briefly discuss X-ray absorption measurements in ferromagnets, and we refer the reader to the contribution by W. Kuch for detailed description of the PEEM technique applied to ferromagnets (Chap. 1 of this book).

2.2.1 X-Ray Absorption Spectroscopy

X-ray absorption spectroscopy (XAS) utilizes the energy-dependent absorption of X-rays to obtain information about the elemental composition of the sample, the chemical environment of its constituents, and its magnetic structure. Core electrons are excited in the absorption process into empty states above the Fermi energy and thereby probe the electronic and magnetic properties of the empty valence levels. X-ray transmission detection and electron yield detection are the most commonly used methods for measuring the absorption coefficient μ as a function of the photon energy E, and can both be used in X-ray microscopes. Transmission microscopes provide direct quantitative access to the spatial distribution of $\mu(E)$, but require samples sufficiently thin to be penetrated by X-rays [34]. The absorption length in $3d$ transition metals in the soft X-ray region is typically about 20–100 nm [30, 35]. An advantage of X-ray detection techniques is their insensitivity to external magnetic fields.

Electron yield detection techniques like PEEM measure the absorption coefficient indirectly, collecting the emitted secondary electrons generated in the electron cascade that follows the creation of the primary core hole in the absorption process. The total electron yield is proportional to the number of absorbed photons in a near surface region of the sample whose depth is given by the mean free path of the low energy secondary electrons. The probing depth of electron yield detection is typically a few nanometers [30]. This is much smaller than the X-ray penetration length, which explains the surface sensitivity of PEEM [36]. After correction of saturation effects caused by the finite X-ray penetration depth, and the application of suitable normalization and background subtraction procedures, the absorption coefficient can be extracted from the yield spectrum [37, 38]. The difficulty of imaging magnetic structures in external applied fields can be overcome by using very localized fields generated by micro-coils.

The $L(2p)$ X-ray absorption spectrum of transition metal oxides such as NiO, CoO, and LaFeO$_3$ is dominated by the large spin-orbit splitting of the core level electronic $2p_{3/2}$ and $2p_{1/2}$ states, which is on the order of 15 eV. An example of an XAS spectrum at the Fe edge of a LaFeO$_3$ sample measured by electron yield detection is shown in Fig. 2.1. Dipole selection rules only permit transitions from the $2p$ states into $4s$ or $3d$ valence states, and the excitation into the narrow $3d$ states is responsible for the strong resonances. The fine structure of the resonance is caused by two effects, (i) the electron correlation due to the electron-electron interaction in the initial and final states of the absorption process and (ii) the crystal field interaction.

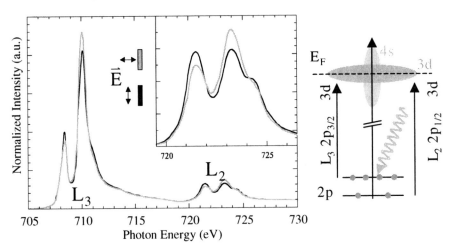

Fig. 2.1. X-ray absorption spectra of antiferromagnetic LaFeO$_3$ (001) measured with the linear X-ray polarization parallel to the sample (black) and perpendicular to the sample (gray). The spectra are corrected for electron yield saturation effects and for the finite linear polarization (80%) of the original spectra. The spectrum for $\vec{E}$ perpendicular to the surface, corresponding to an unphysical grazing X-ray incidence of 0°, has been derived from angle-dependent spectra. The spectra were measured with no spatial resolution on a large sample area. Strong X-ray magnetic linear dichroism (XMLD) appears at the multiplet split L_3 and L_2 edges. The X-ray absorption process is described on the right

Both effects are of comparable size (~ 1 eV), causing the complex structure of oxide spectra. These effects split the absorption resonances into multiplets.

2.2.2 X-Ray Magnetic Linear Dichroism (XMLD)

Magnetic effects become apparent in the absorption spectrum when linearly polarized X-rays are used [22–26]. A change in the X-ray polarization causes a change in the line strength of multiplet lines depending on the angular momentum character of the corresponding initial and final states. Since usually many multiplet lines are superimposed in an absorption spectrum, only a net effect is observable, which depends on the angle between the direction of the X-ray polarization vector and the axis of the magnetization. Thus, the ingredients for a strong magnetic linear dichroism in XAS are a strong spin-orbit interaction of the core level and a well resolved multiplet structure in the absorption resonance. In transition metal oxides, both conditions are fulfilled, leading to a significant polarization-dependent magnetic contribution to the intensity of multiplet lines in XAS. This effect is called X-ray magnetic linear dichroism (XMLD). Linearly polarized X-rays probe the angle α between the linear X-ray polarization vector $\vec{E}$ and the orientation of the magnetization axis $\vec{A}$. The angle and magnetization dependence of the XMLD intensity is then given by: $I_{\text{XLMD}} \sim (1 - 3\cos^2 \alpha)\langle M^2 \rangle_T$. Here $\langle M^2 \rangle_T$ is the statistical average of the squared local magnetization at the temperature T. Contributions linear in M vanish because of the compensated magnetic structure and the zero macroscopic magnetic moment of antiferromagnets. Antiferromagnets, therefore, do not show X-ray magnetic circular dichroism.

The two polarization-dependent absorption spectra in Fig. 2.1 were measured on an antiferromagnetically ordered $LaFeO_3$ thin film, grown on $SrTiO_3(001)$, with the linear X-ray polarization lying parallel to (black) and perpendicular to (gray) the sample surface. The spectra were acquired without spatial resolution and, thus, average over a large sample area. Analysis of the polarization and angular dependence of the XMLD effect in absorption spectra allows the determination of the average direction of the magnetic axis. It is known from multiplet calculations that a higher 2nd peak at the L_3 and the L_2 resonances indicates a more parallel orientation of the magnetic axis $\vec{A}$ and the X-ray polarization $\vec{E}$ in the three antiferromagnetic oxides α-Fe_2O_3 [25], NiO [26], and CoO. The magnetic Fe atom in $LaFeO_3$ sits in an identical local crystallographic environment as in α-Fe_2O_3 and, therefore, has the same multiplet absorption structure and a similar XMLD signature. The spectra show that $\vec{A}$ is closer, on average, to the out-of-plane polarization direction than to the in-plane polarization direction, because the second peak is higher in the out-of-plane spectrum (gray line). By rotating the sample around the surface normal, we can furthermore determine the macroscopic in-plane symmetry of the system, which is fourfold in $LaFeO_3(001)$ [38, 39].

2.2.3 X-Ray Magnetic Circular Dichroism (XMCD)

The interaction of circularly polarized X-rays with a ferromagnetically ordered sample, and its application for quantitative magnetization mapping with X-ray PEEM, are discussed in detail by W. Kuch (Chap. 1 of this book). Circularly polarized X-rays probe the direction of the atomic magnetic moment in a ferromagnet. The angle and magnetization dependence of X-ray magnetic circular dichroism in the total absorption signal is given by $I_{XMCD} \sim \cos\alpha \langle M \rangle_T$, with α denoting the angle between the X-ray helicity vector $\vec{\sigma}$ (parallel to the X-ray propagation direction) and the magnetization $\vec{M}$. Strong XMCD effects of opposite sign appear at the L_3 and L_2 $2p \rightarrow 3d$ resonances of the transition metal ferromagnets Fe, Co, and Ni.

2.2.4 Temperature Dependence of X-Ray Magnetic Dichroism

In contrast to ferromagnets, the spin direction in an antiferromagnet cannot be easily switched or rotated. The characteristic temperature dependence of the XMLD signal is therefore the most direct experimental proof that a feature in the absorption spectrum is of magnetic origin. Let us first consider the simpler case of XMCD. As a first order effect, the contribution of XMCD to the total absorption is directly proportional to the macroscopic magnetization $\langle M \rangle_T$. In mean field approximation the statistical average of the magnetization per atom in units of the electron moment $g\mu_B$ is given by $\langle M \rangle_T = JB_J(x)$, with J the spin moment of the magnetic d-shell and $x = H(M)/kT$. We have neglected the contribution of the orbital moment, which is usually strongly quenched in solids due to the crystal field. $H(M)$ is the molecular field, or Weiss field, generated by the magnetization M, and $B_J(x) = \frac{2J+1}{2J} \coth\left(\frac{2J+1}{2J}x\right) - \frac{1}{2J}\coth\left(\frac{1}{2J}x\right)$ is the Brillouin function. This implicit equation can be numerically evaluated as a function of T. The temperature dependence of XMLD is slightly more complex, because $\langle M^2 \rangle_T$ has to be evaluated [22, 26, 39]. A numerical solution for a given J is possible using the identity $\langle M^2 \rangle = J(J+1) - \langle M \rangle \coth(x/2)$. This formula can again be numerically evaluated using the mean field expression for $\langle M \rangle_T$ and $x = H(M)/kT$. We will later use this expression to describe the temperature dependence of the XMLD contrast in the antiferromagnet $LaFeO_3$.

2.2.5 Experiment

Photoemission electron microscopy utilizes local variations in electron emission, and, therefore, the local X-ray absorption, in order to generate image contrast. The electrons emitted from the sample are accelerated by a strong electric field (typically 15–20 kV) toward the electron optical column, which forms a magnified image of the local electron yield. PEEM is a parallel imaging technique, and the spatial resolution is solely determined by the resolution of the electron optics, while the intensity is proportional to the X-ray flux density. The electron optical column contains typically two or more electrostatic or magnetic electron lenses, corrector elements such as a stigmator and deflector, an angle-limiting aperture in the backfocal plane of one of the lenses, and a detector (a fluorescent screen or a multichannel plate detector that is imaged by a CCD camera). A schematic drawing of the PEEM-2 facility, located at the Advanced Light Source, is shown in Fig. 2.2.

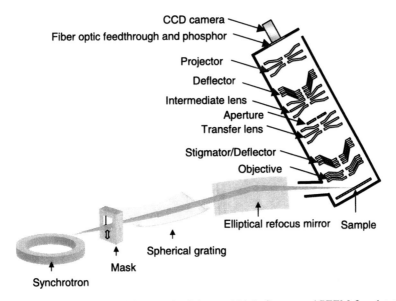

Fig. 2.2. Layout of beamline 7.3.1.1 at the Advanced Light Source and PEEM-2 endstation

An X-ray source providing radiation with tunable energy and variable polarization from left and right circular to linear is essential for the investigation of magnetic materials [29]. The PEEM-2 instrument uses bending magnet radiation. The polarization is selected by moving a mask vertically in the beam. Radiation in the plane of the storage ring is linearly polarized, while above and below the plane the radiation is right and left circularly polarized. The radiation is monochromatized using a spherical grating monochromator and focused onto the sample using an elliptical refocusing mirror. The sample is at high negative potential for this design, and electrons emitted from the sample are imaged using an all-electrostatic four-lens electron optical system [36]. A stigmator-deflector assembly in the backfocal plane of the objective lens and an additional deflector behind the intermediate lens correct for machining and for alignment errors of the lenses. The angle-defining aperture is located in the backfocal plane of the transfer lens.

The best spatial resolution with PEEM instruments is achieved using UV radiation with an excitation energy that just exceeds the work function. The narrow energy distribution of the threshold photoelectrons of 0.1–1 eV minimizes chromatic image errors. A resolution of below 10 nm, close to the theoretical limit of 5 nm, has been

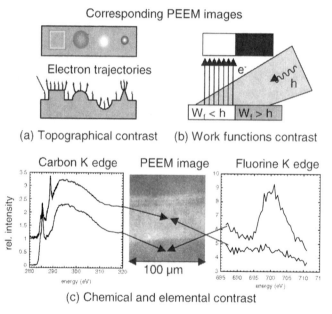

Fig. 2.3. PEEM contrast mechanisms. (**a**) Topographic contrast caused by surface features. (**b**) Work function contrast caused by work function changes that affect the electron yield intensity from different regions. (**c**) Examples for chemical and elemental contrast. The PEEM image in the middle, which was acquired at 280 eV (contrast is mostly topographical), shows a wear track on a computer hard disk lubricated with fluorocarbon [21]. To the left and right, local absorption spectra are shown. The additional peak in the carbon spectrum (288.5 eV) in the wear track indicates oxidation of the carbon. The fluorine spectra show removal of fluorine in the wear track. Both are signatures of lubricant degradation

achieved [40]. If X-rays are used as the excitation source, the emitted electrons are mainly secondary electrons with a substantially wider energy distribution of about 5 eV [41] that leads to a strong increase in chromatic aberrations and a deterioration of the spatial resolution. The highest spatial resolution obtained with X-rays today is in the 20 nm range [36, 42, 43], and for magnetic imaging more typically in the 50–100 nm range.

PEEM relies on several mechanisms for image contrast. Topographic contrast is caused by distortions of the accelerating electric field at topographic surface features. Local changes in work function, for example, caused by different materials, result in changes in the electron yield and, therefore, lead to image contrast. Work function contrast is particularly strong for excitation with UV, depending on whether the UV energy is above or below the work function of the illuminated sample area. These two contrast mechanisms are illustrated in Fig. 2.3a and b. The combination of PEEM and tunable synchrotron radiation adds elemental and chemical contrast by utilizing the sensitivity of the near edge absorption structure to the chemical environment of an atom. Elemental and chemical contrast are explained in Fig. 2.3c. Through the use of XMCD and XMLD magnetic dichroism techniques, it is possible to obtain ferro- and antiferromagnetic contrast, respectively.

2.3 Antiferromagnetic Structure of LaFeO$_3$ Thin Films

We will start our discussion of domain imaging of antiferromagnets with measurements on LaFeO$_3$(001) thin films [39, 44]. These are, to our knowledge, the first measurements showing the magnetic microstructure of an antiferromagnetic thin film. The thin LaFeO$_3$ films were grown by molecular beam epitaxy (MBE) on SrTiO$_3$(001) substrates and were characterized by X-ray diffraction and transmission electron microscopy (TEM). The orthorhombic LaFeO$_3$ layer forms $\sim$ 100-nm-sized crystallographic twins with orthogonal, in-plane c-axes along the cubic [100] and [010] axes of the substrate. The sample geometry is sketched in Fig. 2.4. PEEM images acquired at energies close to the Fe L$_3$ edge show a variation in image contrast, which is a result of the magnetic domain structure of the film. The images were acquired with the linear X-ray polarization vector $\vec{E}$ in the sample plane and parallel to the [100] direction of the cubic SrTiO$_3$ substrate. In this geometry, PEEM is sensitive to the angle between the in-plane projection of the magnetic axis $\vec{A}$ and the polarization $\vec{E}$.

The depicted images represent a subset of a series of images, an image stack, acquired at consecutive X-ray energies. An image stack contains data for a pixel-by-pixel calculation of local X-ray absorption spectra. Exemplary spectra from two regions are shown at the bottom of Fig. 2.4. Two representative regions of reversed contrast are marked in the images by gray and black bordering. The spectra were generated by plotting the local image intensity as a function of the energy of the incident X-rays. A comparison with the spatially averaged, angular resolved spectra in Fig. 2.1 reveals the magnetic origin of the intensity variation in the local PEEM spectra. The higher intensity of the higher-energy L$_2$ and L$_3$ multiplet lines in the dotted-line spectrum indicates a more parallel alignment of $\vec{A}$ and $\vec{E}$ in the gray-bordered area than in the black-bordered area. The PEEM images at energy positions 1 and 7 show no image contrast, demonstrating the chemical homogeneity (no elemental contrast above edge) and smoothness (no topographic contrast below edge) of the sample. However, at three particular energies, corresponding to the main Fe multiplet lines, an image contrast of alternating sign appears that is of magnetic origin (images 2, 4, and 6).

We usually apply a normalization procedure in order to suppress nonmagnetic contrast, such as elemental or topographical contrast, and to correct for the inhomogeneity in the sample illumination. We start with two PEEM images that exhibit strong but opposite XMLD contrast, e.g., images 2 and 4 in Fig. 2.4 or alternatively images A and B in Fig. 2.5, acquired at energies $E_A = 721.5$ eV and $E_B = 723.2$ eV. The approximately equal intensity of the multiplet lines A and B at the Fe L_2 edge usually improves the suppression of nonmagnetic contrast. An image exhibiting enhanced magnetic contrast is derived by calculating the ratio image B/A or the asymmetry image $(B - A)/(B + A)$. If the magnetic contribution is significantly smaller than the total image intensity, then ratio and asymmetry images are equivalent except for a linear transformation. This condition is usually fulfilled (see Fig. 2.4). We refer to the

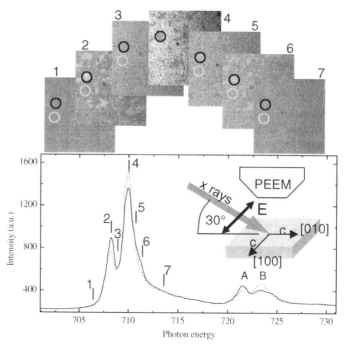

Fig. 2.4. PEEM images (top) and local XMLD spectra (bottom) of a thin LaFeO$_3$(001) film. The measurement geometry is shown in the inset. The photon energies at which PEEM images were acquired are marked by short vertical lines. Circles mark magnetic domains, which appear with opposite contrast in the images

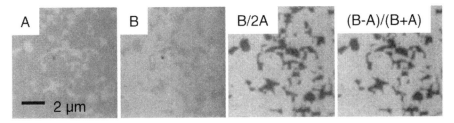

Fig. 2.5. PEEM images for LaFeO$_3$(001) acquired at the Fe L$_2$ edge at $E_A = 721.5$ eV and $E_B = 723.2$ eV. The magnetic contrast is enhanced in the ratio (B/A) and the asymmetry image $(B - A)/(B + A)$

ratio image as the *XMLD image*. Its contrast is a measure for the spatial variation of $\langle M^2 \rangle_T$ and for the local angle α between the magnetic axis and the X-ray polarization vector. On LaFeO$_3$ a higher intensity in the XMLD image indicates a more parallel alignment of $\vec{A}$ and $\vec{E}$ and a lower intensity a more perpendicular alignment. The regions of different brightness, therefore, represent magnetic domains with different orientations of the antiferromagnetic axis, assuming a constant moment.

The magnetic origin of the contrast can be verified by studying its temperature dependence (see above). A selection of images acquired at increasing temperatures is shown in Fig. 2.6. Approaching the Néel temperature of the film, which is 740 K in bulk LaFeO$_3$, the image contrast starts to disappear and, as expected for a magnetic effect, fully reappears after cooling down to room temperature. In Fig. 2.6, the decreasing XMLD contrast, referenced to 0 K, is displayed as a function of temperature. The match between the two temperature cycles and the preserved XMLD signal after the return to room temperature demonstrate the chemical stability of the sample upon heating. The continuous curve in Fig. 2.6 shows the mean field fit for $\langle M^2 \rangle_T$ using $J = 5/2$, appropriately shifted and scaled to fit the data. The curve apparently describes the data accurately, giving further evidence for the magnetic origin of the image contrast. Extrapolation of the curve to zero XMLD signal yields a magnetic transition temperature of 670 K, well below the bulk Néel temperature of 740 K, which is marked by an arrow in the plot. We attribute this reduced transition temperature to epitaxial strain in the film modifying the bonding angles and bonding distances, thereby affecting the strength of the Fe-O-Fe superexchange responsible for the magnetic interaction in the system. The absolute error in the temperature measurement is approximately 10 K, as checked by referencing two thermocouple pairs to each other.

The question arises why domains form in an antiferromagnet. Domain formation in antiferromagnets, as in ferromagnets, is expected to result from a process of energy minimization. The main contributors to the magnetic energy of a magnetic system are the exchange energy (which aligns neighboring moments), the dipolar energy (which favors flux closure), the spin-orbit energy, which gives rise to the magnetocrystalline anisotropy (and favors a spin orientation along an "easy" crystallographic axis), and a magneto-elastic term (which favors a certain spin orientation relative to

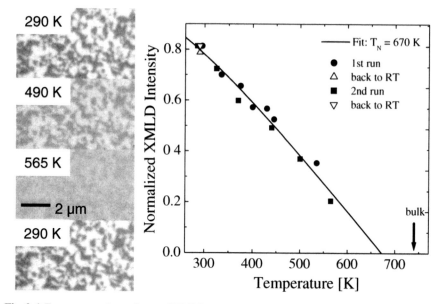

Fig. 2.6. Temperature dependence of XMLD contrast for LaFeO$_3$(001). The contrast disappears approaching the Néel temperature of LaFeO$_3$ (bulk: 740 K). The XMLD intensity measured in two temperature cycles is plotted as a function of temperature and fitted with the expression of in mean field theory using a reduced $T_N = 670$ K

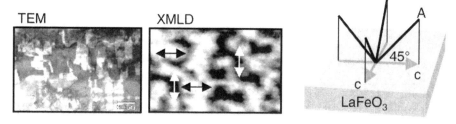

Fig. 2.7. Crystallographic and magnetic structure of LaFeO3(001), in comparison. The contrast in the TEM image (*left*) results from crystallographic twin domains with a horizontal (white domains) and vertical (black domains) c-axis. The XMLD contrast originates from antiferromagnetic domains with horizontal (*white*) and vertical (*black*) in-plane projection of A. Both images were obtained from the same sample, but from different regions, so no one-to-one correspondence exists. The directions of A in relation to the c-axes of the twin domains are shown on the *right*

the strain axis). Unlike in ferromagnets, dipolar effects play a negligible role in *bulk* antiferromagnets, because of their vanishing macroscopic magnetization. However, in two-dimensional antiferromagnets, one may have a competition between dipole-dipole and magnetic anisotropy energies, with the former favoring spin alignment perpendicular to the plane [45].

In practice, the antiferromagnetic domain configuration and the local orientation of the antiferromagnetic axis are typically determined by the local crystallographic structure, including defects, and the local crystal strain. Some domain configurations, while not representing global free-energy minima, may still be meta-stable [46]. A four-wall (four T domains) configuration in NiO is such a configuration [47]. Finally, in a perfect crystal, the lowering of the free energy accompanying an increase in entropy can lead to an equilibrium multidomain structure. An example are S-domain walls in NiO, where a multidomain configuration is thermodynamically favored [48] over a large temperature range [49].

The orthorhombic LaFeO$_3$ film, grown on cubic SrTiO$_3$(001), forms crystallographic twin domains exhibiting orthogonal, in-plane c-axes. These twins are responsible for the black and white contrast in the TEM image, shown on the left in Fig. 2.7. Comparison of the crystallographic structure with XMLD images of the magnetic structure obtained by PEEM (center) reveals a clear similarity in the patterns, which show structures of comparable shape and size. The TEM and XMLD images were acquired on the same sample, but at a different sample position. The correspondence between the patterns demonstrates that the magnetic domain configuration in LaFeO$_3$ is indeed dominated by crystallography. The directions of the antiferromagnetic axes $\bar{A}$ and the crystallographic c-axes of the twin domains, determined from TEM and XMLD microscopy in conjunction with X-ray diffraction and XMLD spectroscopy data, are shown on the right in Fig. 2.7.

2.4 Exchange Coupling at the Co/NiO(001) Interface

The magnetic structure at the interface of magnetic layers influences and controls their magnetic interaction and, therefore, has to be known in order to understand interface phenomena such as exchange bias. PEEM is capable of imaging the antiferromagnetic domain structure, as shown above. Polarization-dependent measurements, furthermore, allow us to quantify the local orientation of the spin directions in ferromagnets and antiferromagnets in three dimensions. In the following, we shall illustrate these capabilities by PEEM studies on 10–20 Å thick Co layers grown on the (001) surface of NiO single crystals.

NiO is often used as the AFM in practical exchange bias structures and, over the years, has served as a valuable model antiferromagnet. We have chosen it for the following reasons:

- The antiferromagnetic bulk structure of NiO has been well established since the 1960s from extensive neutron and optical studies [10, 11, 13, 50].
- Recent studies on thin films have quantitatively established the angle and temperature dependence of the NiO-XMLD effect [26].
- The AFM-ordered (001) surface can be easily prepared by cleaving single crystals.

NiO grows in a rocksalt structure. The Ni spins are ferromagnetically aligned within {111} sheets that are then antiferromagnetically stacked parallel to a ⟨111⟩ axis, creating a completely compensated bulk structure. Each area of the crystal, exhibiting one of the four possible ⟨111⟩ stacking directions – [111], [–111], [1–11], [11–1] – is called a T(win)-domain. In each T-domain the easy spin axis can be one of three different ⟨11–2⟩ directions in the ferromagnetically ordered plane. These domains within a T-domain are called (S)pin-domains. We have studied the (001) surface, which is a natural cleavage plane of NiO and corresponds to a plane of compensated spins in the bulk.

2.4.1 Angular Dependence of Domain Contrast in NiO(001)

The angular dependence of the domain contrast provides information about the magnetic symmetry of the surface. Hillebrecht et al. studied the (001) surface of NiO single crystals at about 1 μm spatial resolution using a commercial PEEM at the new BESSY2 synchrotron facility [51]. The results are shown in Fig. 2.8. The XMLD (asymmetry) image was derived from two PEEM images acquired at the main multiplet lines of the NiO L_2 edge, with linear polarization parallel to the sample plane. The black and white stripes in the XMLD image are several μm wide and several 100 μm long and represent two T-domains that are divided by domain walls along the in-plane [100] direction. The image contrast disappears above the bulk Néel temperature of NiO (525 K), confirming its magnetic origin. Several XMLD images were acquired at different azimuthal sample orientations ϕ varying the angle between the X-ray polarization and the in-plane [100] direction of the sample. The variation in the intensity difference or contrast between the black and white domains in these images shows a $\sin(2\phi)$ dependence, which is easily explained by the $\cos^2 \phi$ dependence of the XMLD intensity within each domain. From the symmetry of the curve with respect to $\phi = 0$, the authors deduced that the antiferromagnetic structure of the studied NiO(001) surface is symmetric with respect to the (100) domain walls. Furthermore, the amplitude of the curve is a measure of the angle between the magnetic axes.

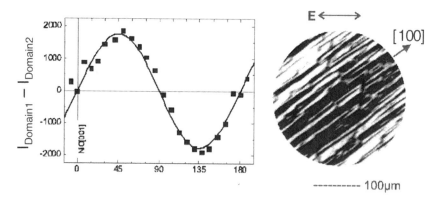

Fig. 2.8. Antiferromagnetic domain pattern on NiO(001) acquired with linear polarization. Black and white areas represent different antiferromagnetic domains, separated by walls parallel to [100]. The intensity difference between black and white domains changes size and sign while rotating the sample around its surface normal, showing a $\sin(2\phi)$ dependence, where ϕ is the angle between the electric field vector and the in-plane [100] direction [51]

2.4.2 Polarization Dependence of Domain Contrast

In the following, we will describe a procedure for the quantitative determination of the orientation of the antiferromagnetic axis within a single domain using spatially resolved spectroscopy (image stacks). This procedure is applicable if the maximum or minimum XMLD effect in an X-ray absorption spectrum of the studied system is known. Since the XMLD effect is superimposed on an isotropic background, which is not known a priori, the experimental spectra need to be compared to reference measurements acquired at perpendicular and parallel alignments of the magnetic axis and the X-ray polarization.

Since the crystallographic symmetry of NiO is known, it is sufficient to search for optimum XMLD contrast with the linear polarization vector aligned with the in-plane projection of the three main symmetry axes of NiO: [100], [110], and [121]. We acquired images and spectra of the same sample area in these three measurement geometries by using linear and "plane" polarization. "Plane" polarization is used synonymously with "circular polarization of either helicity". At the ALS PEEM-2 facility, the degree of circular polarization is about 80%. For plane-polarized radiation, one component of the electric field vector lies in the plane of the sample, and the other lies in the plane of X-ray incidence, tilted by 30° from the surface normal. The second component gives access to the out-of-plane spin component. Note that plane-polarized light can be described by two independent orthogonal electric field vector components. Their relative phase is of no importance in XMLD.

High spatial resolution images for NiO(001), measured with the ALS PEEM-2, are shown in Fig. 2.9. The two images were acquired with linearly (left image) and with plane- (right image) polarized light. The XMLD image acquired with linear polarization shows a distribution of in-plane AFM axes similar to Fig. 2.8, but with improved spatial resolution. Using plane polarization, the out-of-plane component of the spins is revealed. Each antiferromagnetic in-plane domain consists of two classes of domains with different out-of-plane components separated by walls along ⟨110⟩ directions. Because the size of the local XMLD signal is more accurately determined from spectra than from images obtained at a single energy, we acquired image stacks with linear and plane polarization for three different azimuthal sample orientations with the in-plane polarization component parallel to [100], [110], and [120]. The angular dependence of the L_2 XMLD effect then allows the determination of the in-plane antiferromagnetic axis orientation in each domain. In addition, the relative size of the XMLD effect measured with plane and linear polarization is a measure of the inclination angle of the magnetic axis. By comparing these results with spectroscopic measurements by Alders et al. [26] on NiO(001) thin films, we can determine the orientation of the spin axis within each single domain with an accuracy of about $+/-7$ degrees The in-plane component of the orientation of the antiferromagnetic axes is indicated in Fig. 2.9. All four possible T-domains are present, forming a so-called "4-wall" structure. Each T-domain consists of a single ⟨121⟩ S-domain in a configuration that minimizes the magnetic energy of the system. The observed S-domains have their antiferromagnetic axis canted by 35.3° relative to the (001) surface plane. We do not observe any S-domains with a tilt angle of 65.9° from the (001) plane,

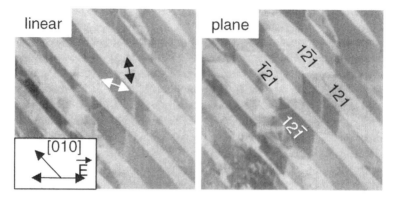

Fig. 2.9. Images of antiferromagnetic domains on NiO(001) acquired with linear (*left*) and plane (*right*) polarization. Using linearly polarized X-rays, two classes of antiferromagnetic domains with different in-plane components of the magnetization are distinguished. Using plane (circularly) polarized X-rays, PEEM becomes sensitive to the out-of-plane component of the magnetic axis, distinguishing four antiferromagnetic T-domains that form a four-wall structure. Arrows and cubic lattice vectors describe the local orientation of the antiferromagnetic axis, derived from a detailed contrast analysis

as are present in the bulk. The breaking of symmetry at the surface, therefore, only supports a subset of all possible bulk antiferromagnetic orientations [10, 11, 13, 50]. Our results demonstrate the capability of polarization and angular-dependent PEEM to determine the orientation of the antiferromagnetic axis at the sample surface in all three dimensions.

2.4.3 Coupling Between Co and NiO–AFM Reorientation

The domain structure of adjacent magnetic layers provides information about the magnetic coupling effects occurring across the interface. We have, therefore, studied the relation of the domain configurations of a 1.5-nm-thick ferromagnetic Co film grown by e-beam evaporation on a NiO(001) substrate [52]. Here, we take advantage of the elemental specificity of X-ray PEEM, which allows the separate investigation of the magnetic properties of the NiO substrate and the Co layer. Taking a close look at the XMLD domain image in Fig. 2.10 (bottom left), which was acquired using plane polarization, we notice that only two AFM domains of the originally four T-domains (top left, before deposition) remain after Co deposition. From this we can conclude that the distribution of the out-of-plane component is now the same in at least two of the four T-domains. A detailed analysis of polarization and angle-dependent images and spectra reveals that the magnetic axis rotates by 35.3° in the (111) plane until it is parallel to either the [110] or [1–10] axis in the (001) surface plane.

The ferromagnetic domain pattern of Co, shown on the right in Fig. 2.10, clearly resembles the antiferromagnetic pattern at the NiO surface. However, we observe four gray levels (black-dark gray-light gray-white), instead of two (darkgray-light gray), because XMCD distinguishes between ferromagnetic spins that have a projection along (lighter) or opposite to (darker) the photon propagation direction (helicity). The domains (gray) that are nearly perpendicular to the X-ray helicity can also be distinguished in our slightly rotated measurement geometry. Arrows in Fig. 2.10 display the local orientations of the antiferromagnetic axes (double arrows) and the ferromagnetic spin directions (arrows). We observe that the Co and NiO spins are coupled parallel, domain by domain, indicating significant exchange coupling at the interface. This coupling is responsible for a strong, induced in-plane anisotropy of the ferromagnetic layer parallel to the ⟨110⟩ antiferromagnetic spin axes.

The importance of these results is twofold: The rotation of the antiferromagnetic axis after Co deposition signifies that the interfacial Co/NiO structure strongly deviates from the NiO bulk structure, emphasizing the need for surface- and interface-sensitive techniques complementing bulk measurements in order to correctly model the interfacial coupling between ferromagnets and antiferromagnets and to conclu-

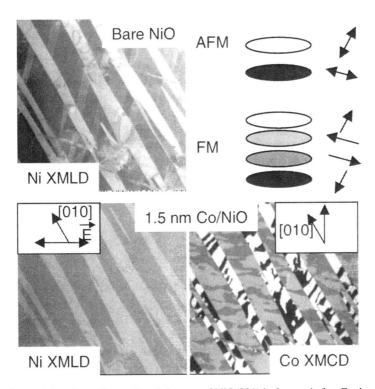

Fig. 2.10. The left column shows XMLD images of NiO(001) before and after Co deposition, recorded with plane polarized X-rays. The original zig-zag pattern of the bare surface visible with plane polarization completely disappears upon Co deposition. The spin axis in the antiferromagnet rotates from [121] toward [110]. Ferromagnetic spins align parallel, domain by domain, to the adjacent antiferromagnetic spins, as indicated in the right image acquired using Co XMCD. Each antiferromagnetic stripe domain breaks up into two ferromagnetic domains with opposite spin directions

sively solve the problem of exchange bias. Furthermore, the measurement reveals that the *parallel* configuration of ferromagnetic and antiferromagnetic spins is the lowest energy state of the system Co/NiO (001). This information is crucial for exchange bias models.

2.4.4 Interfacial Spin Polarization in Co/NiO(001)

After observation of the domain-by-domain coupling across the NiO/Co antiferromagnet-ferromagnet interface, the question arises as to the origin of the coupling. After all, the NiO is supposed to be magnetically neutral! In order to explain exchange coupling and bias, it has long been speculated that the antiferromagnetic interface must contain "uncompensated spins". As described in the Introduction, previous models for their origin involved statistical arguments associated with the termination of bulk antiferromagnetic domains [9], spin-flop canting of antiferromagnetic spins [17], and various defect-based explanations [18, 19]. Here we will discuss PEEM experiments that suggest an altogether different origin of the interfacial spins, namely, the formation of an interfacial layer through chemical reaction. The interfacial layer is shown to create uncompensated spins and to mediate the coupling between the layers [53].

As pointed out in the introduction, the determination of an interface-specific signal is quite challenging, in general. One anticipates only a small concentration of uncompensated interfacial spins of about one monolayer or less. Their signature is superimposed on the large background of the bulk NiO and buried beneath the ferromagnetic Co layer. Our microscopy experiments are aided by recent X-ray absorption experiments on ferromagnet/NiO systems, which revealed chemical oxidation and reduction processes across the interface, leading to the formation of a thin, chemically reacted $CoNiO_x$ layer [37]. We can take advantage of the observed chemical reduction of the NiO layer by tuning the X-ray energy to the maximum of the XMCD effect of Ni metal, thus increasing the sensitivity of PEEM to the interface layer. The moderate surface sensitivity of PEEM, detecting secondary electrons, enhances the strength of the interface signal while retaining the ability to detect a signal from the buried layer.

Our results for the Co/NiO(001) system are shown in Fig. 2.11. Here, the XMCD PEEM image of the ferromagnetic Co film is compared to an XMCD image recorded at the Ni edge. The Ni image was acquired at the metal resonance positions, which are shifted by -0.4 eV compared with the oxide. The Ni XMCD image shows the magnetic structure of *uncompensated* Ni spins that are created at the interface by a chemical reaction of Co with NiO. The images demonstrate a clear correlation of the spin structure of the Co film with that of the $NiCoO_x$ interface layer. This layer is no longer antiferromagnetic, but instead is ferri- or ferromagnetic. The spin contrast in the Ni XMCD image is superimposed on an antiferromagnetic background, caused by the XMLD effect in the antiferromagnet. The thickness of the ferromagnetic layer amounts to 0.5 Å for an as-grown film and increases up to 6 Å after annealing at 600 K, giving rise to an XMCD contrast of 0.3 to 4%. These measurements demonstrate the high sensitivity that can be reached with PEEM, making use of the unique chemical,

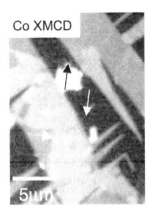 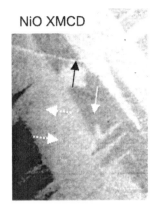

Fig. 2.11. The ferromagnetic domain pattern of Co and the NiO interface are shown using XMCD. Arrows indicate the local direction of the magnetization. The interfacial Ni spin polarization is found to be an exact replica of the ferromagnetic Co domain pattern

elemental, and magnetic specificity of this technique in combination with the surface sensitivity afforded by secondary electron detection. We learn that the chemical reaction at the boundary between an antiferromagnetic oxide and a ferromagnetic metal results in the coupling of the magnetic structures on both sides of the boundary. This first microscopic observation of uncompensated interface spins proves that these spins indeed play an important role in the development of exchange coupling and exchange bias in ferromagnet-antiferromagnet systems.

2.5 Summary

We have demonstrated the power of PEEM to reveal the detailed spin structure near antiferromagnet-ferromagnet interfaces. This capability arises from a unique combination of lateral resolution coupled with elemental, chemical, magnetic, and depth sensitivity. Using this approach, we can pick apart complex magnetic multilayers, one layer at a time, and obtain unprecedented new information on the spin structure near interfaces. The solution of a nearly 50-year-old problem, the microscopic origin of exchange anisotropy, has come within our reach.

The spatial resolution of the PEEM technique can be significantly improved by compensation of the chromatic and spherical image aberrations introduced mostly by the accelerating field and the objective lens. This can be achieved by the introduction of an aberration-correcting electrostatic mirror. Theoretical estimates predict a limit for the spatial resolution of about 1–2 nm [54, 55], which is sufficient for the investigation of typical polycrystalline materials. Two groups, one in the U.S. and one in Germany, are currently designing aberration-corrected instruments that have a design resolution of about 2 nm. These new instruments will be located at BESSY2 (SMART) [54, 55] and the ALS (PEEM-3) [56].

Complementary to the development of higher resolution instruments is the development of higher brightness X-ray sources that provide higher flux and allow tighter focusing of the X-rays. Helical undulators at modern synchrotron facilities are an optimal source for magnetic studies because of their high brightness and control from circular to vertical and horizontal linear polarization. We envision that the combination of ultrahigh resolution PEEM with brilliant insertion device sources will allow access to the full range of nanoscale science, allowing the study of single magnetic grains, clusters, and ultrasmall magnetic patterns. Furthermore, the 50 picosecond X-ray pulse width at modern synchrotron facilities will allow one to explore a time regime in magnetic switching of nanostructures that is presently still a technological dream.

References

1. J.B. Kortright, D.D. Awschalom, J. Stöhr, S.D. Bader, Y.U. Idzerda, S.S.P. Parkin, I.K. Schuller, and H.C. Siegmann, J. Magn. Magn. Mater. **207**, 7 (1999).
2. P.A. Grünberg, Sensors and Actuators A **91**, 153 (2001).
3. G. Prinz, Science **282**, 1660 (1998), G. Prinz, J. Magn. Magn. Mater. **200**, 57 (1999).
4. R. Fiederling, M. Keim, G. Reuscher, W. Ossau, G. Schmidt, A. Waag, and L.W. Molenkamp, Nature **402**, 787 (1999).
5. H. Ohno, Science **281**, 951 (5379).
6. J. Nogues and I.K. Schuller, J. Magn. Magn. Mater. **192**, 203 (1999).
7. A.E. Berkowitz and K.J. Takano, Magn. Magn. Mater. **200**, 552 (1999).
8. W.H. Meiklejohn and C.P. Bean, Phys. Rev. **105**, 904 (1956).
9. K. Takano, R.H. Kodama, A.E. Berkowitz, W. Cao, and G. Thomas, Phys. Rev. Lett. **79**, 1130 (1997).
10. W.L. Roth, J. Appl. Phys. **31**, 2000 (1960).
11. T. Yamada, S. Saito, and Y. Shimomura, J. Phys. Soc. Jpn. **21**, 672 (1966).
12. J. Baruchel, Physica B **192**, 79 (1993).
13. S. Saito, M. Miura, and K. Kurosawa, Journal of Physics. C **13**, 1513 (1980).
14. J. Stöhr, A. Scholl, T.J. Regan, S. Anders, J. Lüning, M.R. Scheinfein, H.A. Padmore, and R.L. White, Phys. Rev. Lett. **83**, 1862 (1999).
15. M. Fiebig, D. Frölich, G. Sluyterman v.L., R.V. Pisarev, Appl. Phys. Lett. **66**, 2906 (1995)
16. M. Fiebig, D. Fröhlich, S. Leute, and R.V. Pisarev, Appl. Phys. B **66**, 265 (1998).
17. N.C. Koon, Phys. Rev. Lett. **78**, 4865 (1997).
18. T.C. Schulthess and W.H. Butler, Phys. Rev. Lett. **81**, 4516 (1998).
19. P. Miltenyi, M. Gierlings, J. Keller, B. Beschoten, G. Güntherodt, U. Nowak, and K.D. Usadel, Phys. Rev. Lett. **84**, 4224 (2000).
20. J. Stöhr, H.A. Padmore, S. Anders, T. Stammler, and M.R. Scheinfein, Surf. Lett. Rev. **5**, 1297 (1998).
21. J. Stöhr and S. Anders, IBM J. Res. Develop. **44**, 535 (2000).
22. B.T. Thole, G. van der Laan, and G.A. Sawatzky, Phys. Rev. Lett. **55**, 2086 (1985).
23. G. van der Laan, B.T. Thole, G.A. Sawatzky, J.B. Goedkoop, J.C. Fuggle, J.-M. Esteva, R. Karnatak, J.P. Remeika, and H.A. Dabkowska, Phys. Rev. B **34**, 6529 (1986).
24. P. Carra, H. Konig, B.T. Thole, and M. Altarelli, Physica B **192**, 182 (1993).
25. P. Kuiper, B.G. Searle, P. Rudolf, L.H. Tjeng, and C.T. Chen, Phys. Rev. Lett. **70**, 1549 (1993).

26. D. Alders, L.H. Tjeng, F.C. Voogt, T. Hibma, G.A. Sawatzky, C.T. Chen, J. Vogel, M. Sacchi, and S. Iacobucci, Phys. Rev. B **57**, 11623 (1998).
27. G. Schütz, W. Wagner, W. Wilhelm, P. Kienle, R. Zeller, R. Frahm, and G. Materlik, Phys. Rev. Lett. **58**, 737 (1987).
28. C.T. Chen, F. Sette, Y. Ma, and S. Modesti, Phys. Rev. B **42**, 7262 (1990).
29. J. Stöhr and R. Nakajima, IBM J. Res. Develop. **42**, 73 (1998).
30. R. Nakajima, J. Stöhr, and Y.U. Idzerda, Phys. Rev. B **59**, 6421 (1999).
31. J. Stöhr, Y. Wu, B.D. Hermsmeier, M.G. Samant, G.R. Harp, S. Koranda, D. Dunham, and B.P. Tonner, Science **259**, 658 (1993).
32. J. Stöhr, "NEXAFS Spectroscopy", Springer Series in Surface Sciences 25, Springer, Berlin (1992).
33. J. Stöhr, J. Magn. Magn. Mat. **200**, 470 (1999).
34. P. Fischer, T. Eimüller, G. Schütz, G. Denbeaux, A. Pearson, L. Johnson, D. Attwood, S. Tsunashima, M. Kumazawa, N. Takagi, M. Kohler, and G. Bayreuther, Rev. Sci. Instrum. **72**, 2322 (2001).
35. D. Attwood, "Soft X-Rays and Extreme Ultraviolet Radiation", Cambridge University Press (2000).
36. S. Anders, H.A. Padmore, R.M. Duarte, T. Renner, Th. Stammler, A. Scholl, M.R. Scheinfein, J. Stöhr, L. Séve, and B. Sinkovic, Rev. Sci. Instrum. **70**, 3973 (1999).
37. T.J. Regan, H. Ohldag, C. Stamm, F. Nolting, J. Lüning, J. Stöhr, and R.L. White, Phys. Rev. B **64**, 214422 (2001).
38. J. Lüning, F. Nolting, H. Ohldag, A. Scholl, E.E. Fullerton, M. Toney, J.W. Seo, J. Fompeyrine, H. Siegwart, J.-P. Locquet, and J. Stöhr, submitted to Phys. Rev. B (2002)
39. A. Scholl, J. Stöhr, J. Lüning, J.W. Seo, J. Fompeyrine, H. Siegwart, J.-P. Locquet, F. Nolting, S. Anders, E.E. Fullerton, M.R. Scheinfein, and H.A. Padmore, Science **287**, 1014 (2000).
40. G. Rempfer and O.H. Griffith, Ultramicroscopy **27**, 273 (1989).
41. B.L. Henke, J.A. Smith, and D.A. Attwood, J. Appl. Phys. **48**, 1852 (1977).
42. E. Bauer, in "Chemical, Structural, and Electronic Analysis of Heterogeneous Surfaces on Nanometer Scale," R. Rosei, Ed., Kluwer Acad. Publ., Dortrecht (1997)
43. G. De Stasio, L. Perfetti, B. Gilbert, O. Fauchoux, M. Capozi, P. Perfetti, G. Margaritondo, and B.P. Tonner, Rev. Sci. Instrum. **70**, 1740 (1999).
44. F. Nolting, A. Scholl, J. Stöhr, J.W. Seo, J. Fompeyrine, H. Siegwart, J.-P. Locquet, S. Anders, J. Lüning, E.E. Fullerton, M.F. Toney, M.R. Scheinfein, and H.A. Padmore, Nature **405**, 767 (2000).
45. D.S. Deng, X.F. Jin and R. Tao, Phys. Rev. B **65**, 132406 (2002).
46. I.M. Lifshitz, Sov. Phys. JETP **15**, 939 (1962).
47. M.M. Farztdinov, Sov. Phys. Usp. **7**, 855 (1965).
48. Y.Y. Li, Phys. Rev. **101**, 1450 (1956).
49. M.M. Farztdinov, Phys. Met. Metall. **19**, 10 (1965).
50. G.A. Slack, J. Appl. Phys. **31**, 1571 (1960).
51. F.U. Hillebrecht, H. Ohldag, N.B. Weber, C. Bethke, U. Mick, M. Weiss, and J. Bahrdt, Phys. Rev. Lett. **86**, 3419 (2001).
52. H. Ohldag, A. Scholl, F. Nolting, S. Anders, F.U. Hillebrecht, and J. Stöhr, Phys. Rev. Lett. **86**, 2878 (2001).
53. H. Ohldag, T.J. Regan, J. Stöhr, A. Scholl, F. Nolting, J. Lüning, C. Stamm, S. Anders, and R.L. White, Phys. Rev. Lett. **87**, 247201 (2001).
54. R. Fink, M.R. Weiss, E. Umbach, D. Preikszas, H. Rose, R. Spehr, P. Hartel, W. Engel, R. Degenhardt, R. Wichtendahl, H. Kuhlenbeck, W. Erlebach, K. Ihmann, R. Schlogl, H.J. Freund, A.M. Bradshaw, G. Lilienkamp, T. Schmidt, E. Bauer, and G. Benner, J. Electron Spectrosc. Relat. Phenom. **84**, 231 (1997).

55. R. Wichtendahl, R. Fink, H. Kuhlenbeck, D. Preikszas, H. Rose, R. Spehr, P. Hartel, W. Engel, R. Schlogl, H.J. Freund, A.M. Bradshaw, G. Lilienkamp, E. Bauer, T. Schmidt, G. Benner, and E. Umbach, Surf. Rev. Lett. **5**, 1249 (1998).
56. J. Feng, H.A. Padmore, D.H. Wei, S. Anders, Y. Wu, A. Scholl, and D. Robin, Rev. Sci. Instrum. **73**, 1514 (2002).

3

Time Domain Optical Imaging of Ferromagnetodynamics

B.C. Choi and M.R. Freeman

Understanding the magnetic properties of small magnetic elements has become a major challenge in fundamental physics, involving both static and dynamic properties. The importance to technological applications such as high-density magnetic storage is also clear. The investigation of such small magnetic structures relies increasingly on magnetic imaging techniques, since the relevant properties can vary over length scales from micrometers to nanometers. This chapter describes an experimental method for measuring the fast magnetic phenomena in lithographically fabricated magnetic elements in the picosecond temporal regime and with sub-micrometer spatial resolution. The method employs stroboscopic scanning microscopy and is capable of measuring simultaneously all three components of the magnetization vector. This technique allows direct insight into the spatiotemporal evolution of dynamic processes, including ferromagnetic resonance and magnetization reversal in small magnetic elements.

3.1 Introduction

Magnetic phenomena on short time scales are different in many aspects from the static case [1, 2]. Moreover, the fast magnetic phenomena in small patterned elements have little relation to that in continuous films, due to the magnetostatics of element edges, which modify the equilibrium states of the elements in terms of the magnetic moment distribution [3–5]. From an application point of view, it has become crucial to understand reversal dynamics on fast (in the nano- and picosecond range) time scales in magnetic elements with dimensions in the micrometer size regime and below. This is owing to the increasing demands on conventional storage technologies and for newer approaches such as magnetic random access memories (MRAM) [6, 7]. Motivated by all of these accumulated interests, magnetization dynamics in micro- and nano-sized magnets are being actively studied by a number of groups [8–15].

In order to elucidate magnetization dynamics, direct observations of the magnetization configuration during reversal processes are most desirable. Imaging of micromagnetic domain structures has been carried out, for example, by magnetic force microscopy (MFM) [16], Lorentz transmission electron microscopy [17] and scanning electron microscope with polarization analysis (SEMPA) [18, 19], in addition to magneto-optic microscopy [3]. These techniques provide a good spatial resolution, but are generally focused on quasi-static magnetic imaging. For the study of magnetization dynamics, it has been demonstrated that very high spatiotemporal resolution can be achieved by employing stroboscopic scanning Kerr microscopy with pulse excitation [20–23]. This technique is a powerful tool for dynamic micromagnetic imaging and is the topic of this chapter.

3.1.1 Historical Background of Time-Resolved Techniques

The development of proper tools to study fast magnetic phenomena has a long history. In this research field, there have been two widely used experimental techniques; i.e., inductive and imaging methods. The former is to infer the magnetic flux changes by measuring the induced voltage in loops around the sample, whereas the latter is to observe the magnetization configuration using the magneto-optic Kerr effect. Kerr microscopy gives a more detailed picture of the magnetization configuration than does the use of pickup loops. The imaging technique, however, has not been extensively used for dynamics studies due to the complex instrumental requirements. A brief description of the development of these two competitive techniques is given below.

In the early 1960s, Dietrich and his coworkers measured magnetization switching speeds of 1 ns in Permalloy ($Ni_{80}Fe_{20}$) films using an inductive technique [24]. This approach used the discharge of a 50 V charge line through a coaxial mercury relay as a pulsed current and, therefore, pulsed magnetic field source. Strip transmission lines were used to deliver the magnetic pulse to the sample, and pickup coils were positioned around the transmission lines to detect the inductive signal from the changing magnetization in the sample. In their fast-switching experiments, free oscillations in the inductive signal, indicative of underdamped magnetic precession, were observed when magnetic pulses were applied transverse to the easy axis of the film. Since Dietrich's experiment, the inductive technique has been improved in many ways [25, 26]; a time-domain transmission geometry is used in order to avoid the need for inductive pickup coils, lithographically patterned coplanar waveguides are used to deliver magnetic pulses to the samples, modern pulse sources are used to create much faster pulses than can be obtained from mercury reed relays, and digital signal processing is employed to improve signal-to-noise ratios. With such improvements in the performance, impulse- and step-response experiments were recently carried out on a 50-nm-thick $Ni_{80}Fe_{20}$ element (1 mm × 50 µm) to infer that free magnetization oscillations occurred with rotation times as short as 200 ps [26].

Contemporary inductive and related electronic approaches exploit magnetic thin film devices for characterizing the high-speed dynamics [15, 27]. Koch et al. [15] probed the magnetization dynamics in thin film elements using spin-dependent tunnel junctions, in which a change in the tunneling resistance between two magnetic layers, i.e., a pinned (Co) and a free ($Ni_{60}Fe_{40}$) magnetic layer separated by an Al_2O_3 tunnel junction, was measured. Using this technique, the time required to reverse the magnetization direction of the soft ferromagnetic $Ni_{60}Fe_{40}$ layer is measured from greater than 10 ns to less than 500 ps as the amplitude of the applied magnetic field pulse is increased up to 8 kA/m. The optically induced modulation experiment is also of interest, which is carried out in exchange biased ferromagnetic/antiferromagnetic (FM/AF) bilayer ($Ni_{81}Fe_{19}$/NiO) structures [22]. In this experiment, large modulation in the unidirectional exchange bias field is induced via photoexcitation of the FM/AF interface with sub-picosecond laser pulses. Consequently, the unpinning of the exchange bias leads to coherent magnetization rotation in the permalloy film on a time scale of 100 psec.

While the inductive techniques described above are being actively pursued, all of these methods suffer from drawbacks that are not shared by the time-resolved imaging technique capable of giving a picture of the magnetization configuration during a magnetization reversal. The first high bandwidth spatially resolved optical experiments also date back to the 1960s, when J.F. Dillon and coworkers "saw" ferromagnetic resonance in a sample of 45-μm-thick $CrBr_3$ using microwave optical technique [28]. In the late 1960s, Kryder et al. first presented spatially and temporally resolved dynamic magnetization configuration in magnetic thin films with "a nanosecond Kerr magneto-optic camera" [29]. This Kerr photo-apparatus enabled one to capture magnetization reversal process in $Ni_{83}Fe_{17}$ films with a 10 μm spatial resolution and a 10 ns temporal resolution. This spatiotemporal resolution was an enormous achievement even by today's standard.

The time-resolved imaging technique, however, has not been used extensively at higher speeds because the instrumentation required is very complex. Revival of interest in fast imaging techniques began to take place in the mid-1980s. Kasiraj et al. reported magnetic domain imaging with a scanning Kerr effect microscope [30], in which a digital imaging technique was used with a spatial resolution of less than 0.5 μm. This allowed the observation of the nucleation and growth of magnetic domains in the pole tip region of a recording head. In this implementation, a time resolution of 50 ns was achieved. Subsequently, ultrafast (in the picosecond range) optical technique was developed by Freeman in 1991 [31]. In this work picosecond-scale magnetic field pulses were launched from a photoconductive switch and applied to samples, which were then probed by time-delayed optical pulses, yielding temporal information in a conventional stroboscopic manner. This pioneering work extended so-called "pump-and-probe" technique to the field of magnetics, providing an experimental tool for the study of picosecond time-resolved magnetization reversal dynamics. Since then, a number of spatiotemporally resolved experiments to directly measure the dynamical evolution of magnetization configuration have been reported, in particular, in microstructured elements [13, 20, 23, 32, 33]. During recent years, a similar experimental method has been used by several groups like Acremann et al. [14] and Hicken et al. [34].

Recently, spatiotemporal magneto-optic imaging technique was combined with second-harmonic magneto-optic Kerr effect (SHMOKE), whereby a magnetic sample is illuminated with light at frequency f and generates light at $2f$ [21]. This new method is a good complement to linear magneto-optics, since SHMOKE offers a unique feature that shows an extreme sensitivity to the magnetization at surfaces and interfaces [35]. The temporal images, however, do not provide a good spatial resolution yet, due to the low second-harmonic ($2f$) signal [36].

A further prospect in the field of magnetization dynamics is a time-resolved technique based on X-ray magnetic circular dichroism (XMCD) [37, Chap. 1 and 2 of this book]. This technique provides element selectivity, which is obtained as the X-ray energy is tuned to the adsorption edge of the desired element and is useful for studying separately each magnetic layer in multilayers like a spin valve system. Recent efforts to use pulsed X-ray beam from synchrotron radiation for the study of time-resolved and element-selective magnetization dynamics have been fruitful [38, 39].

3.2 Instrumentation

This section introduces the magneto-optic Kerr effect and time-resolved magnetic imaging technique for measuring high-speed magnetic phenomena. Such measurements on very short time scales (e.g., picosecond regime) involve many techniques such as the generation of fast magnetic pulses, precise control of time intervals between the pump and probe beam, and detection of magnetic signals by means of high frequency techniques.

3.2.1 Physical Principle of Magneto-Optic Effect

A simple way to dynamically probe the magnetization is by means of interactions of light with a magnetic medium. When linearly polarized light is reflected from a magnetic surface, the incident light is transformed into elliptically polarized light [3]. Thus, the final state of polarization can be characterized by both a rotation of the major axis θ_K and an ellipticity δ_K defined as the ratio between the minor and major axes. Both θ_K and δ_K are proportional to the magnetization of the material (however, not to the applied magnetic field). This effect is known as the magneto-optic Kerr effect (MOKE) [40]. If light transmission through the magnetic film is considered instead of reflection, then it is termed the Faraday effect. Faraday rotations of $\sim 10^{-3}$–10^{-2} degrees/Å are reported for iron, while the Kerr rotation is on the order of $\sim 10^{-4}$–10^{-3} degrees per monolayer [41].

Microscopically, it is known that the magneto-optic effect originates from the coupling of the electric field of the light and the magnetization of the spin system occurring through the light-induced electronic dipole transitions associated with spin-orbit interaction [42]. Macroscopically, the magneto-optic effect can be described by the inequivalent interaction of polarized light with the magnetization of the material by the two oppositely handed circularly polarized components of the incident light, since the right and left circularly polarized light propagates in the magnetic material with different velocities $v_\pm = c/N_\pm$. The signs $+$ and $-$ refer to the right and left circularly polarized light, respectively. $N_\pm$ are the complex refractive indices and are written as $N_\pm = n_\pm - ik_\pm$, where $n_\pm$ are refractive indices and $k_\pm$ the extinction coefficients that are related to the optical absorption. For cubic materials where the incidence angle of light and the direction of the magnetization are arbitrary, the complex refractive index is related to the dielectric tensor $\bar{\bar{\varepsilon}}$ as follows [40]:

$$\overline{\overline{\varepsilon}} = N^2 \begin{pmatrix} 1 & iQm_3 & -iQm_2 \\ -iQm_3 & 1 & iQm_1 \\ iQm_2 & -iQm_1 & 1 \end{pmatrix}, \quad (3.1)$$

where **m** is the magnetization unit vector and Q the material-specific magneto-optic constant, which is proportional to the magnetization. From the equation (3.1) above, it is clearly seen that the off-diagonal terms in the dielectric tensor $\overline{\overline{\varepsilon}}$ represent the magneto-optic effect contribution to ε and that these terms vanish in the absence of the magnetization.

The interaction of the electromagnetic field with the magnetic material can also be described phenomenologically by the Lorentz force, which gives rise to a change in polarization. This relation is deduced from the dielectric tensor in the equation (3.1), which can be expressed as follows:

$$\boldsymbol{D} = \varepsilon \boldsymbol{E} + i\varepsilon Q \boldsymbol{m} \times \boldsymbol{E}. \quad (3.2)$$

As shown in this equation, the origin of the magneto-optical effect can be represented by the magnetization vector **m** and the electric field vector **E** of light. This relation is illustrated in Fig. 3.1, in which a longitudinal Kerr effect measurement is represented. If the linearly polarized light with electric field amplitude E is incident, the electric field induces the oscillation of electrons in its direction. Therefore, without the magneto-optic effect, the reflected light $\boldsymbol{E}_r$ is also linearly polarized parallel to the incident electric field vector $\boldsymbol{E}_i$. If it is supposed that the material is magnetized, then the oscillating electrons will feel the additional Lorentz force v_{Lor}, which induces a small motion perpendicular to **m**. Thus, the reflected light contains a small perpendicular component in the electric field vector $\boldsymbol{E}_k$ compared with the incident field vector $\boldsymbol{E}_i$. This component contributes to the magneto-optic Kerr amplitude.

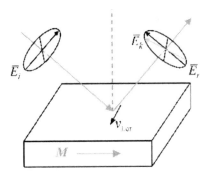

Fig. 3.1. Phenomenological description of magneto-optic Kerr effect by means of Lorentz force v_{Lor}. Illustrated is the longitudinal Kerr configuration, i.e., the magnetization orientation lies in the incident plane

3.2.2 Time-Resolved Experiments

Time-resolved experiments directly measure the dynamic evolution of systems, away from or toward equilibrium, in response to sudden perturbations. For such measurements, a so-called "pump-and-probe" technique is widely used [43]. The method is schematically illustrated in Fig. 3.2. The experimental implementation of time-resolved methods relies, for example, on ultrashort optical pulses. One part of the pulses is used to excite a nonequilibrium state in the system, φ, at the time $t = t_0$ ("pumping"), while the delayed part of light pulses is used to detect the corresponding change in the system at $t = t_0 + \Delta t$ ("probing"). The time interval Δt can be created just by making the optical path of the probe beam longer than that of the pump beam. After setting a time point t, the perturbation of the system, $\varphi(t)$, is detected. The probe beam path is then changed to give a new Δt temporal position, and the probe beam detects the corresponding change in the system again. This procedure is repeated until the entire nonequilibrium response profile of the system is measured. In the study of magnetization reversal dynamics, the nonequilibrium state $\varphi(t)$ corresponds to the magnetization switching process excited by magnetic pulses. Typically, a synchronously triggered transient magnetic pulse is propagated past the sample under study, perturbing the magnetization system, and the subsequent evolution of the magnetization configuration is monitored through its interaction with a time-delayed probe beam. The experimental details of the time-resolved magneto-optical Kerr effect technique are described in the next section.

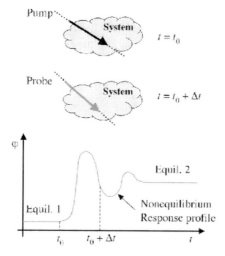

Fig. 3.2. Schematic illustration of the "pump-and-probe" technique. At time $t = t_0$, the system is excited out of its equilibrium with the pump pulse. After a short time interval Δt, the system is probed with the probe pulse. The time delay between the probe and pump beam is then changed to give a new Δt temporal position and the signal built-up again. This procedure is repeated until the entire nonequilibrium profile is measured

3.2.3 Experimental Apparatus

Optical microscopy using the magneto-optic Kerr effect is a well-established tool for imaging magnetic microstructures [44, 45]. In the present experimental setup, a stroboscopic technique is implemented, adding a time-resolving capability down into the deep picosecond range [13]. Experimental arrangements are based on a scanning Kerr microscope, including ultrafast solid-state laser and optics, a piezo-driven flexure stage for rastering the sample, and electronics controlling the time-delay of probe beam and magnetic pulse generation. A schematic diagram for the entire system is shown in Fig. 3.3, including the optical and electronic layouts. Details of the particular parts of the apparatus are discussed below.

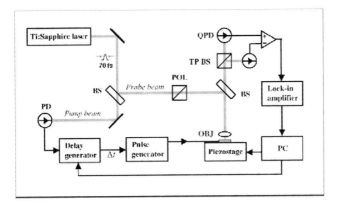

Fig. 3.3. A block diagram of the layout of the time-resolved scanning Kerr microscope

3.2.3.1 Optical Setup and Signal Detection

The light source is a mode-locked Ti:Sapphire laser, which provides 70-fs-long pulses of near-infrared light ($\lambda = 800$ nm) with a repetition rate of 82 MHz. During measurement, the pulsed laser beam is split into two beams (i.e., a pump and a probe beam) with a beam splitter (BS). The probe beam passes through a linear polarizer (POL) and is deflected toward the sample and focused onto the sample using an infinity-corrected microscope objective (OBJ), while the pump beam is directed for triggering magnetic pulses (as will be described in more detail below). The optical power of the probe beam is reduced before being brought to a sharp focus on a sample in order to avoid permanent damage to the sample surface. Typically, an average power of 35 µW is focused onto the sample through a microscope objective. The spatial resolution (d) is determined by the numerical aperture (N.A.) of the objective lens and wavelength (λ) of the laser beam, given by the diffraction limited Rayleigh criterion, $d = (0.82\lambda)/\text{N.A.}$ In our experimental setup, a spatial resolution down to 0.9 µm is yielded using the 0.75 N.A. microscope objective and near-infrared light source.

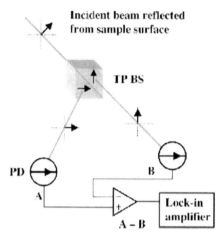

Fig. 3.4. Schematic illustration describing differential detection of polarization rotations induced by the magneto-optical Kerr effect. Both outputs of a polarizing beam splitter (at 45° from incident polarization) are used in subtraction. The intensity in each arm rises or falls respectively with rotations of the plane of polarization. The subtracted signal $(A - B)$ is only non-zero when a polarization rotation occurs

After the probe beam is reflected from the sample, magnetization measurements are accomplished through the polarization analysis of the reflected light in an optical bridge. A particular detection method using quadrant photodiodes (QPD) has been recently developed to allow for simultaneous detection of all three magnetization components (i.e., vector magnetometry) [23]. This approach is adopted from static Kerr imaging [46–48] and works equally well in time-resolved measurements. The principle of the differential detection method is schematically depicted in Fig. 3.4. The probe beam reflected from the sample surface is split into two orthogonal polarization states by the Thomson polarizing beam splitter (TP BS), which is set at 45° to the incident polarization plane. Consequently, equal intensities are sent to the photodiodes. If there is no polarization rotation in the incident beam, each portion of the split beam will be of equal intensity and differential subtraction of the outputs coming from the two photodiodes will result in a zero signal. A small polarization rotation induced by the magnetization in the sample, however, will be turned into an intensity shift by the Thomson analyzer, in which the intensity in one PD increases while decreasing in the other. This differential detection technique, where each PD signal is used as a reference to the other, has the advantage of common mode rejection of laser noise while doubling the signal [47, 50].

Another important issue in the signal detection is the signal-to-noise ratio, which has always been a problem in Kerr imaging due to the weak magneto-optic Kerr effect (see Sect. 3.2.1.). In static Kerr measurements, wide-field imaging with digital enhancement is generally used, in which the image with magnetization pattern is digitally subtracted from an image of magnetization saturation [51]. This technique removes edge defect and optical system polarization artifacts. In our experimental setup, signal enhancement is achieved at each pixel of a raster scan using synchronous modulation of the magnetic excitation and lock-in signal detection technique. The modulated magnetic excitation is accomplished by modulating the trigger electronics, which generates magnetic pulses, at a low frequency (1 ~ 4 kHz), as described below. The modulation on the magnetization in the sample itself isolates the signal from other artifacts, such as depolarizing effects, in the system.

3.2.3.2 Synchronization and Magnetic Pulse Generation

In stroboscopic imaging, the temporal excitation ("pumping") of the system must be repetitive and triggered synchronously with the probe pulses. This is because interactions of many probe pulses and repetitive excitation events are averaged and represented as a single event. Such a synchronization is achieved as a portion of the laser pulse itself triggers magnetic pulses, which are used to excite the nonequilibrium magnetic state of the sample while another portion of the pulse probes (Fig. 3.2).

In our experimental setup (Fig. 3.3), pump pulses are directed to a fast photodiode (e.g., ThorLabs DET210), which creates and sends clock signals to the variable electronic delay generator (Stanford Research Systems DG535). At this stage, the repetition rate of the pulsed beam is reduced from 82 to 0.8 MHz via pulse picking of the mode-locked laser pulse train. This is required since the maximum trigger rate of the delay generator electronics is limited to 1 MHz. On the other hand, the delay electronics creates the propagation delay on the order of 100 ns. Therefore, an additional time delay between the pump and probe beam is required to achieve temporal synchronization. This can be achieved by delaying the pump beam by an equivalent amount by propagation through a length of coaxial cable until the current pulse is actually synchronized with the laser pulse immediately following the one it was triggered by. After setting a time delay Δt, the delay generator sends the clock pulses on to the pulse generator.

The electronic delay method is very convenient, particularly when delay ranges of 10 ns or more are needed, but adds jitter of about 50 ps rms. This trigger jitter is the main limiter of temporal resolution in this case. Alternatively, an optical delay line can be used, in which the travel path of the pump beam with respect to the probe beam is computer controlled using mirrors. This technique is inherently jitter-free and is generally beneficial to measurements for faster (low picosecond regime) processes. However, in practice, the delay range usually spans only a few nanoseconds.

The generation of magnetic pulses relies on the current driver, which is based on the avalanche transistor pulser (Model 2000D Pulse Generator, Picosecond Pulse Labs) using the technique of discharging a transmission line. Pulses from this source have 0.5 ns rise times, 1.5 ns fall times, and pulse widths of 10 ns with the amplitude of 50 V. The current pulses are synchronously triggered by Ti:Sapphire fs laser pulses ($\lambda = 800$ nm, 0.8 MHz repetition rate) and are delivered to micro strip lines, which create magnetic field pulses. The inset in Fig. 3.5 shows an image for such strip lines, on or near which magnetic elements are placed. To excite the sample with an out-of-plane magnetic field pulse, samples are situated between lines and for an in-plane pulse, on top of a line. Strip lines are fabricated using lithographic techniques

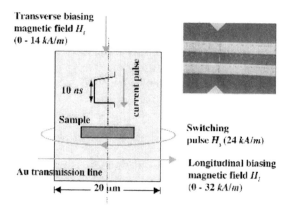

Fig. 3.5. Schematic measurement configuration of a 180° dynamic magnetization reversal experiment for microstructure excitation. H_s, H_l, and H_t indicate the switching field, longitudinal (easy-axis) biasing field, and transverse (hard-axis) biasing field, respectively. In the inset, an image using an optical microscope is given showing magnetic elements on top of or near the gold transmission lines

and have the width of 20 μm and thickness of 300 nm in this case. The strip lines create magnetic pulses as high as 24 kA/m. The temporal shape of field pulses can be measured by a commercially available 2 GHz inductive probe (Tektronix CT-6), or by measuring Faraday rotation in a garnet indicator film [12]. A garnet film allows optical measurement of the current waveforms in a very high bandwidth (over 50 GHz), in addition to providing an absolute time reference for the time-resolved magnetic measurements [23].

3.2.3.3 System Operation

Generally, two operation modes are usually employed in TR-SKM experiments. Temporal-resolving mode: One can obtain a majority of the information very efficiently by measuring the dynamic response of the magnetic state of the sample as a function of time. In this mode, the probe beam is focused on a particular place of the sample surface, and then the time delay Δt is changed. The Kerr signal is detected after each time step, building up the time-dependent profile for selected magnetization components. This mode is suitable for quick local characterization of the magnetic dynamics (see an example in Fig. 3.6). Temporal resolution is ultimately limited by the laser pulse width, but practically limited by the trigger jitter from the delay electronics, as described above.

Spatiotemporal-resolving mode: After the time-dependent profile of the magnetization is measured, the sample surface can be scanned at a particular fixed time delay in order to obtain two-dimensional images mapping the magnetization configuration (see an example in Fig. 3.7). This is required since the dynamic response of most systems studied can be spatially nonuniform [13, 14]. Using this measurement mode, the sample is placed on a computer-controlled piezo-driven flexure stage providing

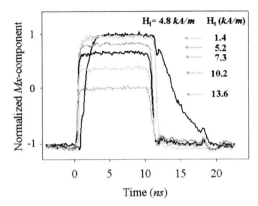

Fig. 3.6. Temporal evolution of the easy-axis magnetization component M_x for several values of the transverse biasing field H_t while longitudinal biasing field H_l is being held fixed at 4.8 kA/m. The thick line indicates the magnetization component measured at $H_t = 0$ kA/m. Small peaks found in the back reversal occur due to electrical reflections in the magnetic pulse line

scanning motion at a typical scan rate of 8 pixels/s (which typically corresponds to about 0.3 μm/s). A time record of a few hundred separate sample measurements of the average magnetization was collected for each pixel in the image. Critical in this operation mode is the stability of the piezostage over long time intervals, since the quality of this stage also contributes to the effective spatial resolution of the system. The ThorLab stages used for our data-acquisition were relatively stable, except for slow drifts on the order of 0.5 μm over two or three hours. Such drifts are probably produced by minor fluctuations in the ambient temperature. For high resolution experiments, a positioning feedback stage (such as the Melles Griot feedback stage) would work better, though the Melles Griot feedback has suffered from its own problems such as ringing.

3.3 Representative Results in Thin Film Microstructures

In this section, two representative results using TR-SKM are discussed. Experiments are carried out in a microstructure of $Ni_{80}Fe_{20}$ (Permalloy), which is a magnetic alloy used throughout the data storage industry for recording heads. The sample used is a 15-nm-thick polycrystalline $Ni_{80}Fe_{20}$ structure (10 μm × 2 μm) patterned by electron beam lithography. The $Ni_{80}Fe_{20}$ structures are made on a 20-μm-wide and 300-nm-thick gold transmission line that carries a fast current pulse (Fig. 3.5).

3.3.1 Picosecond Time-Resolved Magnetization Reversal Dynamics

In the present experiment, a 180° reversal configuration was used (Fig. 3.5). The sample is first saturated in the x-direction by an in-plane bias field H_l, and then a switching pulse H_s is applied antiparallel to H_l in order to flip the magnetization. Additionally, an in-plane transverse bias field H_t is applied along the hard axis of the sample to manipulate the switching time and magnetization reversal processes.

Figure 3.6 represents the time dependence of the magnetization component M_x along the easy axis, measured in the center of the structure for different H_t, while H_l is kept at 4.8 kA/m. For $H_t = 0$, a definite delay in the magnetic response after the beginning of the pulse is observed. The subsequent dynamics are relatively slow with the magnetization fully reversed after $\sim$ 3.5 ns. Furthermore, the asymmetry between front and back reversal is pronounced, owing to different net switching fields H_s^{net} driving the switching in the two cases (i.e., 19.2 kA/m for the forward and 4.8 kA/m for the back reversal). Under these field conditions we find $\tau_s = 1.6$ and 7.3 ns for forward and back switching, respectively, defining switching time τ_s as the interval for 10 to 90% of the total M_x change.

A remarkable difference in the reversal is found by applying a transverse bias field H_t. When H_t is applied, the sample responds earlier to the switching pulse. Correspondingly, the rise time rapidly decreases and the magnetization reverses within 1 ns after the beginning of the pulse. The fast switching can be understood as follows: If H_t is applied, the effective H_c is lower than the case where $H_t = 0$. This situation arises because for a finite H_t the equilibrium position of $\boldsymbol{M}$ is away from the easy axis, the position of minimum anisotropy energy. Thus, lower longitudinal Zeeman energy or smaller H_s is required to overcome the energy barrier.

The effect of applying H_t is directly examined through time domain images taken during the reversal. Figure 3.7 shows a sequence of frames illustrating the change of M_x. For $H_t = 0$, the reversal is mainly governed by a domain nucleation process. In

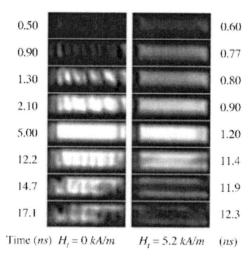

Fig. 3.7. Spatial magnetization profiles of M_x component as a function of time (ns) after the magnetic pulse was applied. $H_l = 4.8$ kA/m, while the transverse field are varied $H_t = 0$ and 5.2 kA/m. The field of view of each frame is 12×4 μm and contains the entire 10×2 μm sample

the beginning ($t = 0.5$–0.9 ns), a stripe-like instability is observed inside the sample. This is followed by expansion of the nucleated domains (1.3 and 2.1 ns), and finally leads to a uniform distribution of fully reversed magnetization, excluding the left and right edge regions (5 ns). These edge regions correspond to free magnetic poles related to the demagnetized areas in a ferromagnet of finite size. On the back reversal, the stripe instability is also pronounced (12.2 ns). From this result, it becomes clear how the switching (Fig. 3.6) evolves spatiotemporally: The finite domain nucleation limits switching time to ~ 3.5 ns.

It is this nucleation-dominated reversal process that is manipulated through the application of H_t. Applying $H_t = 5.2$ kA/m, the 180° domains at the short edges are formed (0.6 ns), but there appears no stripe-like distribution inside. The edge domains expand quickly in the easy-axis direction to form a long, narrow domain parallel to the easy axis (0.77 ns). In the next stage, this elongated domain expands by parallel shifts in the hard direction toward the long edge (0.8 and 0.9 ns) until saturation is reached (1.20 ns). This type of reversal, which is characteristic of domain wall motion, is considerably faster, as revealed in the time dependence of magnetization in Fig. 3.6. The differences in the time domain sequences shown in Fig. 3.7 demonstrate that the formation of nuclei inside the sample is avoided in the presence of H_t, and that the nucleation process is replaced by domain wall motion. Switching occurs over longer times when the stripe domains are involved in the reversal process than if pure domain wall motion occurs.

3.3.2 Precessional Magnetization Reversal and Domain Wall Oscillation

Another aspect of reversal concerns the time scale and mechanism for the removal of the initial excess Zeeman energy from the system. Coherent oscillations of the magnetization M may be observable after the direction has reversed in cases where the damping is not too strong and where the energy has not propagated into a spin wave manifold, which averages incoherently over the spatial resolution of the measurement (so-called "indirect damping" [52]). In these experiments, this regime is encountered when M is pulsed by H_s in the presence of a high H_l and moderate strength H_t. Figure 3.8 shows the time-resolved M_x for different H_l with $H_t = 5.2$ kA/m. Two distinct resonance frequencies are found depending on H_l. First, small oscillations at $f \sim 2$ GHz are found for $H_l = 8.8$–10.4 kA/m. It is also apparent that the amplitudes at $H_l = 8.8$ and 9.6 kA/m rapidly decrease with increasing time. We treat this oscillation as damped ferromagnetic resonance about the new equilibrium direction to infer the saturation magnetization, M_s, through $f = \gamma\mu_0(H_t M_s)^{1/2}/(2\pi)$ (derived from the Landau-Liftshitz equation [15]). Using the previously determined value [20], $\gamma = 1.41 \cdot 10^5$ m/s · A yields $\mu_0 M_s = 820$ kA/m, close to the bulk value 864 kA/m for $Ni_{80}Fe_{20}$.

When H_l increases beyond 10.4 kA/m, an oscillation with another frequency occurs, and well-developed oscillations are found at $H_l = 11.2$ and 12 kA/m. This type of oscillation is characteristically a first "spike" followed by a series of smaller oscillations. The spike is an overshoot associated with the application of H_s, with over-rotation past what will become the new equilibrium magnetization direction. Some

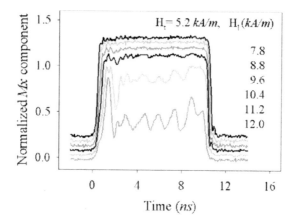

Fig. 3.8. Time-resolved M_x component at the center of the element for different H_l with $H_t = 5.2$ kA/m. The data are offset vertically in order to compare the oscillation behavior directly

of the excess energy associated with this overshoot subsequently devolves to the new oscillation mode, with a typical frequency of ~ 0.8 GHz, much lower than that of the previous oscillation. According to the sequence of spatiotemporal magnetization profiles measured over cycles of low frequency oscillation, time domain images at this mode are characterized as a repetitive expansion and withdrawal of elongated domains in the short axis directions, accompanied by a gradual propagation of the domain along the long axis [13]. Therefore, it is concluded that the observed low frequency oscillation originates from domain wall movement and not from the precession of *M*.

3.4 Conclusion and Outlook

In summary, the current state of TR-SKM technique is reviewed, showing its significance for metrology in fundamental sciences and industrial applications today. The unique combination of temporal and spatial resolution puts it foremost in elucidating the world of ultrafast phenomena.

A challenging question in the present technique is whether the entire magnetization reversal dynamics studied is perfectly repeatable, since the stroboscopic scanning technique does, by its nature, capture only the repetitive part of the process being imaged. Nonrepetitive instabilities, such as thermal fluctuation of the individual spins, will lead to the averaging of the temporal response. In addition to varying scan rate, number of averages, etc., the most sensitive test for underlying stochastic behavior is found to come from spectrum analysis of the noise on the magneto-optic signal [53]. In the discussed experimental results, however, evidence of additional random behavior has not been observed, and if there is a random component present it is too small to detect.

Desirable for the future TR-SKM studies is the improvement of the spatial resolution, which is limited to 0.9 µm in the present experiments. In order to obtain a higher spatial resolution, an oil immersion objective can be employed. With a 1.3 N.A. oil immersion objective and the same light source, for example, a resolution of 0.65 µm can be obtained. This can again be improved by a factor of two just by converting the near-infrared light into the blue region ($\lambda = 400$ nm), refraining from taking the decrease of the detection efficiency into account. Further improvement of spatial resolution requires the incorporation of a solid immersion lens (SIL) [54] into the stroboscopic technique. Incorporating a solid immersion lens, which uses high index material such as $SrTiO_3$ ($n = 2.4$) or GaP ($n = 3.4$), will improve the spatial resolution considerably down to the order of 100 nm. At some point, however, a crossover to near-field techniques becomes essential if one hopes to extend ultrafast optical imaging to the nanometer scale [55].

Acknowledgement. We gratefully acknowledge support from the Natural Sciences and Engineering Research Council of Canada, the Canadian Institute for Advanced Research, and the National Storage Industry Consortium. The samples were produced at the University of Alberta MicroFab, and the experiments performed on equipment donated by Quantum Corporation.

References

1. S.W. Yuan and H.N. Bertram, J. Appl. Phys. **73**, 5992 (1993).
2. B. Heinrich, Canadian Journal of Physics **78**, 161 (2000).
3. A. Hubert and R. Schäfer, "Magnetic Domains," Springer Verlag, (1999).
4. A.F. Popkov, L.L. Savchenko, N.V. Vorotnikova, S. Tehrani, and J. Shi, Appl. Phys. Lett. **77**, 277 (2000).
5. K.J. Kirk, J.N. Chapman, and C.D.W. Wilkinson, Appl. Phys. Lett. **71**, 539 (1997).
6. J.M. Daughton et al., Thin Solid Films **216**, 162 (1992).
7. W.J. Gallagher et al., J. Appl. Phys. **81**, 3741 (1997).
8. C.H. Back, D. Weller, J. Heidmann, D. Mauri, D. Guarisco, E.L. Garwin, and H.C. Siegmann, Phys. Rev. Lett. **81**, 3251 (1998).
9. C. Stamm, F. Marty, A. Vaterlaus, V. Weich, S. Egger, U. Maier, U. Ramsperger, H. Fuhrmann, and D. Pescia, Science **282**, 449 (1998).
10. M. Hehn, K. Ounadjela, J.-P. Bucher, F. Rousseaux, D. Decanini, B. Bartenlian, and C. Chappert, Science **272**, 1782 (1996).
11. R.P. Cowburn, D.K. Koltsov, A.O. Adeyeye, M.E. Welland, and D.M. Tricker, Phys. Rev. Lett. **83**, 1042 (1999).
12. A.Y. Elezzabi, M.R. Freeman, and M. Johnson, Phys. Rev. Lett. **77**, 3220 (1996).
13. B.C. Choi, G.E. Ballentine, M. Belov, W.K. Hiebert, and M.R. Freeman, Phys. Rev. Lett. **86**, 728 (2001).
14. Y. Acremann, C.H. Back, M. Buess, O. Portmann, A. Vaterlaus, D. Pescia, and H. Melchior, Science **290**, 492 (2000).
15. R.H. Koch, J.G. Deak, D.W. Abraham, P.L. Trouilloud, R.A. Altman, Y. Lu, W.J. Gallagher, R.E. Scheuerlein, K.P. Poche, and S.S.P. Parkin, Phys. Rev. Lett. **81**, 4512 (1998).
16. Y. Martin, H.K. Wickramasinghe, Appl. Phys. Lett. **50**, 1455 (1987).
17. J.N. Chapman, J. Phys. D: Appl. Phys. **17**, 623 (1984).

18. K. Koike and K. Hayakawa, Jpn. J. Appl. Phys. **23**, L187 (1984).
19. J. Unguris, G. Hembree, R.J. Cellota, and D.T. Pierce, J. Microscopy **139**, RP1 (1985).
20. W.K. Hiebert, A. Stankiewicz, and M.R. Freeman, Phys. Rev. Lett. **79**, 1134 (1997).
21. T.M. Crawford, T.J. Silva, C.W. Teplin, and C.T. Rogers, Appl. Phys. Lett. **74**, 3386 (1999).
22. G. Ju, A.V. Nurmikko, R.F.C. Farrow, R.F. Marks, M.J. Carey, and B.A. Gurney, Phys. Rev. Lett. **82**, 3705 (1999).
23. G.E. Ballentine, W.K. Hiebert, A. Stankiewicz, and M.R. Freeman, J. Appl. Phys. **87**, 6830 (2000).
24. W. Dietrich and W.E. Proebster, J. Appl. Phys. **31**, 281S (1960).
25. G.M. Sandler, H.N. Bertram, T.J. Silva, and T.M. Crawford, J. Appl. Phys. **85**, 5080 (1999).
26. T.J. Silva, C.S. Lee, T.M. Crawford, and C.T. Rogers, J. Appl. Phys. **85**, 7849 (1999).
27. S.E. Russek, J.O. Oti, S. Kaka, and E.Y. Chen, J. Appl. Phys. **85**, 4773 (1999).
28. J.F. Dillon, Jr., H. Kamimura, and J.P. Remeika, J. Appl. Phys. **34**, 1240 (1963).
29. M.H. Kryder and F.B. Humphrey, Rev. Sci. Instr. **40**, 829 (1969).
30. P. Kasiraj, D.E. Horne, and J.S. Best, IEEE Trans. Magn. **MAG-23**, 2161 (1987).
31. M.R. Freeman, R.R. Ruf, and R.J. Gambino, IEEE Trans. Magn. **27**, 4840 (1991).
32. A. Stankiewicz, W.K. Hiebert, G.E. Ballentine, K.W. Marsh, and M.R. Freeman, IEEE Trans. Magn. **134**, 1003 (1998).
33. M.R. Freeman, W.K. Hiebert, and A. Stankiewicz, J. Appl. Phys. **83**, 6217 (1998).
34. R.J. Hicken and J. Wu, J. Appl. Phys. **85**, 4580 (1999).
35. J. Reif, J.C. Zink, C.M. Schneider, and J. Kirschner, Phys. Rev. Lett. **67**, 2878 (1991).
36. C.T. Rogers, NSIC (National Storage Industry Consortium) 2000 Symposium, June 26–29 (2000) Monterey, CA, U.S.A. G. Schutz, W. Wagner, W. Wilhelm, P. Kienle, R. Zeller, R. Frahm, and G. Materlik, Phys. Rev. Lett. **58**, 737 (1987).
37. F. Sirotti, R. Bosshard, P. Prieto, G. Panaccione, L. Floreano, A. Jucha, J.D. Bellier, and G. Rossi, J. Appl. Phys. **83**, 1563 (1998).
38. M. Bonfim, G. Ghiringhelli, F. Montaigne, S. Pizzini, N.B. Brookes, F. Petroff, J. Vogel, J. Camarero, and A. Fontaine, Phys. Rev. Lett. **86**, 3646 (2001).
39. S.D. Bader, J. Magn. Magn. Mater. **100**, 440 (1991).
40. S.D. Bader, E.R. Moog, and P. Grünberg, J. Mag. Mag. Mater. **53**, L295 (1986).
41. P.N. Argyres, Phys. Rev. **97**, 334 (1955).
42. D.D. Awschalom, J.-M. Halbout, S. von Molnar, T. Siegrist, and F. Holtzberg, Phys. Rev. Lett. **55**, 1128 (1985).
43. H.J. Williams, F.G. Foster, E.A. Wood, Phys. Rev. **82**, 119 (1951).
44. C. A. Fowler. Jr., E.M. Fryer, Phys. Rev. **86**, 426 (1951).
45. W.W. Clegg, N.A.E. Heyes, E.W. Hill, and C.D. Wright, J. Mag. Mag. Mater **83**, 535 (1990).
46. W.W. Clegg, N.A.E. Heyes, E.W. Hill, and C.D. Wright, J. Mag. Mag. Mater. **95**, 49 (1991).
47. T.J. Silva and A.B. Kos, J. Appl. Phys. **81**, 5015 (1997).
48. B. Petek, P.L. Trouilloud, and B.E. Argyle, IEEE Trans. Magn. **26**, 1328 (1990).
49. P. Kasiraj, R.M. Shelby, J.S. Best, and D.E. Horne, IEEE Trans. Magn. **22**, 837 (1986).
50. P.L. Trouilloud, B. Petek, and B.E. Argyle, IEEE Trans. Magn. **30**, 4494 (1994).
51. M. Ramesh, E. Jedryka, P.E. Wigen, and M. Shone, J. Appl. Phys. **57**, 3701 (1985).
52. M.R. Freeman, R.W. Hunt, and G.M. Steeves, Appl. Phys. Lett. **77**, 717 (2000).
53. J.A.H. Stotz, M.R. Freeman, Rev. Sci. Instrum. **68**, 4468 (1997). E. Betzig, J.K. Trautman, J.S. Weiner, T.D. Harris, and R. Wolfe, Appl. Optics **31**, 4563 (1992).

4
Lorentz Microscopy

A.K. Petford-Long and J.N. Chapman

Lorentz microscopy has been used extensively for the past 40 years to study magnetic domain structure and magnetization reversal mechanisms in magnetic thin films and elements. In this chapter, the principal imaging and diffraction modes are reviewed, both qualitative and quantitative. In addition, a description of the instrumental and specimen requirements is included, and in the final section, the application of the various techniques to the study of spin-valve and spin-tunnel junction layered structures is discussed as a means of illustrating the type of information that can be obtained.

4.1 Introduction

Transmission electron microscopy (TEM) was one of the earliest techniques to be used for the imaging of magnetic domain structures in magnetic materials (see, for example, [1, 2]). A principal motivation for the use of TEM for the analysis of magnetic domain structure is that many applicable magnetic properties are extrinsic rather than intrinsic to the materials themselves; in other words, they depend critically on the local microstructure and composition of the material. A detailed knowledge of both the physical and magnetic microstructure is, therefore, essential if the structure-property relation is to be understood and materials with optimized properties are to be produced. Some of the materials of interest are markedly inhomogeneous, with features on a sub-50 nm scale playing an important role in influencing the magnetization reversal mechanism. For example, the grain size in a spin-valve structure is typically on the order of 10 nm, and variations in magnetic properties between grains in the pinning, antiferromagnetic layer can greatly influence the way in which magnetization reversal occurs in a spin-valve device. Hence, the attraction of TEM is twofold: Firstly, it offers very high spatial resolution and the ability to image the local magnetic domain structure, and secondly, because of the large number of interactions that take place when a beam of fast electrons hits a thin solid specimen, detailed insight into compositional, electronic, and structural properties can be obtained from the same region of the sample at the nanometer scale. The resolution that is achievable depends largely on the information sought and may well be limited by the specimen itself. Typical achievable resolutions are 0.2 nm for structural imaging, 0.5–2.0 nm for extraction of compositional information, and 2–20 nm for magnetic imaging. A considerable amount of work has already been carried out using Lorentz microscopy to study the magnetic structure of magnetic materials, and reviews of the techniques can be found, for example, in [3–7].

There are a number of modes of magnetic TEM (Lorentz microscopy) that can be used to obtain information about magnetic domain structure, some of which primarily provide qualitative information and some of which provide more quantitative information. These are discussed in detail below. In addition to the use of TEM, an imaging mode that uses a scanning TEM (STEM) to obtain quantitative maps of the magnetic induction distribution is also discussed. A further advantage of the TEM technique is that in-situ magnetic fields can be applied, which enables the local magnetization reversal of a sample to be followed in real-time (although limitations of image capture and signal mean that the shortest time interval that can be achieved is typically 40 ms).

In summary, studies can be made of specimens in their as-grown state, in remanent states, in the presence of applied fields or currents, and as a function of temperature. From these can be derived basic micromagnetic information, the nature of domain walls, where nucleation occurs, the importance or otherwise of domain wall pinning, and many other related phenomena. In addition, the effects of materials processing (for example, patterning into small elements) on magnetic domain structure and magnetization reversal mechanisms can be analyzed. Section 5 provides a "case study" of the application of the various modes of Lorentz microscopy to the study of the magnetic domain structure and magnetization reversal mechanisms in spin-valves and spin-tunnel junction structures.

4.2 Experimental Requirements

4.2.1 Basic Instrumental Requirements

The principal difficulty encountered when using a TEM to study magnetic materials is that the specimen is usually immersed in the high magnetic field (typically 0.6–1.2 T) of the objective lens [8]. This is sufficient to completely eradicate or severely distort most domain structures of interest. A number of strategies have been devised to overcome the problem of the high field in the specimen region [9]. These include (i) simply switching off the standard objective lens, (ii) changing the position of the specimen so that it is no longer immersed in the objective lens field, (iii) retaining the specimen in its standard position, but changing the pole-pieces [10, 11], once again, to provide a non-immersion environment, or (iv) adding super mini-lenses in addition to the standard objective lens, which is, once again, switched off [12, 13].

In addition to ensuring that the specimen is sitting in a field-free region, further constraints have to be taken into consideration – for example, in order to obtain good quality Foucault mode images (see Sect. 4.3 below) it is important for the objective aperture to sit as close to the back-focal plane of the objective lens as

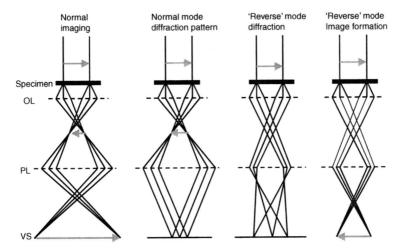

Fig. 4.1. Ray diagram illustrating the way in which the image and diffraction planes are interchanged in a TEM when the objective lens is weakened: the reverse mode configuration. OL – objective lens, PL – projector lens, VS – viewing screen

possible, and this is not always easy to achieve, especially with nonstandard pole-pieces. For example, in [10] the use of a modified aperture mechanism is described in order to make the back-focal plane and objective aperture co-planar. A further requirement is the need for highly coherent illumination. This is best achieved by using a field-emission gun (FEG) TEM, as, for example, in [14]. But if this is not possible, then the use of a small spot size can be used to increase the coherency of the illumination. If the low-angle diffraction (LAD) technique is to be used, as described in Sect. 4.4 below, then it is important to be able to see highly magnified images of the diffraction pattern. One strategy for doing this is to defocus the objective lens to such a degree that the diffraction and image planes are interchanged so that the range of magnifications available for imaging can be applied to the diffraction pattern (the so-called "reverse mode" of operation). Figure 4.1 shows a series of raydiagrams indicating the difference between "standard" and "reverse" modes for imaging and diffraction.

4.2.2 Specimen Requirements

A major consideration that must be borne in mind is that TEM is a *transmission* technique and is, therefore, limited to analysis of thin film samples with a maximum thickness normally on the order of 100 nm, although the limitation on thickness is much more stringent for high atomic number materials. In the case of bulk material, a TEM sample must be prepared by thinning the material to a suitable thickness using a technique such as electropolishing or ion-beam milling (for details of specimen preparation see, for example, [15]), In this case, care must be taken to ensure that the resulting magnetic microstructure is representative of the bulk. However, many technological applications require magnetic material in the form of thin films, and, in this case, the entire film thickness can be analyzed. The problem that remains, however, may be that of removing the substrate. The options are then, firstly, to deposit the film of interest on a suitable electron-transparent substrate such as a carbon-coated TEM support grid or a Si_3N_4 support membrane on a Si wafer substrate, and, secondly, to use some technique to remove the thin magnetic film from the substrate. A focused ion-beam (FIB) system can be used to achieve this, or, in some cases, a suitable chemical etch may be used.

4.3 Basic Theory

In a TEM, a high energy (100–1000 keV) electron beam is incident on a thin specimen. The interaction of the electrons passing through a region of magnetic induction in the specimen results in magnetic contrast, because of the deflection experienced by the electrons [3]. This type of imaging is referred to as Lorentz transmission electron microscopy (LTEM), and there are several LTEM modes that can be used. If the electrons are considered particles, then on passing through the magnetic induction in the specimen, they are deflected by the Lorentz force:

$$F = |e|(v \times B), \qquad (4.1)$$

where e and v are the charge and velocity of the electrons, and B is the magnetic induction averaged along the electron trajectory. Note that only components of the magnetic induction normal to the electron beam give rise to a deflection; the stray fields above and below the specimen also contribute to the image, because Lorentz microscopy is a transmission technique; and the deflection direction depends on the magnetization direction within the domain being imaged and is perpendicular to it.

The Lorentz deflection angle, β_L, is given by:

$$\beta_L = \frac{e\lambda t(B \times n)}{h}, \qquad (4.2)$$

where n is a unit vector parallel to the incident beam, t is the specimen thickness, λ is the electron wavelength, and h is Planck's constant. The deflection is thus proportional to the product of the averaged magnetic induction and the specimen thickness. Substituting typical values into Eq. 4.2 suggests that β_L rarely exceeds 100 μrad. Given the small magnitude of β_L, there is no danger of confusing magnetic scattering with the more familiar Bragg scattering, where angles are typically in the 1–10 mrad range.

The description given so far is essentially classical in nature and much of Lorentz imaging can be understood in these terms. However, for certain imaging modes and, more generally, if a full quantitative description of the spatial variation of induction is sought, a quantum mechanical description of the beam-specimen interaction must be employed [16]. Using this approach, the magnetic film should be considered a phase modulator of the incident electron wave, the phase gradient $\nabla \phi$ of the specimen transmittance being given by:

$$\nabla \phi = \frac{2\pi\, et(\boldsymbol{B} \times \boldsymbol{n})}{h}.\qquad(4.3)$$

Substituting typical numerical values shows that magnetic films should normally be regarded as strong, albeit slowly varying, phase objects [3]. For example, the phase change involved in crossing a domain wall usually exceeds π rad.

4.4 Imaging Modes in Lorentz Microscopy

The most commonly used techniques for revealing magnetic domain structures are the Fresnel (or defocus) and Foucault imaging modes, together with low-angle electron diffraction (LAD) [3]. These modes are normally practiced in a fixed-beam (or conventional) TEM, and schematics of how magnetic contrast is generated are shown in Fig. 4.2.

For the purpose of illustration, a simple specimen comprising domains separated by 180° domain walls is assumed. The Lorentz force present as a result of the magnetic specimen acts on the electrons passing through the specimen and splits each of the diffraction spots into two (as shown schematically in the first ray diagram). One split spot contains information from domains with magnetization lying in one direction, and the other spot contains information from the antiparallel domains.

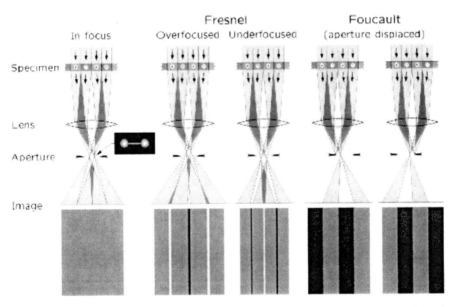

Fig. 4.2. Schematic of ray diagram indicating the paths followed by electrons passing through a magnetic specimen, together with the contrast that would be seen in the image for the Fresnel and Foucault modes of LTEM

4.4.1 Fresnel Mode

For the Fresnel imaging mode, the objective lens is defocused so that an out-of-focus image of the specimen is formed (see Fig. 4.2). (Note that the term "objective lens" is used to refer to the primary imaging lens, although in some modified instruments this will not be a standard objective lens.) Under these conditions the magnetic domain walls are imaged as alternate bright (convergent) and dark (divergent) lines. The bright lines occur when the domain walls are positioned such that the magnetization on either side deflects the electrons toward the wall. If a coherent electron source is used, the convergent wall images consist of sets of fringes running parallel to the wall. Detailed analysis and simulation of the fringe patterns can give information about the domain wall structure, but this is not easy to interpret. A Fresnel mode image of a Co thin film can be seen in Fig. 4.3(a). Information about the magnetization direction within the magnetic domains can also be obtained from the magnetization ripple visible in Fresnel images of polycrystalline specimens as a result of small fluctuations in the magnetization direction. The ripple is always oriented perpendicular to the magnetization direction. A typical image of a NiFe thin film showing magnetization ripple together with domain walls can be seen in Fig. 4.3(b).

The Fresnel mode is useful for real-time studies of magnetization reversal, as it is relatively easy to implement. However, it should be noted that the spatial resolution is not as high as for the Foucault mode, because the images must be recorded at a relatively high value of objective lens defocus if the images of the walls are to be clearly visible (this is of particular significance for very thin specimens or for ones in which the magnetic induction is low).

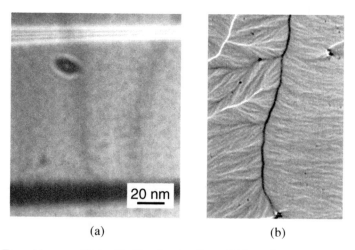

(a)   (b)

Fig. 4.3. Fresnel images of (**a**) a Co film showing Fresnel fringes in a domain wall (seen as series of bright and dark lines near top of image; (**b**) NiFe film showing magnetization ripple and wall contrast (strong contrast dark and bright lines)

4.4.2 Foucault Mode

To image magnetic domains using the Foucault mode, the objective lens is kept in focus, but one of the split spots in the diffraction pattern is blocked by displacing an aperture in the same plane as the diffraction pattern. Under normal conditions, this aperture would be the objective aperture, but if the reverse mode is used, the back-focal plane of the objective lens is co-planar with the selected area aperture. The contrast then results from the magnetization within the domains. Bright areas correspond to domains where the magnetization orientation is such that electrons are deflected through the aperture and dark areas to those where the orientation of magnetization is aligned antiparallel to this direction. By knowing the relative direction of the aperture and image, the direction of magnetization within the various domains can be qualitatively determined. A Foucault image of the magnetization distribution in the NiFe sense layer in a spin-valve structure can be seen in Fig. 4.4. Note that, in general, the splitting of the central spot is more complex than for the simple case considered here.

The Foucault mode is somewhat more difficult to implement than the Fresnel mode, because the contrast observed in the images depends critically on the position of the blocking aperture used. To obtain good quality Foucault mode images, the back-focal plane of the objective lens and the blocking aperture must be as near co-planar as possible. In addition, problems can arise in carrying out real-time imaging of magnetization reversal using a magnetizing stage, because of the fact that the electron beam shifts when a magnetic field is applied, resulting in a change in the relative position of the aperture and the split spots. The use of correction coils that bring the electron beam back onto the optic axis when a field is applied can help reduce this problem.

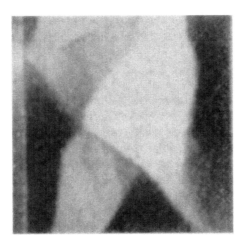

Fig. 4.4. Foucault image of a 10 μm × 10 μm NiFe element. Component of magnetization being imaged runs from top to bottom of the image. Intermediate gray levels indicate that the magnetization lies at an angle to the vertical direction

The principal advantages of both Fresnel and Foucault microscopy are that they are generally fairly simple to implement and provide a clear picture of the overall domain geometry and a useful indication of the directions of magnetization in (at least) the larger domains. Such attributes make them the preferred techniques for in-situ experimentation; this theme is developed further in Sect. 4.6. However, a significant drawback of the Fresnel mode is that no information is directly available about the direction of magnetization within any single domain, whilst reproducible positioning of the contrast-forming aperture in the Foucault mode is difficult. Moreover, both imaging modes suffer from the disadvantage that the relation between image contrast and the spatial variation of magnetic induction is usually nonlinear [17]. Thus, extraction of reliable quantitative data, especially from regions where the induction varies rapidly is problematic.

4.4.2.1 Coherent Foucault Microscopy

A technique that allows the Foucault imaging to be made quantitative is the coherent Foucault imaging mode [13, 18]. This requires the use of a fieldemission gun (FEG) TEM for successful implementation. In this technique, the opaque blocking aperture that is normally used is replaced by a thin phase-shifting aperture such as an amorphous film with a small hole, the edge of which is located on the optic axis. The thickness of the aperture is chosen so that the phase shift experienced is π radians. The technique can only be used near the edge of a specimen, as a reference beam passing through free space is required. The image formed is a magnetic interferogram similar to that produced by electron holography. The fringes seen in the image map lines of constant flux, and adjacent fringes are separated by a flux of h/e and, thus, the image provides quantitative data. An advantage of the technique over electron holography is that no processing of the recorded image intensity distribution is required. Coherent Foucault images of NiFe elements are shown in Fig. 4.5. Note that the technique is not so applicable to ultrathin films, because of the excessive separation of the fringes.

4.4.3 Low-Angle Electron Diffraction

Quantitative data can be obtained from the TEM by direct observation of the structure in the split central diffraction spot. The main requirement for low-angle electron diffraction (LAD) is that the magnification of the intermediate and projector lenses of the microscope is sufficient to render visible the small Lorentz deflections. Camera constants (defined as the ratio of the displacement of the beam in the observation plane to the deflection angle itself) in the range 10–100 m are typically required. One method of achieving this is the use of the "reverse mode", as described in Sect. 2.1 above.

Furthermore, high spatial coherence in the illumination system is essential if the detailed form of the low-angle diffraction pattern is not to be obscured. In practice, this necessitates the angle subtended by the illuminating radiation at the specimen to be considerably smaller than the Lorentz angle of interest. The latter condition is particularly easy to fulfill in a TEM equipped with a field emission gun (FEG).

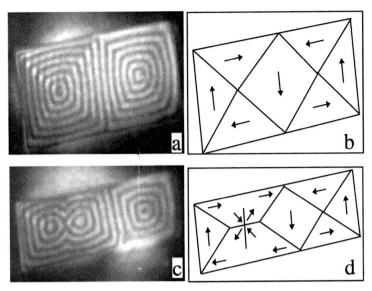

Fig. 4.5. Coherent Foucault images of NiFe elements (**a**) $3.0 \times 1.5\,\mu m^2$ and (**c**) $3.0 \times 1.0\,\mu m^2$ together with schematics of the magnetization distributions (courtesy of Dr. A.B. Johnston)

LAD can be used very effectively to follow magnetization processes, and a sequence of LAD images, recorded during in-situ magnetizing of a Co/Cu multilayer film, is shown in Fig. 4.6. The spots appear as arcs rather than as spots as a result of the dispersion of the magnetization direction in the polycrystalline film (the same effect that gives rise to magnetization ripple in Fresnel images).

LAD can be used to provide some quantitative information, but it must be borne in mind that it provides global information from the whole of the illuminated specimen

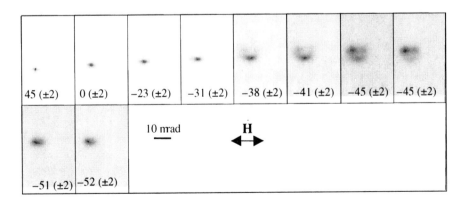

Fig. 4.6. Sequence of low-angle diffraction patterns following a magnetization reversal process in a Co/Cu multilayer film (courtesy of M. Hermann). The numbers indicate the field values in Oe

area rather than local information. It should, therefore, be considered a supplementary technique that can be used to good effect in combination with the qualitative Fresnel and Foucault imaging modes. One important point to note is that it is necessary to be careful that the high intensity in the diffraction spots does not cause damage to the imaging system used, for example, a silicon intensified tube (SIT) or CCD TV camera.

4.4.4 Differential Phase Contrast (DPC) Imaging

The imaging modes discussed so far can only be used to a limited extent to obtain quantitative information except under the rather exacting conditions required for coherent Foucault microscopy. Differential phase contrast (DPC) microscopy [4, 19] overcomes many of the deficiencies of the imaging modes discussed above. Its normal implementation requires a STEM instrument. It can usefully be thought of as a local area LAD technique using a focused probe, and Fig. 4.7 shows a schematic of the way in which DPC contrast is generated.

In DPC imaging, the local Lorentz deflection at the position about which the probe is centered is determined using a segmented detector sited in the far-field. Of specific interest are the difference signals from opposite segments of a quadrant detector, as these provide a direct measure of the two components of the Lorentz deflection, β_L. By monitoring the difference signals as the probe is scanned in a regular raster

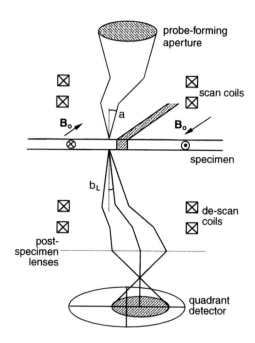

Fig. 4.7. Schematic of the DPC imaging technique

across the specimen, directly quantifiable images with a resolution approximately equal to the electron probe size are obtained. Provided a FEG microscope is used, probe sizes of less than 10 nm can be achieved with the specimen located in field-free space [20]. The validity of the simple geometric optics argument outlined above has been confirmed by a full wave-optical analysis of the image formation process, for experimentally realizable conditions [21]. The main disadvantage of the DPC technique is that there is an increase in instrumental complexity and operational difficulty compared with the fixed-beam imaging modes. The relatively long time needed to record each image means that recording real-time series of images during a magnetization reversal process is not possible. However, provided that time-dependent effects do not contribute significantly to the reversal process, the DPC technique can still be used to record images during a magnetization reversal cycle.

The two difference-signal images are collected simultaneously and are, therefore, in perfect registration. From them a map of the component of magnetic induction perpendicular to the electron beam can be constructed. In addition, a third image formed by the total signal falling on the detector can be formed. Such an image contains no magnetic information (the latter being dependent only on variations in the position of the bright field diffraction disc in the detector plane) but is a standard incoherent bright field image, as would be obtained using an undivided spot detector. Thus, a perfectly registered structural image can be built up at the same time as the two magnetic images, which is a further advantage of DPC imaging. However, at this point it is important to recognize one of the primary difficulties encountered in *all* Lorentz microscopy modes. The simple analyses given above have assumed that image contrast arises solely as a result of the magnetic induction/electron beam interaction, whereas, in reality, contrast arising from the physical microstructure is frequently present as well. In fact, the microstructural contrast is frequently stronger than the magnetic contrast, with the result that the resolution at which useful magnetic information is extracted is often limited by the specimen, rather than by the inherent instrumental capability.

Although the presence of unwanted microstructural contrast is serious, its effect can be reduced to some extent by a suitable choice of operating conditions or by the modification of the techniques themselves. It is frequently (but not always!) the case that the physical microstructure is on a significantly smaller scale than its magnetic counterpart. If this is so, the influence of high spatial frequency components in the image can be reduced by the suitable choice of defocus in the Fresnel mode. In DPC imaging, an annular quadrant detector can be used in place of the solid quadrant detector – so-called "modified DPC" imaging [22]. Indeed, in the latter case, not only is the unwanted signal component suppressed considerably, but also the signal-to-noise ratio in the magnetic component is significantly enhanced. A pair of modified DPC images of a small lithographically defined NiFe element are shown in Fig. 4.8. Although it is not as possible to follow magnetization reversals in real-time using DPC as it is using Fresnel imaging, the DPC technique still provides a powerful technique for imaging the effect that an applied field has on a magnetic domain configuration. An example is shown in Fig. 4.9, in which the interaction between a domain wall in a NiFe film and an inclusion is seen.

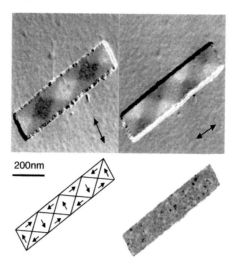

Fig. 4.8. DPC images, arrow map indicating magnetization configuration and bright field image of 40-nm-thick NiFe element (images courtesy of Dr.S. McVitie)

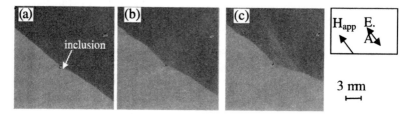

Fig. 4.9. Sequence of DPC images recorded at different values of in-situ applied magnetic field showing the interaction between a domain wall in a NiFe film and an inclusion (courtesy of Dr. M.F. Gillies)

4.4.4.1 Summed Image DPC (SIDPC)

More recently, a technique known as the summed image differential phase contrast (SIDPC) technique has been developed, which allows quantitative maps of the magnetic induction in a thin film specimen to be obtained from summed series of Foucault images [23]. The contrast that is produced is equivalent to that resulting from the DPC technique, but the SIDPC technique does not require there to be scanning coils or a quadrant detector present in the TEM. The SIDPC magnetic induction maps are collected in the following way: Four series of Foucault images are recorded with small incremental tilts (about 3×10^{-6} rad) of the incident illumination. Each series is summed to produce a single Foucault sum image. The directions of tilt are such that the four images are recorded in orthogonal directions ($+x$, $-x$, $+y$, and $-y$). The $+x$ and $-x$ images are then subtracted one from the other to produce a map of the y-component of the magnetic induction, $\boldsymbol{B}_{yi}(x, y)$, integrated along the electron trajectory. The magnitude of the component at each point is determined by the pixel value at that point. The x-component map can be produced in a similar way. The component maps are then combined to produce a map of $B_i(x, y)$. A schematic of the image collection process is shown in Fig. 4.10. Figure 4.11 shows a pair of SIDPC images of the x- and y-components of the magnetization in a 30 μm × 10 μm NiFe element, together with an arrow map of the magnetic induction produced from the x- and y-component maps. Note that for both the DPC and SIDPC images, it is not possible to show a map of B_i in a monochrome image.

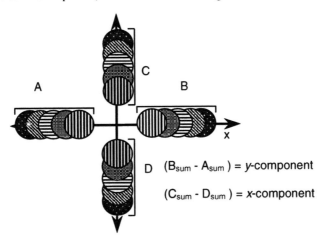

Fig. 4.10. Schematic illustrating the SIDPC technique for obtaining quantitative magnetization maps using Lorentz microscopy

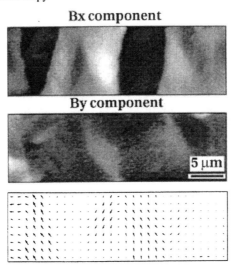

Fig. 4.11. SIDPC images showing (**a**) the x- and (**b**) the y-components of the in-plane magnetic induction of a 10 μm × 30 μm spin-valve element. (**c**) Arrow map of the magnetization determined from the images shown in (**a**) and (**b**)

4.4.5 Electron Holography

While DPC imaging can provide high spatial resolution induction maps and direct information on domain wall structures, difficulties still remain when absolute values of β_L, rather than relative variations, are sought. A final class of techniques, involving electron holography, go some way to overcoming this problem. Here, the electron wave, phase-shifted after passing through the specimen in accordance with Eq. 4.3, is mixed with a reference wave, and the resulting interference pattern is analyzed to yield a full quantitative description of the averaged induction perpendicular to the electron trajectory. However, this will not be discussed further, and the reader is referred to Chap. 5 for a fuller discussion of the electron holography technique.

4.4.6 In-situ Magnetizing Experiments – Use of the TEM as a Laboratory

One of the major attributes of TEM is the ability to change the magnetic state of the specimen in-situ. To observe how magnetic structures evolve as a function of temperature is particularly straightforward and simply involves mounting the specimen in a variable temperature holder. The use of a small furnace built into a holder allows temperatures to be raised from room temperature to temperatures in excess of 1200 K, which is above the Curie temperature of essentially all magnetic materials of interest. Alternatively, cooling holders can be used, in which case the temperatures attained depend on whether liquid nitrogen or helium is the coolant employed. In these cases, the minimum temperatures at the points of observation are usually significantly greater than the liquefaction temperatures of the coolants involved. Realistic temperatures achievable at the center of a thin film specimen are 110 K and 15 K for liquid nitrogen and liquid helium, respectively. Lower temperatures are difficult to achieve mainly as a result of poor thermal paths, rather than because of significant heating in the electron beam itself. The energy deposited in a thin film sample, and, hence, the consequent temperature rise, is generally very small for a thin sample.

However, of greater interest for the study of magnetic materials is the ability to perform in-situ observations of the magnetization reversal process itself. The ability to observe domains during the process is crucial if hysteresis is to be understood at a microscopic level. There are two different approaches to subjecting the specimen to varying magnetic fields. In the first, a magnetizing stage is used that generates a horizontal field in the plane of the specimen, as, for example, described in [24]. Two variants are possible: A magnetizing stage can be built into the specimen mounting rod, as shown in Fig. 4.12, or, if spare ports are available at appropriate positions in the microscope column, magnetizing coils can be introduced as separate entities. As the maximum available field relates strongly to the volume of the exciting coils and the nature of the magnetic circuit, greater fields can usually be generated using an independent stage. However, irrespective of which variation is adopted, the presence of a horizontal field over a vertical distance many times greater than the specimen thickness results in the electron beam being deflected through an angle orders of

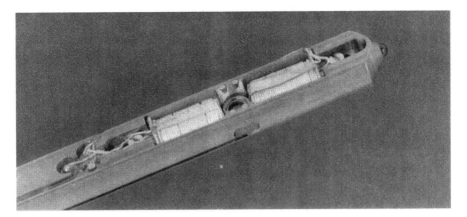

Fig. 4.12. Image of TEM holder for in-situ magnetizing experiments

magnitude greater than typical Lorentz angles. Under these conditions the illumination normally disappears from the field of view and can only be restored using compensating coils. This introduces an additional element of complexity, and the inevitable result is that the field range over which "real-time" magnetization reversal events can be followed is relatively limited (e.g., a few hundred Oe). The inclusion of a magnetizing stage mounted on the specimen holder, combined with a second set of coils at the same height in the microscope column, allows a rotatable field to be applied, or two perpendicular fields to be applied either simultaneously or in succession.

An alternative approach, which can be used in instruments where super mini-lenses are present as well as the standard objective lens, is to use the latter lens as a source of magnetic field, given that the specimen remains located somewhere close to its center [25, 26]. The objective lens now acts as a source of vertical field whose excitation is under the control of the experimenter. Variation in the excitation of the mini-lenses compensates for the additional focusing effect introduced. As the field of the objective lens is parallel to the optic axis, it is clearly suitable for studying magnetization processes in perpendicular magnetic materials. However, by tilting the specimen, a component of field in the specimen plane can also be introduced. Furthermore, given that demagnetizing effects perpendicular to the plane of a thin film are very large, the presence of even moderately large perpendicular fields can often be ignored, and it is the much smaller fields in the plane of the specimen that are of interest. When this is the case, the objective is set to an appropriate fixed excitation giving a constant vertical field, and the sample is simply tilted from a positive angle to a negative one and back again to take it through a magnetization cycle. The principal attraction of the second approach is that the electron optical conditions do not vary during the experiment. Hence, the specimen can be observed throughout, and the experimenter can devote full attention to the changing magnetization distribution. Note, however, that when small magnetic elements are being imaged the effect of the vertical field does need to be considered, as shown in [27].

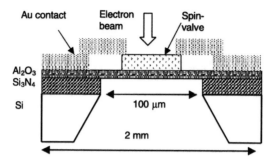

Fig. 4.13. Schematic of sample configuration for active element analysis

A further issue that can be addressed is the effect that an applied current has on the magnetization reversal in a device such as a spin-valve [28]. In this case, design of a special magnetizing stage specimen holder is required that enables the specimen to be connected via Au contacts to an external circuit so that a current can be passed through the specimen in-situ in the TEM whilst magnetizing. This has the added advantage of enabling a direct correlation to be made between the magnetic domain structure and the giant magnetoresistance (GMR) curve. Standard lithographic processing techniques can be used to fabricate suitable spin-valve element specimens with Au contacts at either end. Figure 4.13 shows a schematic of the specimen used for such experiments.

4.5 Application to Spin-Valves and Spin Tunnel Junctions

One of the areas to which Lorentz microscopy has been very successfully applied (see, for example, review article [29]) is to understanding the effect of various parameters on the magnetization reversal mechanisms in spin-valve (SV) structures [30] being developed for use as read-heads in magnetic recording systems and as potential magnetoresistive random access memory (MRAM) elements. A schematic of the structure is shown in Fig. 4.14.

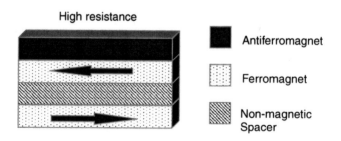

Fig. 4.14. Schematic of a simple SV structure

In their simplest form, SV structures consist of two ferromagnetic (FM) layers separated by a nonmagnetic layer (spacer layer). The magnetization direction (easy axis) of one FM layer is fixed by an adjacent antiferromagnetic (AF) layer (pinning layer) through exchange coupling. The magnetization direction of the other FM layer (sense layer) can be rotated by applying a low external magnetic field (a few Oe). The relative directions of the magnetization in the pinned and sense layers lead to a change in resistance – the so-called giant magnetoresistance (GMR) effect. Two main configurations are widely used for applications. When applying the external field perpendicular to the easy axis, the change in resistance is linear with low coercivity and a magnetization reversal characterized by coherent rotation of the sense layer magnetization. This configuration is used for sensors. A second scheme is to apply the external field along the easy axis direction. Two remanent states can then appear with either parallel (P) or antiparallel (AP) alignment of the magnetizations of the pinned and sense layers. The magnetization reversal is then a hysteretic reversal process, with the P and AP states corresponding to the minimum and maximum resistance values, respectively. Such a configuration is suitable for MRAM devices. This section highlights some of the Lorentz TEM studies that have been carried out on these materials, together with an explanation of the way in which they have increased our understanding of the operation of SV structures.

Magnetostatic coupling occurs at the edges of an SV element via the demagnetizing field between the pinned and sense layers [31, 32]. This phenomenon becomes of primary importance as the size of the elements is continuously reduced. Practically, this magnetostatic coupling favors the opposite magnetic poles between the pinned and sense layers at the free edges of the element and may lead to the formation of domains in the AP state in the sense layer at the very beginning of the P to AP transition. As a result, depending on the strength of the demagnetizing field, an offset of the GMR curve with respect to remanence is expected. Two GMR curves recorded in-situ in the TEM are shown in Fig. 4.15(a) with the easy axis and applied field, $\mathbf{H}_a$, parallel (Δ) or perpendicular ($\circ$) to the free edges of a 10 μm × 10 μm SV element. The current used was 6 mA. For the former curve (Δ) the offset of the curve from zero is induced only by the ferromagnetic "orange peel" coupling between the pinned and sense layers. However, when the easy axis and $\mathbf{H}_a$ are perpendicular to the free edges, the curve is centered around zero because the stray-field coupling at the edges introduces a demagnetizing effect that favors the AP state and compensates the orange peel coupling. Corresponding series of Foucault mode images taken during the $P-AP$ reversal of the element are also shown for the easy axis imposed by the AFM layer and the applied field parallel (b) or perpendicular (c) to the free edges of the element. The reversal shown in Fig. 4.15(b) proceeds via domain nucleation and domain wall motion with the last domains to reverse to the AP state being at the free edges of the element. The reversal shown in Fig. 4.15(c) is very different – edge domains are clearly visible at the free edges of the element from the start of the reversal, confirming the effect of the stray field magnetostatic coupling. This feature is very sensitive to the element size and aspect ratio.

A variation on the SV structure is the spin tunnel junction (STJ) structure [33], in which the spacer layer is not a metal but an insulator such as aluminum oxide. In this

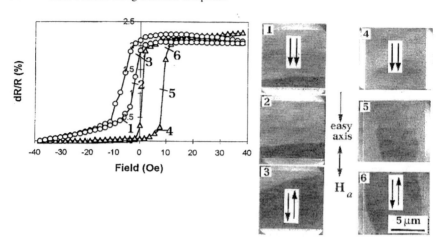

Fig. 4.15. GMR curves for a 10 μm × 10 μm spin-valve element with the easy axis and applied field parallel (△) or perpendicular (○) to the free edges of the element (applied current 6 mA). Associated Foucault images showing the P–AP magnetization reversal for a 10 μm × 10 μm spin-valve element with the easy axis and applied field perpendicular (**1–3**) or parallel (**4–6**) to the free edges of the element (applied current 6 mA). Arrows show the sense layer magnetization directions. The numbers on the images correspond to those shown on the hysteresis curves to indicate the position in the magnetization cycle at which the images were recorded

case, the current tunnels across the insulating barrier layer and the magnitude of the tunnel current is a function of the relative alignment of the magnetization in the two ferromagnetic layers. As for spin-valve structures, there is interest in the way in which the magnetization in the sense layer reverses. Figure 4.16 shows a series of Foucault mode images of the reversal of the sense layer in a NiFe/Al$_2$O$_3$/NiFe/MnFe STJ. The top series of images shows the formation of a 360° wall formed by the combination of two 180° domain walls (imaged as adjacent bright and dark lines) when the two end domains meet. The 360° wall is pinned at a surface feature. The field was then increased to +60 Oe and decreased (center row of images from right to left). The 360° wall defect acted as a domain nucleation site. When the magnetization cycle was then repeated, but with the positive field increased to +80 Oe prior to reversal, the 360° wall was removed and reversal of the sense layer magnetization occurred at a higher negative field (bottom row of images, from right to left). This experiment shows the problems involved in producing large arrays of STJ elements for memory applications, in which the magnetization reversal field of all the element must be the same. Further developments are to heat or cool the specimen (with or without an external applied field) and to apply fields in more than one direction in-situ in the TEM.

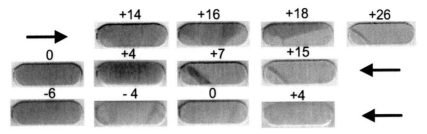

Fig. 4.16. Series of Foucault mode images of a 2 μm wide NiFe/MnFe/NiFe STJ element showing the effect of a 360° wall defect on the magnetization reversal. Arrows indicate the direction of the applied field (and the direction in which the sequence of images should be viewed). The numbers above each image correspond to the applied field value in Oe. (Images courtesy of Dr. P. Shang)

4.6 Summary and Conclusions

The techniques of Lorentz microscopy provide a powerful and direct means of obtaining experimental micromagnetic information from thin magnetic films, mesostructures, and nanostructures. Many of the techniques possess complementary advantages. The experimenter can choose between an imaging mode that is easy to implement and that is suitable for rapid specimen assessment and real-time studies or one that yields a detailed quantitative description of the distribution of induction at very high spatial resolution. It is noteworthy that the phenomenon of magnetization reversal varies markedly in different material systems. Hence, the ability to observe directly how reversal proceeds, and to study the nature of domain walls involved, provides invaluable insight into the overall magnetic behavior. Understanding at a microscopic level is important if material properties are to be optimized and superior materials designed with specific applications in mind.

Acknowledgement. We are grateful to the EPSRC and to the Royal Society for funding some of the research described in this chapter. We are also grateful to all those involved in the work presented here and those who allowed their images to be included.

References

1. M.E. Hale, H.W. Fuller and H. Rubenstein, J. Appl. Phys. **30**, 789 (1959).
2. H. Boersch and H. Raith, Naturwissenschaften **46**, 574 (1959).
3. J.N. Chapman, J.Phys. D: Appl. Phys. **17**, 623 (1984).
4. H. Rose, Ultramicroscopy **2**, 251 (1977).
5. J.P. Jakubovics, in *Handbook of Microscopy: Applications in Materials Science, Solid State Physics and Chemistry*, eds. S. Amelinckx et al., (VCH: Weinheim, New York, 1997) p. 505.
6. A. Hubert and R. Schäfer, *Magnetic Domains* (Springer-Verlag, Berlin, 1998).
7. J.N. Chapman and M.R. Scheinfein, J. Magn. Magn. Mater. **200**, 729 (1999).

8. L. Reimer, Transmission electron microscopy: the physics of image formation and microanalysis, *Springer Series in Optical Sciences* **36** (Springer-Verlag, Berlin, 1994).
9. I.R. McFadyen and J.N. Chapman, EMSA Bulletin **22**, 64 (1992).
10. R.C. Doole, A.K. Petford-Long and J.P. Jakubovics, Rev. Sci. Instrum. **64(4)**, 1038 (1993).
11. K. Tsuno and T. Taoka, Jap. J. Appl. Phys. **22**, 1041 (1983).
12. K. Tsuno and M. Inoue, Optik **67**, 363 (1984).
13. J.N. Chapman, R.P. Ferrier, L.J. Heyderman, S. McVitie, W.A.P. Nicholson, and B. Bormans, in *Electron 15. Microscopy and Analysis* (ed. A.J. Craven, IOPP, Bristol, 1993), 1.
14. J.N. Chapman, A.B. Johnston, L.J. Heyderman, S. McVitie, W.A.P. Nicholson and B. Bormans, IEEE Trans. Mag. **30**, 4479 (1994).
15. D.B. Williams and C.B. Carter, *Transmission Electron Microscopy* (Plenum Press, New York, 1996).
16. Y. Aharanov and D. Bohm, Phys.Rev.**115**, 485 (1959).
17. J.N. Chapman, G.R. Morrison, J.P. Jakubovics, and R.A. Taylor, IOP Conf. Ser. **68**, 197 (1984).
18. A.B. Johnston and J.N. Chapman, J. Microsc. **179**, 119 (1995).
19. N.H. Dekkers and H. de Lang, Optik **41**, 452 (1974).
20. G.R. Morrison, H. Gong, J.N. Chapman, and V. Hrnciar, J. Appl. Phys. **64**, 1338 (1988).
21. G.R. Morrison and J.N. Chapman, Optik **64**, 1 (1983).
22. J.N. Chapman, I.R. McFadyen and S. McVitie, IEEE Trans. Magn. **26**, 1506 (1990).
23. A.C. Daykin and A.K. Petford-Long, Ultramicrosc. **58**, 365 (1995).
24. S.J. Hefferman, J.N. Chapman and S. McVitie, J. Magn. Magn. Mat. **83**, 223 (1990).
25. J.N. Chapman, L.J. Heyderman, S. McVitie and W.A.P. Nicholson, in Advanced Materials '95, Proc. 2nd NIRIM International Symposium on Advanced Materials (eds. Y. Bando, M. Kamo, H. Haneda, T. Aizaw – NIRIM, Japan), 23 (1995).
26. S. McVitie, J.N. Chapman, L. Zhou, L.J. Heyderman, and W.A.P. Nicholson, J. Magn. Magn. Mat. **148**, 232 (1995).
27. K.J. Kirk, M.R. Scheinfein, J.N. Chapman, S. McVitie, M.F. Gillies, B.R. Ward, and J.G. Tennant, J. Phys. D: Appl. Phys. **34**, 160 (2001).
28. X. Portier, A.K. Petford-Long, T.C. Anthony and J.A. Brug, IEEE Trans. Mag. **33(5)**, 3574 (1997).
29. X. Portier and A.K. Petford-Long, J. Phys. D.: Appl. Phys. **32**, R89 (1999).
30. B. Dieny, V.S. Speriosu, S. Metin, S.S.P. Parkin, B.A. Gurney, P. Baumgart and D.R. Wilhoit, J. Appl. Phys. Rev. **69(8)**, 4774 (1991).
31. J.N. Chapman, P.R. Aitchison, K.J. Kirk, S. McVitie, J.C.S. Kools, and M.F. Gillies, J. Appl. Phys. **83**, 5321 (1998).
32. X. Portier, A.K. Petford-Long, T.C. Anthony and J.A. Brug, J. Appl. Phys. **85 (8)**, 4120–4126 (1999).
33. W.J. Gallagher, S.S.P. Parkin, Y. Lu, X.P. Bian, A. Marley, K.P. Roche, R.P. Altman, S.A. Rishton, C. Jahnes, T.M. Shaw and G. Xiao, J. Appl. Phys. **81**, 3741 (1997).

5

Electron Holography of Magnetic Nanostructures

M.R. McCartney, R.E. Dunin-Borkowski, and D.J. Smith

Electron holography is an electron microscope imaging technique that permits quantitative measurement of magnetic fields with spatial resolution approaching the nanometer scale. The theoretical background and usual experimental setup for electron holography are first briefly described. Applications of the technique to magnetic materials and nanostructures are then discussed in more detail. Future prospects are summarized.

5.1 Introduction

The transmission electron microscope (TEM) is an essential tool that is in widespread use for microstructural characterization. Although there are many different TEM imaging modes, most cannot be used to characterize magnetic fields within or surrounding a sample because they are insensitive to changes in the phase of the electron beam. Lorentz microscopy, as described in Chap. 4, uses defocused imaging to distinguish magnetic features such as domain walls [1, 2]. In this approach, which has several closely related variants [3], high-energy electrons of the incident beam are deflected sideways by the magnetic field of the sample to produce bands of light and dark contrast that correspond to local changes in the magnetic field strength or orientation. Compared with electron holography, these Lorentz imaging modes have the advantages that real-time observation of changing domains is possible and no vacuum reference wave is required. However, longer-range magnetic fields are not easily mapped, and compositional contributions to the contrast from the edges of nanostructured elements may be significant, so that magnetic nanostructures become more difficult to characterize as their dimensions are reduced.

Electron holography offers an alternative and powerful approach for characterizing magnetic microstructure. In this technique, access to both the phase and the amplitude of the electron wave can be obtained after the electrons have traveled through the sample [4, 5]. Since the phase change of the electron wave can be related directly to the magnetic (and electric) fields in the sample, magnetic materials can be studied at high spatial resolution and sensitivity using electron holography. The technique is intrinsically capable of achieving a spatial resolution of better than 1 nm for magnetic materials, but this resolution level has yet to be demonstrated on real samples, due primarily to practical limitations. These limitations are associated with the recording process and/or the subsequent hologram processing, with the available signal-to-noise ratio in the hologram being a major restriction. Moreover, because the electrons are transmitted through the specimen, the sample thickness for electron holography is limited to about 500 nm to avoid degradation of the hologram due to multiple scattering effects. One particular attraction of electron holography relative to most other magnetic imaging techniques, which measure the first or second differential of the phase, is that much smaller sample regions are accessible for analysis because unwanted effects arising from local variations in composition and sample thickness can be removed more easily. With the ongoing downscaling of dimensions for magnetic storage devices, holographic approaches thus offer much potential for solving important industrial problems, as well as contributing toward the advancement of fundamental scientific knowledge.

The technique of electron holography is based on the interference of two (or more) coherent electron waves to produce an interferogram or "hologram." This interference pattern must then be processed in order to retrieve, or reconstruct, the complex electron wavefunction, which carries the desired phase and amplitude information about the sample. At least 20 different forms of electron holography have been identified [6], many of which have been demonstrated in practice. Off-axis (or sideband) electron holography, as illustrated in Fig. 5.1, is the mode most commonly used. In this approach, the electrostatic biprism, as developed originally by Möllenstedt and Düker [7], is used to overlap the electron wave scattered by the sample with a vacuum reference (Fig. 5.1).

The earliest attempts at electron holography [8] were severely restricted because the tungsten hairpin filament used as their electron source had limited brightness, in turn limiting the available coherence of the incident beam. The development of the high-brightness, field-emission electron gun (FEG) for the TEM [9] made possible the practical implementation of electron holography. All electron holography applications subsequently reported in the scientific literature have used an FEG as the electron source.

Off-line optical methods have traditionally been used to achieve wavefunction reconstruction from electron holograms [10], but digital processing of electron holograms has become widespread in recent years [11] due to the advent of the slow-scan charge-coupled-device (CCD) camera for digital recording [12]. Coupled with the recent rapid growth in computer speed and memory, quantitative digital electron holography has become a reality [13, 14]. Digital recording with the CCD camera also provides linear output over a large dynamic range, so that correction for nonlinearity of photographic-plate optical density is no longer needed. The speed, accuracy, and reliability of the reconstruction process are greatly enhanced, and accurate registration of sample and reference holograms is easily achieved (see below).

In this chapter, we first outline the basic principles of electron holography, from both theoretical and practical points of view. We briefly describe some of the more general applications of the technique, in particular to magnetic materials such as

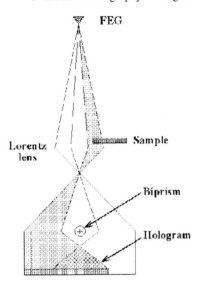

Fig. 5.1. Schematic ray diagram showing setup used for off-axis electron holography in the TEM. Essential components are the field emission electron source (FEG) used to provide coherent illumination and the electrostatic biprism, which causes overlap of object and (vacuum) reference waves

recording media and hard magnets. Results for magnetic nanostructures are then presented in more detail. Finally, possible opportunities for future developments are discussed. These include the in situ application of variable external fields and real-time viewing of dynamic events.

5.2 Basis of Electron Holography

5.2.1 Theoretical Background

The electron wave incident on a TEM sample undergoes phase shifts due to the mean inner potential (i.e., the composition and density) of the sample, as well as the in-plane component of the magnetic field integrated along the incident beam direction. Neglecting dynamical diffraction, which can have a significant effect on crystalline materials oriented close to major zone axes [15], the electron phase change after passing through the sample, relative to a wave that has passed only through vacuum, is given (in one dimension) by the expression

$$\phi(x) = C_E \int V(x, z)\,dz - \frac{e}{\hbar} \iint B_\perp(x, z)\,dx\,dz , \qquad (5.1)$$

where z is the incident beam direction, x is a direction in the plane of the sample, V is the mean inner potential, and **B** is the component of the magnetic induction perpendicular to both x and z. The constant C_E is given by the expression

$$C_E = \frac{2\pi}{\lambda E} \frac{E + E_0}{E + 2E_0},\tag{5.2}$$

where λ is the wavelength, and E and E_0 are the kinetic and rest mass energies, respectively, of the incident electron.

When neither V nor B vary with z within the sample thickness t, and neglecting any electric and magnetic fringing fields outside the sample, this expression can be simplified to

$$\phi(x) = C_E V(x)t(x) - \frac{e}{\hbar} \int B_\perp(x)t(x)\,dx.\tag{5.3}$$

Differentiation with respect to x leads to

$$\frac{d\phi(x)}{dx} = C_E \frac{d}{dx}\{V(x)t(x)\} - \frac{e}{\hbar} B_\perp(x)t(x).\tag{5.4}$$

When the sample has uniform thickness and composition, the first term in Eq. 5.4 is zero, and the phase gradient can be interpreted directly in terms of the in-plane magnetic induction. However, in many cases, the projected thickness or the composition of the magnetic sample may vary rapidly. In such cases, the first mean inner potential term $V(x)t(x)$ is likely to dominate both the phase and the phase gradient. Attempts to quantify the magnetization in the sample then become complicated, and additional processing is required before the in-plane magnetization can be extracted.

When two coherent objects and reference waves interfere to produce a hologram, the intensity distribution in the holographic interference fringe pattern takes the form

$$\begin{aligned}I(x,y) &= |\Psi_1(x,y)|^2 + |\Psi_2(x,y)|^2 + |\Psi_1(x,y)||\Psi_2(x,y)| \\ &\quad \times \left(e^{i(\Phi_1-\Phi_2)} + e^{-i(\Phi_1-\Phi_2)}\right) \\ &= A_1^2 + A_2^2 + 2A_1 A_2 \cos(\Delta\phi),\end{aligned}\tag{5.5}$$

where Ψ is the electron wavefunction, $\Delta\phi$ is the phase change of the electron wave, and the subscripts refer to the reference and object waves. Thus, a series of cosinusoidal fringes is superimposed onto a normal TEM bright field image. The relative phase shift of the electron wave after passing through the sample is represented by shifts in the positions and spacings of the interference fringes. Finite beam divergence (effective source size) and energy spread (temporal coherence) cause loss of contrast in the interference fringes, which, in turn, affects the precision and accuracy of the reconstructed holographic phase image.

Figure 5.2 provides an illustration of the recording and processing steps. The off-axis electron hologram at the top left (Fig. 5.2a) corresponds to a chain of magnetite nanocrystals originating from a single magnetotactic bacterial cell, which is supported on a holey carbon film. A region of vacuum outside the specimen is located in the upper left-hand corner of the hologram. Large deviations in the spacing and the angle of the fringes occur as they cross the crystallites, and these variations can be interpreted in terms of the phase shift of the electron wavefunction relative to the reference vacuum wave. By Fourier transforming the hologram, a two-dimensional frequency

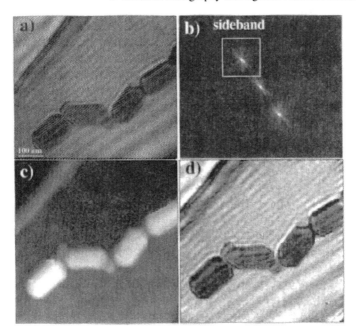

Fig. 5.2. (a) Off-axis electron hologram showing chain of magnetite nanocrystals; (b) Fourier transform of (a), indicating sideband used in phase reconstruction; (c) reconstructed phase image; (d) reconstructed amplitude image

map is obtained, as shown in Fig. 5.2b. The two strong sideband spots correspond to the fundamental cosine frequency. Their separation is due to the relative tilt of the object and reference waves, which depends on the voltage applied to the biprism. The intensity variations around the spots reflect the local phase shifts of the electron wavefunction caused by the sample. The phase of the image wave cannot be obtained from the central autocorrelation function, but the sidebands, which are complex conjugates of each other, contain this information. The reconstruction process makes use of either one of these sidebands, hence the term "off-axis" holography. This off-axis approach provides a convenient solution to the problem of overlapping twin images, which is a serious issue for in-line holographic techniques [6].

Hologram reconstruction involves the extraction and re-centering of one of the sidebands, followed by the calculation of its inverse Fourier transform. The amplitude and phase of the resulting complex image are then

$$\phi = \arctan(i/r)$$
$$A = \text{sqrt}\left(r^2 + i^2\right), \tag{5.6}$$

where r and i are the real and imaginary parts of the wavefunction, respectively.

Figures 5.2c and 5.2d show the corresponding reconstructed phase and amplitude, respectively, of the hologram shown in Fig. 5.2a. The phase is usually calculated modulo 2π, which means that 2π phase discontinuities that are unrelated to particular specimen features may appear at positions in the phase image where the phase shift exceeds this amount. Suitable phase-unwrapping algorithms must then be used to unwrap the phase and to ensure reliable interpretation of the image features [14]. In addition, reference holograms are usually recorded with the sample removed. Any artifacts associated with local imperfections or irregularities of the imaging and recording systems are then excluded by dividing the sample wavefunction by the reference wavefunction before recovering the amplitude and phase.

5.2.2 Experimental Setup

Our attention here is focused on the off-axis mode of electron holography since this mode has been used almost exclusively in electron microscope studies of magnetic materials. The microscope geometry for off-axis electron holography in the TEM is illustrated schematically in Fig. 5.1. Equivalent configurations can also be achieved in the scanning TEM using a stationary defocused probe [6,16]. The sample is examined using defocused, coherent illumination, usually from a FEG electron source, and it is positioned so that it covers approximately half the field of view. The electrostatic biprism is usually a thin ($< 1\mu m$) metallic wire or quartz fiber coated with gold or platinum, which is biased by means of an external dc power supply or battery. The application of a voltage to the biprism causes overlap between the object wave and the vacuum or reference wave, resulting in the formation of the holographic interference pattern on the final viewing screen or detector. Voltages of between 50 and 200 V are typically used for the medium-resolution examples shown below, but higher voltages are required for higher-resolution applications [17]. A rotatable biprism is highly useful since it is often necessary to align the direction of the interference fringes with particular sample features of interest.

Although the biprism may be located in one of several positions along the beam path, the usual position is in place of one of the selected-area apertures. In this configuration, the holographic interference pattern that is formed at the first image plane must be translated electron-optically below the selected area plane. This is achieved by increasing the excitation of the diffraction or intermediate lens so that the image is located below the biprism. The interference fringe spacing and the width of the fringe overlap region are usually then referenced to the image magnification in this plane [18], which, in turn, depends on the method used to ensure that the magnetic sample is located in a field-free region. For example, a low magnification and a correspondingly large field of view are obtained when the normal objective lens is turned off. A post-column imaging filter can then be used to obtain additional magnification, but the image resolution is relatively poor because the imaging intermediate lens is only weakly excited. An alternative approach is to use a modified specimen holder, with the sample located just outside the field of the immersion objective lens. A far-out-of-focus image is then obtained, but the phase-shifting effects due to the objective lens defocus can be corrected during the reconstruction process [16]. Yet another approach involves the use of a weak imaging lens below the normal objective lens. The so-called "Lorentz" minilens of the Philips CM200 FEG-TEM allows image magnifications of up to 70,000x to be obtained, and the reconstructed phase image can have a spatial resolution equal to the 1.4 nm line resolution of the lens [19]. A special low-field ($\sim$ 5.5 G) objective lens for studying small magnetic particles has been reported to provide a maximum magnification of 500 000x [20].

Alternatively, by placing the biprism in one of the condenser aperture positions preceding the sample, a very different configuration is obtained that is equivalent to creating two closely spaced, overlapping plane waves incident on the sample [21]. An equivalent mode can be obtained by using the stationary focused probe of the STEM [16]. By defocusing the observation plane relative to the sample, differences in phase shift between adjacent areas are recorded in the interference pattern, resulting in a differential phase contrast (DPC) image. Either a rotating biprism or a rotating sample holder must be used in this configuration to enable both components of the in-plane magnetic field to be characterized. This approach has the disadvantages that any one hologram only contains information about one component of the magnetization and the spatial resolution of the recovered phase is limited because the image is defocused. However, this DPC imaging mode is particularly attractive for low magnification holography applications because the need for a vacuum reference wave is eliminated, thus making it easier to prepare suitable specimens.

5.2.3 Practical Considerations

The coherence of the electron beam is all-important for practical electron holography. The beam convergence angle must be minimized as far as practicable to avoid loss of phase detail in the reconstructed hologram stemming from poor interference fringe contrast. Even with a FEG source, it is common practice to employ illumination that is deliberately made to be highly elliptical using appropriate condenser lens stigmator settings. Aspect ratios of 100:1 are not uncommon [18]. The major axis of the elliptical illumination patch must also be aligned perpendicular to the biprism wire direction to maximize the coherence and fringe contrast.

Further experimental factors impact the reconstruction process. Higher biprism voltages result in smaller fringe spacings (needed for small object dimensions) and larger fringe overlap width (greater field of view). However, fringe contrast typically drops off as the biprism voltage is increased, and higher magnifications are required to avoid insufficient sampling of the finely spaced interference fringes by the recording system. A common rule of thumb is that four effective pixels per hologram fringe is the minimum acceptable sampling [22], although for sensitive phase measurements even greater sampling rates are recommended [18].

When quantitative analysis of phase shifts is required, holograms should be recorded using electron detectors that have linear output over a large dynamic range. The CCD camera is ideal for this purpose, and subsequent computer processing is also facilitated by digital acquisition [13, 23]. Note that the sampling density of the recovered amplitude and phase is determined by the effective pixel size of the hologram. The size of the sideband extracted during hologram reconstruction can also affect the final resolution of the phase image. In situations where a Lorentz lens has been used for holographic imaging, aberrations of the lens do not usually limit the spatial resolution of the final reconstructed image. The strength of the magnetic signal and the phase sensitivity of the hologram (i.e., the signal-to-noise ratio in the phase image) are more often the factors that limit the resolution for magnetic materials. Phase sensitivities of $2\pi/100$ can be achieved routinely during holographic studies using quantitative recording and processing [13]. It is then possible to detect magnetic details in thin films on a scale of about 5 nm [24]. Thicker films, larger magnetic fields, or longer acquisition times are necessary for recording smaller features using instrumentation that is currently available.

5.2.4 Applications

Off-axis electron holography has been used successfully in a variety of applications. Electron holography was originally proposed by Gabor [25] as a means of correcting electron microscope lens aberrations. Instrumental factors (primarily the coherence of the electron source) prevented his goal from being realized for many years. Optical methods [26,27] and computer reconstruction (e.g., [17,28]) have been used to surpass conventional microscope resolution limits.

Off-axis electron holography has also been used in many studies related to electrostatic fields. For example, reverse-biased p-n junctions were studied [29] at low magnification ($\sim$ 2500x), and electrostatic fields associated with charged latex spheres were also investigated [30]. Depletion region potentials at a Si/Si *p-n* junction were reported [31], and the two-dimensional electrostatic potential associated with deep-submicron transistors has been mapped at a resolution of $\sim$ 10 nm and a sensitivity approaching 0.1 V [32]. An improvement in spatial resolution to $\sim$ 5 nm has been achieved in studies of 0.13 µm device structures [33]. The electrostatic potential and associated space charge across grain boundaries in $SrTiO_3$ has also been reported [34]. Recently, attention has turned to the (Ga, In, Al)N system. Examples include the observation of piezoelectric fields in GaN/InGaN/GaN strained quantum wells [35] and the mapping of electrostatic potentials across an AlGaN/InGaN/AlGaN heterojunction diode [36].

Investigations of magnetic materials by electron holography have historically been limited to spatial resolutions of considerably more than 10 nm due to the obvious requirement for the sample to be located in a region of very low external field to prevent magnetization saturation. Typically, the strong objective lens has been switched off, so that imaging was restricted to the diffraction/projector system of lenses below the sample. Significant applications have included experimental confirmation of the Aharonov-Bohm effect at low temperature [37], studies of vortex "lattices" in a superconductor [38], and observations of thin magnetized Co film used for magnetic recording applications [39]. Further applications of electron holography to magnetic materials are described in the following sections.

5.3 Applications to Magnetic Materials

5.3.1 FePt Thin Films

Anisotropic magnetic films that have either longitudinal (in-plane) or perpendicular (out-of-plane) magnetization are of central importance to the magnetic recording industry. Ordered alloys of Fe-Pt and Co-Pt with the $L1_0$ crystal structure have this desired magnetocrystalline anisotropy and large magnetic moments. Spontaneous ordering of Fe-Pt can occur during deposition, causing development of the ordered phase [40]. The tetragonal c-axis is then predominantly either out of the plane or in-plane, depending on the particular substrate used for growth. Figure 5.3 shows results from a study of an epitaxial $Fe_{0.5}Pt_{0.5}$ ordered alloy [19] that had been deposited using molecular beam epitaxy onto an MgO (110) single-crystal substrate. In this growth direction, the anisotropic $L1_0$ ordered alloy phase has the (001) easy axis parallel to the film normal. The reconstructed phase image in Fig. 5.3b clearly shows the presence of magnetic fringing fields outside the material, and the induction map in Fig. 5.3c confirms that adjacent domains within the FePt film have opposite polarity.

This study highlights some of the difficulties associated with using electron holography to quantify the distribution of magnetic flux in vacuum. The technique will always be insensitive to any magnetic field component that is parallel to the electron beam direction. Moreover, the information contained in a recorded hologram represents a two-dimensional projection of a three-dimensional field. Extensive calculations were thus required, for example, in order to quantify the magnetic field from the tip of a magnetic force microscope [41], and further tomographic experiments were needed to confirm the cylindrical symmetry that was assumed in the calculations [42]. Full three-dimensional simulations will always be needed for quantitative interpretation in situations where the fringing fields extend relatively large distances into vacuum [43, 44].

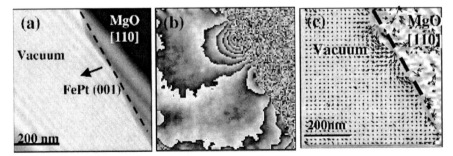

Fig. 5.3. (a) Off-axis electron hologram obtained from epitaxial FePt/MgO with (001) easy axis parallel to film normal; (b) contoured phase image showing magnetic flux extending into vacuum; (c) induction (arrow) map confirming opposite polarity of adjacent FePt domains

5.3.2 NdFeB Hard Magnets

Hard magnetic materials such as NdFeB alloys have high magnetic remanence and coercivity that lead to many practical applications. Off-axis electron holography has been used to image the magnetic induction in $Nd_2Fe_{14}B$ with nanometer-scale resolution and high signal-to-noise [45–47]. As a representative example, Fig. 5.4 shows reconstructed phase images obtained from a sample of die-upset $Nd_2Fe_{14}B$ that was prepared for electron microscopy observation by standard dimpling and ion-milling. A variety of serpentine and "Y"-shaped domains that extend to the sample edge are visible in Fig. 5.4a: These domain shapes are similar to those previously observed at grain boundaries in sintered NdFeB [48]. After heating the sample to a nominal temperature of 300 °C and subsequent cooling to room temperature, rearrangement of these domains occurred, as visible in Fig. 5.4b. The well-defined domains on the left-hand side have disappeared, and reduced phase gradients indicate that the remaining structure has a large out-of-plane component. Domain walls have been released from the thin edge of the sample, although some remaining domains were found to have interacted with structural features that were identified as planar defects and grain boundaries. Further heating to 400 °C, which is above the Curie temperature of 312 °C [49], resulted in the complete disappearance of the domain structure. A single pixel line scan across the central part of Fig. 5.4a put an upper limit of 10 nm on the domain wall width. This result agreed well with theoretical values [50], particularly bearing in mind that no special care was taken to ensure that the walls were parallel to the electron beam path.

In order to map the magnetic induction within the sample, simple gradients of the phase image at ±45° angles were calculated according to Eq. 5.4. These two gradient images were then combined to form a vector map, thereby producing a direct image of the magnetic induction within the domains, as shown in Fig. 5.5. The vector map in Fig. 5.5a is divided into 20 nm × 20 nm squares, and a low contrast image of the x-gradient has been superimposed for reference. The minimum vector length is zero (which is indicative of out-of-plane induction), while the maximum vector length, as calculated from the modulus of the gradients near the center of the image, corresponds to $4\pi M_s = B = 1.0$ T for an estimated sample thickness of 90 nm. The

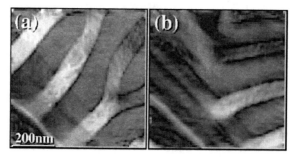

Fig. 5.4. Reconstructed phase images from $Nd_2Fe_{14}B$ sample: (**a**) before, and (**b**) after heating to 300 °C

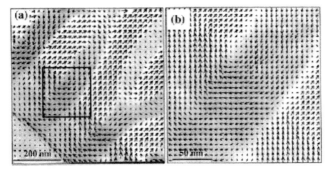

Fig. 5.5. (a) Induction map derived from gradients of phase image shown in Fig. 5.4; (b) enlargement of area indicated in (a) showing singularities and domain wall character

vector map shows that the magnetizations of the domains in this sample region are oriented at approximately 90° to each other, rather than 180°, as might be expected for a material having such strong uniaxial anisotropy. The domains near the sample edge close the in-plane induction more or less parallel to the thin edge. Note also the pairs of singularities at the intersection with the striped domains, which are reminiscent of cross-tie walls that contain periodic arrays of Bloch-lines of alternating polarity [50].

A vector map of the area outlined in Fig. 5.5a is shown at higher magnification in Fig. 5.5b. This map is divided into 6.7 nm × 6.7 nm squares with a maximum vector length again corresponding to 1.0 T. Portions of the domain walls that run perpendicular to the sample edge show distinct out-of-plane character, and there is an apparent tendency for the induction in the center domain to rotate toward a 180° orientation near the vortices. The vortices show Bloch-like character with vanishingly small vector length, indicating small in-plane components. The majority of the 90° walls show large in-plane components, as expected for a thin sample. However, it is emphasized that care is needed when interpreting fine details in such induction maps, due to the undetermined effects of the fringing fields immediately above and below the sample surfaces. Moreover, caution is needed when extrapolating the magnetization states of thin films to bulk materials. The need for an electron-transparent sample should always raise concerns about reliability whenever a bulk magnetic material is thinned for TEM examination.

5.4 Magnetic Nanostructures

5.4.1 Co Spheres

In small magnetic particles, the energy associated with the formation of domain walls becomes an increasingly large contribution to the total magnetostatic energy of the entire particle. Domain walls are then less likely to be observed within small particles when they are in their remanent state. Moreover, because of the limited resolution of most magnetic imaging techniques, it becomes increasingly difficult to characterize micromagnetic structure. Electron holography thus has an opportunity to provide unique experimental information unobtainable with other techniques. For example, isolated polyhedral particles of barium ferrite, with sizes ranging from about 0.1 to 1.0 μm, were studied using electron holography [51]. On the basis of the magnetic-flux-line geometry surrounding the particles, it was concluded that these existed as single magnetic domains. However, flux lines *within* the ferrite particles, and also in smaller-sized iron particles, were not identified in this early study, presumably because the dominant influence of the mean-inner-potential contribution to the phase and phase gradients could not be extracted (see Eqs. 5.3 and 5.4). It is our experience that information about the thickness profile of small particles must be obtained before their internal magnetization can be determined [52]. In some special geometries, such as for CrO_2 needles [53], the particle cross sections may be constant, thus allowing thickness variations to be neglected.

In some magnetic nanostructures, large variations in the projected thickness of the sample are unavoidable. Figure 5.6 shows an off-axis electron hologram obtained from a chain of carbon-coated spherical Co particles [54], and Fig. 5.6b shows the corresponding reconstructed phase image. Holographic analysis shows that the magnetic contribution to the phase shift of the Co particle along the line labeled "1" results in a difference in the value of the phase in vacuum on each side of the sphere. However, it is clear that the mean inner potential makes the dominant contribution to the phase profile. In this particular study, extensive modeling taking account of the

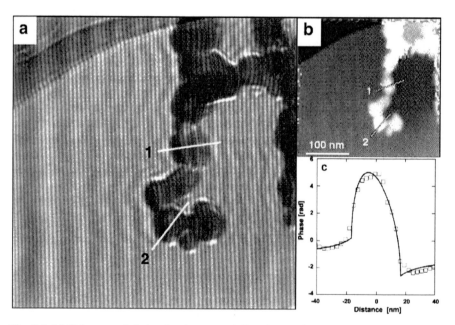

Fig. 5.6. (**a**) Hologram of chain of carbon-coated Co spheres; (**b**) reconstructed phase image; (**c**) phase profile along line in phase image in (**b**) (open squares). Solid line shows fitted phase shift calculated for an isolated sphere

particle shape was required before quantitative analysis of the magnetization within the crystalline nanoparticles could be completed.

5.4.2 Magnetotactic Bacteria

Certain sample geometries lend themselves to unidirectional remanent states, and phase shifts due to sample thickness and electrostatic effects can then be easily separated from those due to magnetostatic effects. Magnetization reversal can be accomplished *in situ* within the microscope by tilting the sample in the applied field of the conventional microscope objective lens. The external field is then removed, and separate holograms are recorded with the induction in the sample in two opposite directions [52]. The sum of the phases of these two holograms provides twice the mean inner potential contribution to the phase if the magnetization has reversed exactly (See Eqn. 5.3), while the difference between the phases provides twice the magnetic contribution. As an example of this procedure, Fig. 5.7a shows a hologram of a chain of magnetite crystals from a single cell of the aquatic magnetotactic bacterium *Magnetospirillum magnetotacticum*. Following magnetization reversal and addition of the phases, the electrostatic contribution is extracted as shown in Fig. 5.7b: the thickness contours reveal the crystallites to be cubocathedral in shape [55]. The magnetic contribution to the phase is shown in Fig. 5.7c. The phase contours are parallel to lines of constant magnetic induction integrated in the incident beam direction and have been overlaid onto the mean inner potential contribution to the phase so that the positions of the crystals and the magnetic contours can be correlated.

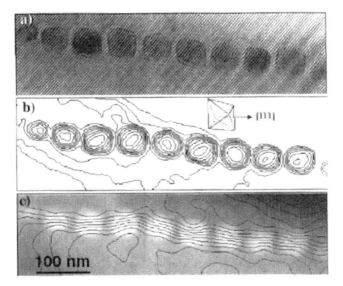

Fig. 5.7. (a) Hologram of chain of magnetite crystals in Magnetospirillum Magnetotacticum; (b) mean inner potential contribution to phase of reconstructed hologram. Thickness contours indicate cuboctahedral shape; (c) magnetic contribution to phase

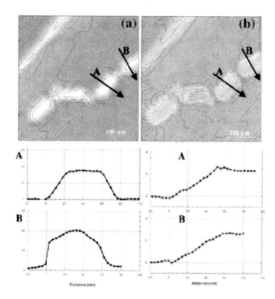

Fig. 5.8. (a) Thickness map and (b) phase contour map from part of a chain of magnetite magnetosomes from bacterial strain Itaipu 3. Line profiles along lines "A" and "B" are shown below the images

In this way, the magnetic flux both within and in between the crystallites becomes visible, and the total magnetic dipole moment of the particle chain can also be determined.

Further holography studies of magnetotactic bacteria have been reported [56,57]. Single cells from two different bacterial strains, designated as MV-1 and MS-1, consisted of magnetite crystals that were $\sim$ 50 nm in length and arranged in chains. Analysis revealed that the individual crystals existed as single magnetic domains [56], as anticipated from numerical micromagnetic modeling, which predicted that the transition to a multi-domain state should occur at a size of $\sim$ 70 nm for cube-shaped particles [58]. Furthermore, by careful monitoring of the phase shifts measured across the chain as a function of applied field, the coercive field was determined to be between 300 and 450 Oe. Similar analysis has been applied to chains of larger magnetosomes identified as Itaipu 1 and Itaipu 3 [57]. Figures 5.8a and b, respectively, show thickness and phase contour maps from part of a chain of Itaipu 3 magnetosomes. The corresponding line profiles along the lines designated "A" and "B" are shown directly underneath these two images. Such thickness profiles can be used to determine particle morphologies, and their magnetizations can then be determined from the phase maps.

5.4.3 Patterned Nanostructures

Because of their potential utilization for information storage, there is much current interest in the magnetization reversal of individual magnetic nanostructures. Due to their small size, magnetostatic energy contributions have a major influence in determining their overall magnetic response. Further important factors include individual particle geometries, as well as their proximity to other particles. We have used off-axis electron holography to investigate the micromagnetic behavior of a wide range of nanostructured elements. This approach enables the magnetization state to be visualized during hysteresis cycling [59–63].

Elements of various shapes, sizes, and separations were prepared by electron-beam evaporation onto self-supporting 55-nm-thick silicon nitride membranes using standard electron-beam lithography and lift-off processes [61]. Magnetic fields were applied by tilting the sample in situ within the microscope and then using the conventional microscope objective lens to obtain the desired field. The hysteresis loops of individual elements could thus be determined, and the extent of any interactions between closely-spaced elements could also be assessed.

The application of electron holography enabled flux lines both inside and outside the elements to be visualized even though substantial loss of interference-fringe contrast occurred due to the presence of the underlying silicon nitride support. The results of holographic reconstruction for two closely spaced Co rectangles over a complete hysteresis cycle, working in a counter-clockwise direction, are shown in Fig. 5.9. (The in-plane components of the applied field are indicated in each image.) The separation of the phase contours is proportional to the magnetic induction integrated in the incident beam direction. The fringing fields extending between elements are only minimized when the field lines are located entirely within both elements. The characteristic solenoidal shape then observed is indicative of flux closure associated with a vortex state.

Micromagnetic simulations based on the Landau-Lifshitz-Gilbert equations were made to aid in the interpretation of the phase profiles [60]. Reasonable agreement with the experimental results was obtained, although further analysis revealed subtle but important differences. Simulated vortices matched the experimental results except that they were formed at higher fields in the simulations, presumably because of local defects or inhomogeneities in the real elements. The simulated results were also affected by the squareness of the element corners, and slight changes in the initial state had a strong influence on subsequent domain formation during switching. Simulations also confirmed that the strength and direction of the applied field had a marked impact on the observed domain structure [60], thus emphasizing the need to correlate experimental measurements with micromagnetic simulations. Figure 5.10 compares simulations for two cells simulated as a pair (left set) and simulated separately (although displayed as a pair). The different domain configurations visible for the smaller cell in this figure demonstrate convincingly that the magnetization behavior of the adjacent Co elements is affected by their mutual proximity and interaction. Intercell coupling is clearly an important factor that must not be overlooked when designing magnetic storage devices with very high density.

A further important consideration for the design of magnetic elements for device applications is the formation of remanent states from different stages in a magnetization reversal loop. In normal operation, the storage element must first be magnetized and the external field then removed, the objective being to retain a non-solenoidal dis-

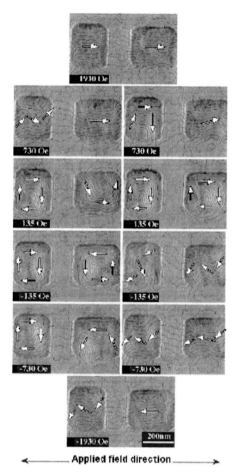

Fig. 5.9. Magnetic contributions to phase for 30-nm-thick Co elements over hysteresis cycle. Phase contours are separated by 0.21π radians. Field was applied in horizontal direction of figure, and loop should be followed counter-clockwise. Average out-of-plane field of 3600 Oe directed into the page

tribution. As shown by the representative examples in Fig. 5.11, none of the different remanent states of the two Co elements showed the non-solenoidal domain structures visible at the extreme ends of the hysteresis loop. Indeed, new and unexpected domain configurations were sometimes observed such as the double-vortex structure visible in the larger rectangle. It is possible that the presence of the external vertical field caused these states to be stable, Nevertheless, the value of experimental observations, and the need for reproducibility in the formation of remanent states, are emphasized.

Further studies of patterned nanostructures have focused on submicron Co (10 nm)/Au (5 nm)/Ni (10 nm) "spin-valve" (SV) elements shaped as rectangles, diamonds, ellipses, and bars with lateral dimensions on the 100-nm scale [61,63]. Similar

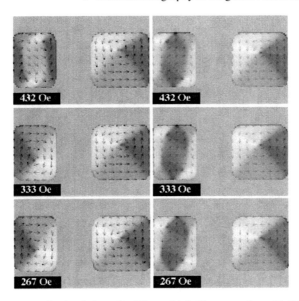

Fig. 5.10. Micromagnetic simulations for 30-nm-thick Co rectangles with 3600 Oe field directed into page, rounded bit corners, and applied fields as indicated. Left set simulated for two cells together. Right set simulated for cells separately (but shown together)

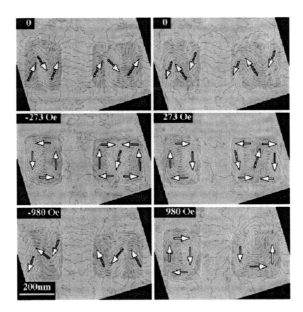

Fig. 5.11. Remanent states for 30-nm-thick Co elements. Left column: in-plane fields applied directly after large positive in-plane field. Right column: starting with large negative in-plane

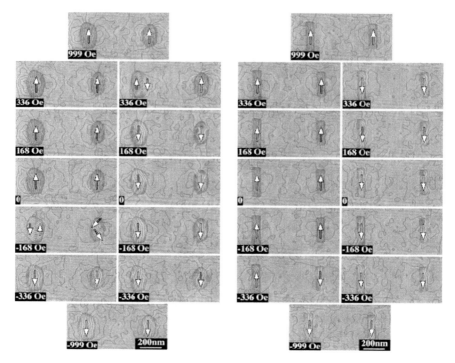

Fig. 5.12. Magnetic contributions to phase for elliptical and bar-shaped Co(10nm)/Au (5nm)/Ni(10nm) spin-valve elements over complete hysteresis cycle. Phase contours of 0.064π radians. Field applied in vertical direction. Loop should be followed counter-clockwise. Average out-of-plane of 3600 Oe directed into the page

SV layered structures are the subject of much current research and development activity because of the large differences in resistance obtained when the magnetization directions of the two magnetic layers are aligned either parallel or anti-parallel. The understanding and control of magnetization reversal in such layered combinations is essential for future information storage applications. Representative results from a complete hysteresis cycle for elliptical and bar-shaped elements are shown in Fig. 5.12. The direction of the in-plane component of the applied field is aligned along the long axis of each element, and the arrows within each element indicate the field direction. It was notable in additional studies of rectangular SV elements that, unlike some of the magnetization states during hysteresis cycling, vortex states were not visible in any of the remanent states. Micromagnetic simulations later suggested that this difference in behavior was most likely due to the presence of strong coupling between the Co and Ni layers within each element. It was also interesting that the smaller diamond-shaped elements had a substantially larger switching field, which could possibly be attributed to an increased difficulty in nucleating end domains that initiate magnetization reversal [64].

A significant observation made in these SV studies was the occurrence of two different contour spacings within each element at different applied fields, with corresponding steps in the hysteresis loops. Micromagnetic simulations, described in detail elsewhere [63], indicate that coupling between the Co and Ni layers accounted for this behavior. The Ni layer in each element reverses its magnetization direction well before the external field reaches 0 Oe. Thus, an antiferromagnetically coupled state, caused by the strong demagnetization field of the closely adjacent and magnetically more massive Co layer, must be the remanent state of these Co/Au/Ni SV elements. Flux closure associated with an antiferromagnetic state could also contribute to the absence of end domains, which are commonly observed in thicker single layer films of larger lateral dimensions. Solenoidal states were observed experimentally for both elliptical and diamond-shaped elements, but they could not be reproduced in the simulations. Structural imperfections such as crystal grain size or orientation are factors that could contribute to this difference in behavior.

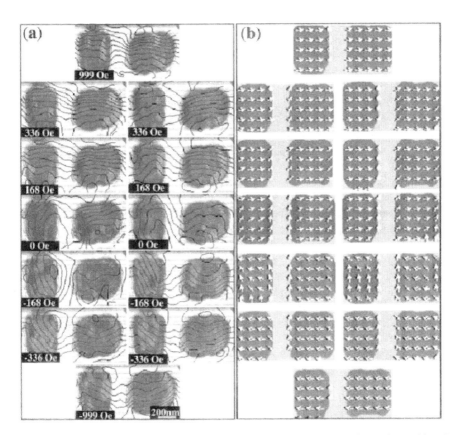

Fig. 5.13. Comparison of experimental and simulated magnetization states for exchange-biased CoFe/FeMn patterned nanostructures over complete magnetization reversal cycle. Applied in-plane field as shown directed along horizontal axis. Phase contours of 0.085 radians

In practical SV applications, it is usual to pin, or "exchange bias," the magnetization direction of one of the magnetic layers using an adjacent antiferromagnetic layer such as FeMn [65]. Off-axis electron holography has been combined with micromagnetic simulations to investigate magnetization reversal mechanisms and remanent states in exchange-biased submicron CoFe/FeMn nanostructures [61,62]. Figure 5.13 compares a typical experimental hysteresis loop with micromagnetic simulations, proceeding in a counter-clockwise direction with the in-plane applied field directed along the horizontal direction. No solenoidal or vortex states are observed, and complete reversal is not achieved, suggesting that shape anisotropy has a controlling influence on the response. Smaller and larger elements behaved quite differently, again implying the likely dominant role of shape anisotropy.

5.5 Outlook

The results described here convincingly demonstrate that off-axis electron holography has a valuable role to play in understanding the complex behavior of real magnetic nanostructures. The technique enables both visualization and quantification of magnetic fields with high spatial resolution and sensitivity. The theoretical basis for the technique is well established, and experimental applications to important materials are starting to be explored. As field-emission-gun TEMs become increasingly available, it can be anticipated that electron holography will develop into a widely used tool for micromagnetic characterization.

A major challenge for the technique is to provide direct visualization of dynamic effects induced by variations of the magnetic field, which are needed for a full understanding of micromagnetic behavior. The problem for the microscopist is that when changes are made to the applied field, the electron trajectories through the objective lens are seriously affected, and the holographic interference process is also likely to be impacted. Our experiments have involved tilting the sample in situ within the field of the weakly excited objective lens to change the in-plane field component [59]. However, the out-of-plane component of the applied field has been shown to impose switching asymmetries during hysteresis cycling of Co nanostructures [60]. An alternative, and obviously preferable, solution would be to provide three sets of auxiliary coils: One set would be used to apply the field at the sample level, and the subsequent sets would be used to steer the beam back onto the optic axis [66]. This possibility is not yet generally available.

The technique of off-axis electron holography is not well-suited to real-time observations because the holograms are usually processed off-line. One approach used to circumvent this problem has been to feed the signal from an intensified TV camera to a liquid crystal (LC) panel, which was then used to provide the input for a light-optical reconstruction [67]. However, geometric distortions may result from the TV camera and the LC panel, and reference holograms are not conveniently available to correct these distortions, unlike the situation for digital recording and processing. The dynamic behavior of magnetic domains in a thin permalloy film has been observed in real time using this type of system [51]. As faster computers and better CCD cameras become available, real-time viewing of reconstructed phase images may soon become possible.

Acknowledgement. Much of the electron holography described here was carried out at the Center for High Resolution Electron Microscopy at Arizona State University. We are grateful to Drs. R.F.C. Farrow, R.B. Frankel, B. Kardynal, S.S.P. Parkin, M. Posfai, M.R. Scheinfein, and Y. Zhu for provision of samples and ongoing collaborations.

References

1. J.P. Jakubovics (1976) In: Lorentz Microscopy and Applications (TEM and SEM). U. Valdre and E. Ruedl (eds) Electron Microscopy in Materials Science, Part IV, Commission of the European Communities, Brussels, p. 1303.
2. J.N. Chapman, J. Phys. D: Appl. Phys. **17**, 623 (1984).
3. S. McVitie and J.N. Chapman, Microscopy and Microanalysis **3**, 146 (1997).
4. H. Lichte, Adv. Opt. Electron Micros. **12**, 25 (1991).
5. A. Tonomura (1993) Electron Holography, Springer Series in Optical Sciences, Vol. **70**. Springer, Heidelberg.
6. J.M. Cowley, Ultramicroscopy **41**, 335 (1992).
7. G. Möllenstedt and H. Düker, Naturwissen. **42**, 41 (1955).
8. M.E. Haine and T. Mulvey, J. Opt. Soc. Amer. **42**, 763 (1952).
9. A. Tonomura, T. Matsuda, and J. Endo, Jpn. J. Appl. Phys. **18**, 1373 (1979).
10. A. Tonomura, Rev. Mod. Phys. **59**, 639 (1987).
11. E. Völkl and M. Lehmann, The reconstruction of off-axis electron holograms, In: E. Völkl, L.F. Allard, and D.C. Joy (eds.), Introduction to Electron Holography, Kluwer Academic, New York. pp. 125–151.
12. W.J. de Ruijter, Micron **26**, 247 (1995).
13. W.J. de Ruijter and J.K. Weiss, Ultramicroscopy **50**, 269 (1993).
14. D.J. Smith, W.J. de Ruijter, J.K. Weiss and M.R. McCartney (1999) Quantitative electron holography, In: E. Völkl, L.F. Allard, and D.C. Joy (eds.), Introduction to Electron Holography. Kluwer Academic, New York. pp. 107–124.
15. M. Gajdardziska-Josifovska, M.R. McCartney, W.J. de Ruijter, D.J. Smith, J.K. Weiss, and J.M. Zuo, Ultramicroscopy **50**, 285 (1993).
16. M. Mankos, A.A. Higgs, M.R. Scheinfein and J.M. Cowley, Ultramicroscopy **58**, 87 (1995).
17. A. Orchowski, W.D. Rau, and H. Lichte, Phys. Rev. Lett. **74**, 399 (1995).
18. D.J. Smith and M.R. McCartney (1999) Practical electron holography, In: E. Völkl, L.F. Allard and D.C. Joy (eds) Introduction to Electron Holography. Kluwer Academic, New York. pp. 87–106.
19. M.R. McCartney, D.J. Smith, R.F.C. Farrow, and R.F. Marks, J. Appl. Phys. **82**, 2461 (1997).
20. T. Hirayama, J. Chen, Q. Ru, K. Ishizuka, T. Tanji, and A. Tonomura, J. Electron Microsc. **43**, 190 (1994).
21. M.R. McCartney, P. Kruit, A.H. Buist, and M.R. Scheinfein, Ultramicroscopy **65**, 179 (1996).
22. D.C. Joy, Y.-S. Zhang, X. Zhang, T. Hashimoto, R.D. Bunn, L.F. Allard, and T.A. Nolan, Ultramicroscopy **51**, 1 (1993).

23. J.M. Zuo, M.R. McCartney, and J.C.H. Spence, Ultramicroscopy **66**, 35 (1997).
24. H. Lichte, H. Banzhof, and R. Huhle, In: Electron Microscopy 98, Vol. 1, H.A. Calderon Benavides and M.J. Yacaman (eds.), (IOP, Bristol, 1998) pp. 559–560.
25. D. Gabor, Proc. Roy. Soc. London, A**197**, 454 (1949).
26. A. Tonomura, T. Matsuda, J. Endo, H. Todokoro, and T. Komoda, J. Electron Micr. **28**, 1 (1979).
27. H. Lichte, Ultramicroscopy **20**, 293 (1986).
28. W.D. Rau and H. Lichte (1999) High resolution off-axis electron holography, In: E. Völkl, L.F. Allard, and D.C. Joy (eds.). Introduction to Electron Holography, Kluwer Academic, New York, pp. 201–229.
29. S. Frabboni, G. Matteucci, G. Pozzi, and M. Vanzi, Phys. Rev. Lett. **55**, 2196 (1985).
30. B.G. Frost, L.F. Allard, E. Völkl and D.C. Joy (1995), Holography of electrostatic fields, In: A. Tonomura, L.F. Allard, G. Pozzi, D.C. Joy, and Y.A. Ono (eds.) Electron Holography. Elsevier, Amsterdam, pp. 169–179.
31. M.R. McCartney, D.J. Smith, R. Hull, J.C. Bean, E. Voelkl, and B. Frost, Appl. Phys. Lett. **65**, 2603 (1994).
32. W.D. Rau, P. Schwander, F.H. Baumann, W. Höppner and A. Ourmazd, Phys. Rev. Lett. **82**, 2614 (1999).
33. M.A. Gribelyuk, M.R. McCartney, J. Li, C.S. Murthy, P. Ronsheim, B. Doris, J.S. McMurray, S. Hedge, and D.J. Smith, Phys. Rev. Lett. **89**, 022502 (2002).
34. V. Ravikumar, R.P. Rodrigues, and V.P. Dravid, Phys. Rev. Lett. **75**, 4063 (1995).
35. J.S. Barnard and D. Cherns, J. Electron Microscopy **49**, 281 (2000).
36. M.R. McCartney, F.A. Ponce, J. Cai, and D.P. Bour, Appl. Phys. Lett. **76**, 3055 (2001).
37. A. Tonomura, N. Osakabe, T. Matsuda, T. Kawasaki, J. Endo, S. Yano, and H. Yamada, Phys. Rev. Lett. **56**, 792 (1986).
38. J. Bonevich, K. Harada, T. Matsuda, H. Kasai, T. Yoshida, G. Pozzi, and A. Tonomura, Phys. Rev. Lett. **70**, 2952 (1993).
39. N. Osakabe, K. Yoshida, S. Horiuchi, T. Matsuda, H. Tanabe, T. Okuwaki, J. Endo, H. Fujiwara, and A. Tonomura, Appl. Phys. Lett. **42**, 792 (1983).
40. R.F.C. Farrow, D. Weller, R.F. Marks, M.F. Toney, A. Cebollada, and G.R. Harp, J. Appl. Phys. **79**, 5330 (1996).
41. D.G. Streblechenko, M.R. Scheinfein, M. Mankos, and K. Babcock, IEEE Trans. Magn. **32**, 4124 (1996).
42. D.G. Streblechenko, Ph.D. dissertation, Arizona State University (1999).
43. G. Matteucci, G. Missiroli, E. Nichelatti, A. Migliori, M. Vanzi, and G. Pozzi, J. Appl. Phys. **69**, 1853 (1991).
44. G. Lai, T. Hirayama, A. Fukuhara, K. Ishizuka, T. Tanji, and A. Tonomura, J. Appl. Phys. **75**, 4593 (1994).
45. M.R. McCartney and Y. Zhu, Appl. Phys. Lett. **72**, 1380 (1998).
46. M.R. McCartney and Y. Zhu, J. Appl. Phys. **83**, 6414 (1998).
47. Y. Zhu and M.R. McCartney, J. Appl. Phys. 84, 3267 (1998).
48. H. Kronmüller (1990) In: Science and Technology of Nanostructured Materials, G.C. Hadjipanayis and G. Prinz (eds.). Plenum Press, New York. p. 657.
49. J.F. Herbst and J.J. Croat, J. Magn. Magn. Mater. **100**, 57 (1991).
50. B.O. Cullity (1992) Introduction to Magnetic Materials. Addison-Wesley, New York.
51. T. Hirayama, J. Chen, T. Tanji, and A. Tonomura, Ultramicroscopy **54**, 9 (1994).
52. R.E. Dunin-Borkowski, M.R. McCartney, D.J. Smith, and S.S.P. Parkin, Ultramicroscopy **74**, 61 (1998).
53. M. Mankos, J.M. Cowley, and M.R. Scheinfein, Phys. Stat. Sol. (a) **154**, 469 (1996).

54. M. de Graef, T. Nuhfer, and M.R. McCartney, J. Microscopy **194**, 84 (1999).
55. R.E. Dunin-Borkowski, M.R. McCartney, R.B. Frankel, D.A. Bazylinski, M. Posfai, and P.R. Buseck, Science **282**, 1868 (1998).
56. R.E. Dunin-Borkowski, M.R. McCartney, M. Posfai, R.B. Frankel, D.A. Bazylinski, and P.R. Buseck, Eur. J. Mineral. **13**, 671 (2001).
57. M.R. McCartney, U. Lins, M. Farina, P.R. Buseck, and R.B. Frankel, Eur. J. Mineral **13**, 685 (2001).
58. K.A. Fabian, A. Kirchner, W. Williams, F. Heider, T. Leibel, and A. Hubert, Geophys. J. Int. **124**, 89 (1996).
59. R.E. Dunin-Borkowski, M.R. McCartney, B. Kardynal, and D.J. Smith, J. Appl. Phys. **84**, 374 (1998).
60. R.E. Dunin-Borkowski, M.R. McCartney, B. Kardynal, D.J. Smith, and M.R. Scheinfein, Appl. Phys. Lett. **75**, 2641 (1999).
61. R.E. Dunin-Borkowski, M.R. McCartney, B. Kardynal, S.S.P. Parkin, M.R. Scheinfein, and D.J. Smith, J. Microscopy **200**, 187 (2000).
62. R.E. Dunin-Borkowski, M.R. McCartney, B. Kardynal, M.R. Scheinfein, D.J. Smith, and S.S.P. Parkin, J. Appl. Phys. **90**, 2899 (2001).
63. D.J. Smith, R.E. Dunin-Borkowski, M.R. McCartney, B. Kardynal, and M.R. Scheinfein, J. Appl. Phys. **87**, 7400 (2000).
64. M. Rührig, B. Khamsehpour, K.J. Kirk, J.N. Chapman, P. Aitchison, S. McVitie, and C.D.W. Wilkinson, IEEE Trans. Magn. 32, 4452 (1996).
65. J.C.S. Kools, IEEE Trans. Magn. **32**, 3165 (1996).
66. J. Bonevich, G. Pozzi, and A. Tonomura (1999) Electron holography of electromagnetic fields In: Introduction to Electron Holography, E. Völkl, L.F. Allard, and D.C. Joy (eds.). Kluwer Academic, New York. pp. 153–181.
67. J. Chen, T. Hirayama, G. Lai, T. Tanji, K. Ishizuka, and A. Tonomura, Opt. Rev. **2**, 304 (1994).

6

SPLEEM

E. Bauer

Spin-polarized low energy electron microscopy is one of several methods for the study of the magnetic microstructure of surfaces and thin films on surfaces. It is a non-scanning, full-field imaging method that allows much faster image aquisition than scanning methods, provided that the electrons are elastically backscattered along or close to the optical axis of the instrument. This is the case in single crystals and epitaxial films or films with strong fiber texture. After a brief introduction (6.1), this chapter first discusses the physics of the electron beam–specimen interaction that is the basis of SPLEEM (6.2). This is followed by a brief description of the experimental aspects of the method (6.3). The remaining part is devoted to the applications of SPLEEM mainly in the study of thin film systems (6.4). The final section (6.5) briefly summarizes the possibilities and limitations of the method.

6.1 Introduction

SPLEEM (spin-polarized low energy electron microscopy) is an imaging method that is based on the spin dependence of the elastic backscattering of slow electrons from ferromagnetic surfaces. It is related to SPLEED (spin-polarized low energy electron diffraction) in the same manner as LEEM is to LEED. SPLEEM differs from LEEM in that the incident beam is partially spin-polarized, always normal to the surface, and only the specularly reflected beam and its close environment is used for imaging, while in LEEM sometimes tilted illumination or other diffracted beams are used. The normal incidence and reflection ensures, at least in the absence of multiple scattering, that spin-orbit interaction is not contributing to the signal, so that the magnetic contribution to the signal results from only the exchange scattering.

The first demonstration of the ability of spin-polarized slow electrons to provide information on the state of magnetization of a magnetic material was performed by Celotta et al. [1]. They used the specular reflection of 125 eV electrons incident at 12° from the normal of an in-plane magnetized thin Ni(110) crystal to measure the hysteresis curve and the temperature dependence of the magnetic contribution to the reflected intensity. When LEEM had demonstrated its possibilities in the late 1980s, one of the pioneers in the field, D. Pierce, suggested adding a spin-polarized gun to the LEEM system and looking at magnetic properties with high lateral resolution. It took, however, some strong stimulation-provided by H. Poppa – and financial support from a research laboratory strong in thin film magnetics, IBM Almaden Research Center, to follow up this suggestion more than 20 years after the first SPLEED experiments. The first results were reported in 1991 [2], and since then, the method has developed very slowly, in part due to the complexity of the systems, in part due to relocation and personnel changes.

SPLEEM has many of the possibilities and limitations that LEEM and SPLEED have. With LEEM, for example, it shares the resolution limitations, with SPLEED the fact that quantitative analysis is complicated by spin-dependent multiple scattering and attenuation. The main strengths are its high surface sensitivity and rapid image acquisition. The possibilities and limitations are in part determined by the beam-specimen interaction, in part by the instrument, i.e., the electron source, the electron optics, and the image detection.

6.2 Physical Basis of Beam-Specimen Interactions

As mentioned in the introduction, it is the exchange interaction between the spin-polarized beam electrons and the spin-polarized electrons in the ferromagnetic material – a consequence of the Pauli principle – that makes SPLEEM magnetization sensitive. Slater [3] has shown that this interaction can be represented by an exchange potential V_x that has to be added to the Coulomb potential V_c. V_c describes the interaction of an electron with the nuclei and all electrons, including itself. The original form of V_x was $-3[(3/8\pi)\rho]^{1/3}$, where ρ is the charge density. Hammerling et al. [4] and Kivel [5] were the first to use a potential of this form to calculate scattering cross sections of free atoms for very slow electrons ($E < 2$ eV) successfully. Calculations over a wider energy range [6], however, gave acceptable agreement with experiment only when different $V_{x\ell}$ were used for different orbitals. The agreement could be further improved [7] when the energy dependence

$$F(\eta) = 1/2 + [(1-\eta^2)/4\eta]\ln[(1+\eta)/(1-\eta)] \tag{6.1}$$

of the exchange potential of the free electron gas [8] was taken into account, so that

$$V_x(\eta) = -4[(3/8\pi)\rho]^{1/3} F(\eta), \tag{6.2}$$

where $\eta = k/k_F$, k being the wave number of an electron with kinetic energy $\hbar^2 k^2/2m$ and $k_F = (3\pi^2 \rho)^{1/3}$ the Fermi wave number. Experience with the application of this potential to atomic calculations has shown that the original expression overestimates the influence of the "exchange hole", which can be taken into account by a factor of about 2/3 [9]. The scattering calculations mentioned above were made without regard to spin, that is for unpolarized beam and target. When the spin polarization of the electron or of the target is of interest, a more sophisticated treatment is needed, as described in Chap. 3 and 4 of [10]. However, for atoms for which the Thomas-Fermi-Dirac model of the electronic structure is a reasonable approximation, the exchange interaction may be approximated again with an energy-dependent Slater-type potential for spin up and spin down charge densities separately. The good agreement that Slater et al. [9] obtained with various forms of V_x for the various ground state quantities of the closed shell ion Cu^+ suggests that $V_x^\uparrow(\eta)$ and $V_x^\downarrow(\eta)$ should also be good approximations of the exchange interactions with spin up and spin down electrons, respectively, in electron scattering.

In the elastic scattering of slow spin-polarized electrons from crystals, that is, in SPLEED, not only is the energy- and spin-dependent exchange potential of importance, but also proper inclusion of multiple scattering. This is due to the fact that the wave field incident on a given atom is not only the plane wave incident from the outside, but consists also of the waves scattered by the surrounding atoms. The multiple scattering ("dynamical") theory required in this case has reached a high level of sophistication in LEED and has also been developed for SPLEED mainly by Feder and has been well reviewed by him [10–12]. In this theory, the diffracted intensity is calculated for positive and negative V_x contribution corresponding to the scattering of electrons with spin parallel and anti-parallel to the spins in the magnetic material. The magnetic signal, the so-called scattering asymmetry, is obtained from the normalized intensity difference

$$A = (I_+ - I_-)/(I_+ + I_-). \tag{6.3}$$

A discussion of the SPLEED theory is beyond the scope of this review, particularly since very little theoretical work has been done for the conditions used in SPLEEM (180° scattering, very low energies). In the energy range of SPLEEM (< 10 eV) fewer partial waves are needed and the exchange potential is significantly stronger than at the usual SPLEED energies. For a typical SPLEEM energy (2 eV) V_x is about four times larger than at a typical SPLEED energy (50 eV) for Fe, Co, and Ni. In the scattering from the solid state potentials of these atoms – whose energy zero is the muffin tin zero – the phase of the $l = 2$ partial wave passes through $\pi/2$ near the vacuum level. This causes strong backward scattering that is different for spin-up and spin-down electrons because of the different V_x contributions $(+/-V_x)$. As a consequence, A is large at these energies. Of course, the spin-independent part of the backward scattering is also large, which facilitates focusing and astigmatism correction. As a consequence of multiple scattering, the connection between scattering asymmetry and magnetization is not straightforward. It is generally accepted that the magnetization is layer-dependent. First-principle calculations give enhancements ranging from a few percent to about 50% for the topmost layer that decrease rapidly to the bulk value within a few atomic layers. SPLEED calculations for various layer profiles show [13]:

i) the top layer magnetization M_s determines A if all other layers have bulk magnetization M_b, but A is not proportional to M_s or some average magnetization;
ii) if all layers have the same magnetization M, then A is proportional to $M = M_s = M_b$;

iii) if the layer dependence of the magnetization M_n is homogeneous, then A is proportional to M_s.

These results are for a rigid lattice ($T = 0$). Thermal vibrations reduce the spatial coherence in multiple scattering and relax these limitations somewhat. The temperature dependence of the magnetic correlation length ξ plays a more important role. If ξ is large compared with the thickness contributing to the scattering, then the M_n scale is like M_s. ξ becomes large in the temperature range close to the Curie temperature T_C, in which the asymptotic power laws of the critical behavior are valid.

These theoretical considerations for SPLEED, wich are at least to some extent also valid for SPLEEM, suggest that SPLEED and SPLEEM give information on the surface magnetization and its critical behavior if used with proper caution. To which extent this is the case can only be decided by a comparison between theory and experiment. A test case for the ability of SPLEED to determine the magnetization profile M_n ($n = 1 = s, 2, 3, 4$) is the Fe(110) surface. Theory [14] predicts enhancement factors M_n/M_b of 1.194, 1.068, 1.027, and 1.014 in the first four layers over the bulk magnetization. The experimental SPLEED data for thin Fe(110) films on W(110) of Waller and Gradmann [15] were analyzed by two groups. Tamura et al. [16] analyzed the experimental data for 62 eV and concluded $M_1/M_b = 1.35$ and $M_2/M_b = 0.85$, confirming an earlier, less detailed analysis [17]. Ormeci et al. [18,19] simulated the theoretical M_n/M_b data [14] in their model calculations and obtained equally good agreement with the experimental data for 62 eV, but found little sensitivity to changes of M_1/M_b at other energies. On the other hand, a large surface enhancement is also deduced from magnetometric measurements [20]. Thus, it appears difficult to extract magnetization profile information from SPLEED measurements. A similar conclusion was drawn by Plihal et al. [21] from a comparison of SPLEED calculations with experimental results for the Fe double layer on W(100). On the Ni(100) surface, an enhancement in the range from 0 to 10% was found [22], in agreement with theoretical values of 5.8–6.8%. On the Ni(111) surface $M_s/M_b = 0.9-1.3$ was deduced from a theory-experiment comparison [23] in agreement with the theoretical value 1.16.

These comparisons show that there is considerable uncertainty regarding the relation between A and M. Two observations are of interest in this connection. Kirschner [24] found in his study of the temperature dependence of A of a Fe(110) surface strong deviations from the expected behavior in the 10 eV range, while the $A(T)$, measured at 3.5 eV, fitted the bulk behavior quite well. Elmers and Hauschild [25] found characteristic deviations from the asymptotic power law $A(T) \propto (T_C - T)^\beta$ for Fe films on W(100) that were thicker than two monolayers. These observations suggest the use of either very slow electrons, that is to work in the SPLEEM energy range or to restrict the measurements to thicknesses below two monolayers if reliable information on M is to be derived from A, or to use A only as an indicator of the magnetic behavior. The latter philosophy was adopted in a study of Fe films on W(100) [26], the second has been followed with considerable success in Gradmann's group, and the results have been well reviewed by Elmers [27]. More work of this kind, all on Fe films on W(110) and W(100) can be found in references [28–30]. The first philosophy is the basis of recent SPLEEM studies of the magnetization of thin Fe films on Cu(100) [31] and of Co films on W(111) [32], which will be discussed below.

As pointed out earlier, at very low energies, the phase of the $l = 2$ partial wave passes through $\pi/2$, causing strong spin-dependent backscattering and, thus, a significant A. This simple picture is applicable in the case of amorphous or polycrystalline samples. In single crystals, the band structure also plays an important role, as illustrated in Fig. 6.1 for a Co(0001) surface. The band structure in the ΓA direction ([0001] direction) is exchange-split with an exchange splitting between 1.0 and 1.6 eV, depending upon the computational method. Below the two bands there is a 6–7 eV wide band gap in which the wave number k is imaginary. The material acts as a reactive medium with (theoretically) 100% reflectivity. When the electron energy is increased from zero to the onset of the spin-up band, the reflectivity for spin-up electrons suddenly decreases, while the spin-down electrons are still 100% reflected until the onset of the spin-down band is reached. Therefore, A is negative between the two band onsets. With further increasing energy, A is also nonzero, because of the slightly different density of states at a given energy. This picture has to be refined below by the inclusion of inelastic processes. Figure 6.1 also shows that the k-values of spin-up and spin-down electrons at a given energy differ. This has important consequences for thin film studies in which "quantum size effects" occur due to the interference effects in thin films well known from optics: Whenever the wave length $\lambda = 2t/n$ (n integer), the reflectivity has a maximum that occurs at a different thickness for spin-up and spin-down electrons because of $\lambda_\uparrow \neq \lambda_\downarrow$. A also oscillates with energy because the interference condition is fulfilled for different energies for spin-up and spin-down electrons. Fig. 6.2 illustrates this for a 6 monolayer thick Co film on W(110) [33]. The oscillations are clearly seen despite the fact that the film is a three-level system consisting of 5, 6 and 7 monolayer thick regions. At the first quantum size resonance the asymmetry is much larger than upon reflection from the surface of a thick crystal that can be utilized for very efficient polarization detection. The experimental data could be well fitted with SPLEED calculations [34].

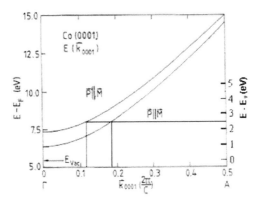

Fig. 6.1. Band structure of Co in the ΓA direction ([0001]) above the vacuum level

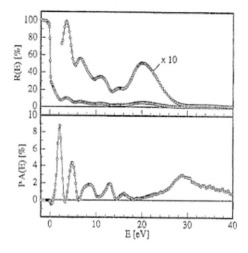

Fig. 6.2. Intensity (top) and asymmetry (bottom) of the specular beam from a six-mono-layer-thick epitaxial Co layer with (0001) orientation on a W(110) surface as a function of electron energy

The discussion up to now did not take into account inelastic scattering. Except for the elastic attenuation of the incident wave at energies in the band gaps, inelastic scattering is the main factor that determines the sampling depth of slow electrons. At the low energies generally used in SPLEEM, $3p \rightarrow 3d$ excitations, which have been considered important in the SPLEED energy range [22,23], and plasma excitations do not occur, only electron-hole pair creation, phonon, and magnon excitations occur. The last two processes involve small energy losses ("quasi-elastic scattering") and cannot be separated from elastic scattering in SPLEEM because of insufficient energy resolution. Phonon excitation attenuates the reflected beam irrespective of its polarization because of the large momentum change associated with them and the small transverse momentum acceptance of the SPLEEM instrument dictated by the large aberrations of the objective lens. Magnon excitation is possible only by spin-down electrons. It involves a spin flip and thus increases the number of spin-up electrons upon reflection, that is, it changes the polarization of the beam. The cross section for magnon excitation has long been considered to be orders of magnitudes smaller than that for the (spin-conserving) phonon and elastic scattering [35,36]. Recent detailed calculations for Fe and Ni by Hong and Mills [37] have confirmed this for electron energies above the vacuum level, but at the same time have shown that magnon excitation is an important damping mechanism in Fe for electrons close to the Fermi level. In the energy range of SPLEEM, magnon excitation may be neglected.

Electron-hole pair creation can occur with spin-flip (Stoner excitations) or without spin-flip (direct excitations). Both excitations attenuate the signal when the momentum change is larger than that accepted by the angle-limiting aperture, or when the energy change is so large that the electron is deflected outside the angle-limiting aperture by the dispersion of the beam separator. In practice, the second attenuation effect is unimportant because of the small energy change, the weak dispersion, and the aperture sizes used. In addition to the attenuation, Stoner excitations with sufficiently small momentum and energy transfer (so that the reflected electron is accepted by the angle-limiting aperture) change the polarization of the reflected beam. In these excitations, a spin-down (spin-up) electron drops into an empty minority (majority) spin state and excites via an exchange process a majority (minority) spin electron to the final state with a lower energy than that of the incident electron. Because the density of unoccupied minority spin states is much larger than that of the majority spin states, this process effectively increases (decreases) the number of spin-up (spin-down) electrons in the reflected beam. This is best seen in the case of Ni, in which the majority spin band is completely filled [38].

As a consequence, the total inelastic scattering cross section σ is spin-dependent and is given for spin-up ($\sigma_\uparrow$) and spin-down ($\sigma_\downarrow$) electrons to a first approximation by [39]

$$\sigma_{\uparrow\downarrow} = \sigma_o + \sigma_d(5 - n_{\uparrow\downarrow}). \tag{6.4}$$

Here σ_d is the cross section for scattering into an unoccupied d-orbital, σ_o the cross section for scattering into all other non-d-orbital, and $n_\uparrow$, $n_\downarrow$ are the numbers of occupied majority and minority spin states, respectively. These cross sections may differ by a factor as large as three [40]. Recent model calculations for Fe, Co, and Ni [41] and explicit calculations for Fe and Ni [37] confirm the general trend with the number $5 - n_{\uparrow\downarrow}$ of d-holes, but show that the relationship is not linear. Figure 6.3 [37] shows the inelastic mean free paths $l_\uparrow$ and $l_\downarrow$ for spin-up and spin-down electrons in Fe and Ni derived from the imaginary part of the self-energy of an electron with energy E above the Fermi level. The work functions of Fe(110) and Ni(111) surfaces are approximately 5.1 and 5.55 eV. Although the largest differences between $l_\uparrow$ and $l_\downarrow$ occur below the vacuum level, they are still significant in the SPLEEM energy range: In Fe $l_\uparrow/l_\downarrow \approx 3.2$, in Ni $l_\uparrow/l_\downarrow \approx 1.8$ at 1 eV above vacuum level. 5 eV above the Fermi level, this ratio is still about 2.0 for Fe, while in Ni it has already decreased to about 1, that is, the IMFP is already spin-independent [37]. From experiment, however, a ratio of about 3 has been deduced for Ni at a still higher energy (12 eV above the vacuum level) [38]. Thus, there is still some uncertainty in the magnitude of the difference of the two IMFPs, but there is no doubt that spin-down electrons are damped more than spin-up electrons.

A comparison of the IMFPs with the elastic attenuation lengths in band gaps suggests that the importance of band gaps for the magnetic contrast, as determined by the asymmetry A, has been overestimated in the past. As an example we consider the Co(0001) surface using the model of Feibelman and Eastman (Sect. IV.B in [42]) for the calculation of the elastic attenuation lengths $a_{\uparrow\downarrow}$ and an interpolation between Fe and Ni based on the calculations of Hong and Mills [37,41] for the IMFPs of Co [43]. At the vacuum level $a_\uparrow \approx 0.35$ nm, $a_\downarrow \approx 0.30$ nm, at the spin-up band onset at 0.9 eV (see Fig. 6.1) $a_\uparrow \to \infty$ while $a_\downarrow \approx 0.38$ nm vs. $l_\downarrow \approx 0.31$ nm. At the spin-down band onset at 2.0 eV, where also $a_\downarrow \to \infty$, $l_\downarrow \approx 0.33$ nm. This shows that between the band onsets the elastic attenuation length for spin-down electrons

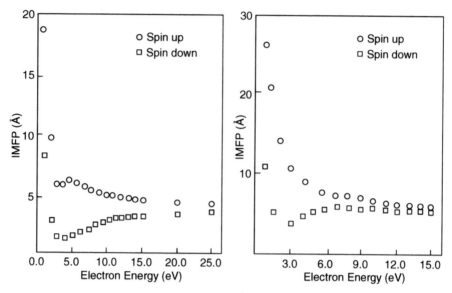

Fig. 6.3. Inelastic mean free paths for spin-up and spin-down electrons in Fe (left) and Ni (right) as a function of electron energy

is sufficiently large, so that significant inelastic damping, that is, reduction of $l_\downarrow$, can occur, resulting in a smaller asymmetry than that in the absence of (inelastic) damping, which is determined by $\boldsymbol{P} \bullet \boldsymbol{M}$. This reduction is, however, apparently insignificant, as several transmission and reflection experiments suggest [44–46]. For example, in Co the contribution of Stoner excitations – which are responsible for the difference of the IMFPs – to the polarization is below 5% for electrons between 5 and 15 eV above the vacuum level [45]. For this reason, inelastic damping is presently taken into account mainly to estimate the sampling depth and the various layer contributions to the signal. The asymmetry is attributed mainly to the exchange potential V_x in the elastic scattering from the ion cores that contain most of the spin density (see for example Fig. 6.1 in [47]). Band gaps caused by the periodic array of the ion cores are relegated into a second place because their influence is largely suppressed by inelastic scattering, as discussed above. In films whose thickness is less than the elastic attenuation length and the IMFP, e.g., in monolayers and double layers, these considerations do not play an important role.

To summarize this section, SPLEEM may be used as a qualitative measure of the magnetization of thin films or of the near-surface region of crystals. Before quantitative conclusions about the magnitude and depth distribution of the magnetization can be drawn from the measured asymmetry, a much better understanding of the elementary interaction processes is necessary. What SPLEEM can do at present is determine the lateral magnetization distribution; in particular, it can image magnetic domains and domain walls.

6.3 Experimental Setup and Procedure

A SPLEEM instrument is a LEEM instrument with a spin-polarized illumination system. The principles and realization of LEEM have been described repeatedly (see, for example [48,49]). Therefore, they will only be sketched briefly. In a LEEM instrument (Fig. 6.4), the object (1) is part of a "cathode" objective lens (2), in which the incident electrons from the illumination system (3) are decelerated from high energies, typically 5–20 keV, to the desired low energy with which they enter the object at (near) normal incidence. After reflection they are reaccelerated in the cathode lens to their original energy – minus possible energy losses in the object. Incident and reflected beams are separated by a magnetic sector field (4) with deflection angles ranging from 20 to 90° in the various instruments. The incident electrons are focused into the back focal plane of the objective lens, so that the illumination of the object is (nearly) parallel. As a consequence, the diffracted electrons produce in this plane the diffraction pattern that is transferred further downstream with a "transfer lens" (5). The angle-limiting aperture (6) is placed into this image of the diffraction pattern. The objective lens images the first image of the object into the center of the beam separator. An intermediate lens (7) allows one to image either the first image of the object or the image of the diffraction pattern in front of the projective lens (8), which produces the final image or diffraction pattern on the microchannel plate – fluorescent screen image detection system (9). The image on the fluorescent screen is recorded with a CCD camera.

In an ordinary LEEM instrument, the illumination system uses a LaB_6 or field emission cathode whose crossover is imaged with 2–3 condenser lenses into the back focal plane of the objective lens. In a SPLEEM system, the electron source is a GaAs(100) surface with high p-doping and negative electron affinity, which is obtained by Cs and oxygen deposition. Spin-polarized electron emission is stimulated

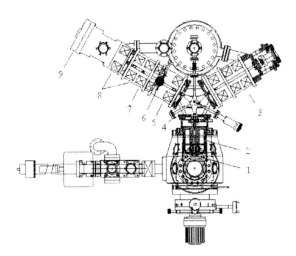

Fig. 6.4. Schematic of a LEEM instrument. For explanation, see text (courtesy C. Koziol)

by illumination with circularly polarized light (σ^+, σ^-), usually from a semiconductor diode laser with a wavelength close to the band gap energy. The conduction band has predominantly s-character and is twofold degenerate ($m_j = -1/2, +1/2$), the valence band has predominantly p-character and is split by spin-orbit interaction into a fourfold degenerate $p_{3/2}$ band ($m_j = -3/2, -1/2, +1/2, +3/2$) and a twofold degenerate $p_{1/2}$ band ($m_j = -1/2, +1/2$). The selection rules and relative transition probabilities give a maximum polarization P_{max} of 50% of the excited electrons in the crystal. The polarization of the emitted electrons depends on the doping level and profile, on the surface cleaning, in particular, the removal of carbon, and on the activation. 25% to 30% polarization is usual, but P values of close to 50% have also been reported. With strained lattice and quantum well superlattice emitters, P values up to 80% and more have been obtained, though with much lower quantum yield, typically less than 0.1% at the wavelength of maximum P [50]. In contrast, ordinary GaAs emitters may have quantum yields as high as several percent. The vacuum around the emitter is crucial to its lifetime. Less than 1×10^{-10} Torr are needed for lifetimes longer than one day. The science and technology of spin-polarized electron sources has been well reviewed by Pierce [51].

A Pockels cell allows one to switch rapidly between σ^+ and σ^- light and thus between the sign of the polarization of the emitted electrons ($\boldsymbol{P} \rightarrow -\boldsymbol{P}$). For SPLEEM experiments, one wants not only to change between $\boldsymbol{P}$ and $-\boldsymbol{P}$, but also to be able to orient $\boldsymbol{P}$ in any direction, usually into two orthogonal directions in the surface and one perpendicular to it. This can be done with a spin manipulator that is schematically shown in Fig. 6.5 [52]. The combined electrostatic-magnetic 90° deflector, together with the Pockels cell, allow one to orient $\boldsymbol{P}$ in any direction in the plane of deflection by changing the ratio of electrostatic to magnetic deflection. Whenever $\boldsymbol{P}$ has a nonzero transverse component, it can be rotated around the optical axis by changing the magnetic field of the axial-symmetric "rotator" lens. In addition, the spin manipulator contains condenser lenses, deflectors, and a stigmator, as is usual in electron-optical illumination systems. Both presently existing SPLEEM instruments use this principle, but differ in that in one (system A), which is described briefly in [53], the emitter is at high negative potential (-15 kV), while in the other (system B) [54, 55] it is at ground potential, which requires different design in detail.

The next instrument component is the beam separator. In system A, it is a single magnetic deflection field with circular boundary and a semi-circular cutout that acts as a nearly non-focusing 60° deflector. In system B, the beam separator is split into three segments, each deflecting 45° in order to fit the instrument onto a 6-inch flange. Stigmators are needed in both systems to correct for the residual astigmatism. The fields are weak enough, so that the spin precession in them may be neglected. The heart of the instrument is the objective lens that produces the first image and the diffraction pattern, in addition to decelerating and reaccelerating the electrons in front of the object. Both instruments use an electrostatic tetrode lens that has only slightly larger aberrations than magnetic lenses, but significantly smaller ones than the electrostatic triode lens originally used. The aberrations, in particular, the chromatic aberration, determine the resolution limits, which are shown in Fig. 6.6, as a function of the start energy of the electrons for an energy half-width of 0.5 eV. At a typical

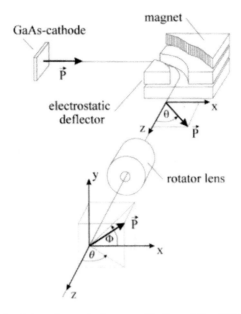

Fig. 6.5. Schematic of the spin manipulator used in the existing SPLEEM instruments

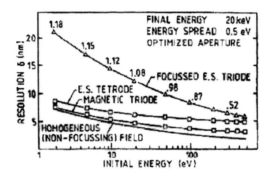

Fig. 6.6. Resolution of various cathode lenses as a function of electron energy. The numbers are typical voltages on the center electrode needed for focusing

SPLEEM energy of 2 eV, a resolution of 8 nm could be expected with the tetrode lens, but in practice it is usually larger than 10 nm.

The magnetic contrast is proportional to $\vec{P} \cdot \vec{M}$. Separation of the magnetic contrast from the structural contrast requires the acquisition of two images with opposite polarization ($I_\uparrow$, $I_\downarrow$) and their subtraction from each other. In order to normalize these images, the difference image is divided by the sum image. This results in an image that is proportional to the asymmetry A : $PA = P(I_\uparrow - I_\downarrow)/(I_\uparrow + I_\downarrow)$. When $I_\downarrow > I_\uparrow$, A is negative. Furthermore, the magnetic contrast is usually small compared with the structural contrast. Therefore, the final magnetic image is displayed in the form

$$A^* = N/2 + C(I_\uparrow - I_\downarrow)/(I_\uparrow + I_\downarrow). \tag{6.5}$$

Here, N is the number of gray levels of the CCD camera and C is a contrast enhancement factor chosen in such a manner so that no pixel in the final (A^*) image is saturated. During routine studies, image subtraction is done on-line, which sometimes causes loss of resolution or artificial contrast due to specimen shifts between two images. Typical image acquisition times range from 1 to 5 sec per image. Image shifts can be eliminated by off-line processing with suitable algorithms. A purely structural image is obtained from $I_\uparrow + I_\downarrow$.

In view of the fact that the magnetic contrast is proportional to $\boldsymbol{P} \bullet \boldsymbol{M}$, $\boldsymbol{P}$ is chosen to be parallel and anti-parallel to $\boldsymbol{M}$ for easy analysis if there is a strong in-plane magnetic anisotropy and similarly perpendicular to the surface in the case of pure out-of-plane magnetization. In general, $\boldsymbol{M}$ shows a more complicated angular distribution, so images with three $\boldsymbol{P}$ directions have to be taken for complete analysis. In the case of uniaxial in-plane anisotropy, for example, Néel domain walls can be imaged with $\boldsymbol{P} \perp \boldsymbol{M}$. These procedures are illustrated in Fig. 6.7 for an eight-monolayer-thick epitaxial Co film on W(110), which has a strong uniaxial in-plane anisotropy with the easy axis in the [−110] direction. Whenever the direction of $\boldsymbol{P}$ is changed, the illumination conditions are changed somewhat due to imperfections

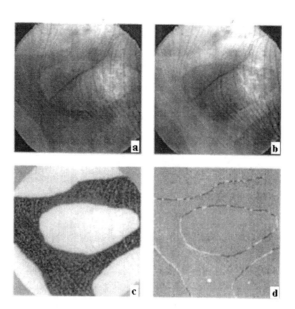

Fig. 6.7. SPLEEM images of a 6 monolayer thick Co layer on W(110). (a) and (b) are the two images taken with opposite polarization parallel to the easy axis, (c) the contrast-enhanced difference image, and (d) the difference image, of images taken with in-plane polarization perpendicular to the easy axis. Field of view 13 μm

in the illumination system and in the alignment of the laser beam (Pockels cell switching) and the electron beam. These can be largely corrected with deflectors, but usually cause some intensity changes that are eliminated in the final magnetic image by the normalization procedure described above.

As discussed in Sect. 6.2, SPLEEM gives at least a relative measure of the magnetization averaged with an exponential damping factor over the sampling depth. In thin films that show a spin-dependent quantum size effect, it is in principle possible to obtain information on the perpendicular magnetization distribution by shifting the nodes and antinodes of the waves via changing the energy once the necessary theory is developed.

6.4 Applications

6.4.1 Single Layers

Single layers were studied with SPLEEM mainly in order to understand i) how the magnetic domain structure and the magnetization direction evolve with increasing film thickness, ii) how they are influenced by substrate defects, and iii) how they are modified by segregated and deposited thin nonmagnetic overlayers. Most of the work has been done on Co on W(110), followed by Co on Au(111) and W(111). Fe on Cu(100) has also been a subject of recent studies. In the early Co/W(110) studies, two phases of Co were observed: fcc Co with the (100) plane parallel to the substrate in two equivalent azimuthal orientations and hcp Co with the epitaxial orientation $(0001)_{Co}$ ∥ $(110)_W$, $[-1100]_{Co}$ ∥ $[-110]_W$ [56]. The former developed from a poorly ordered film deposited at room temperature on carbide-covered surfaces upon annealing to 530 K and showed magnetic domains with several gray levels, that is, no preferred magnetization direction. The hcp phase was obtained on the clean surface, was also well-ordered when deposited at room temperature and showed a pronounced in-plane anisotropy with the easy axis along the W[-110] direction. In this direction, the misfit is small, so the layer is dilated by 2.4% into a one-dimensional pseudomorphy, while in the W[001] direction, the misfit is large, so the layer is one-dimensionally floating. The strong magnetic anisotropy was attributed to the strain in the W[-110] direction. In both phases, magnetic contrast appeared at 3 monolayers (ML) and increased up to at least 6 ML. In this early work, P could be rotated only parallel to the surface. Once the spin manipulator was also available, the out-of-plane component of M became accessible. This led to the discovery that Co layers on clean W(110) also had an out-of-plane M component (Fig. 6.8) [57]. The deviation of the resulting M direction from in-plane magnetization decreased from about 36° at 3 ML to about 6° at 8 ML in an apparently oscillatory manner and was practically zero at 10 ML. As also seen in Fig. 6.8, the domain boundaries of the in-plane M images are little influenced by the monatomic substrate steps (Fig. 6.8a), while the domain boundaries of the out-of-plane M images in general coincide with substrate steps (Fig. 6.8b). Neither of them is influenced by the thickness variation in the three atomic level system consisting of 4, 5, and 6 ML thick regions, which

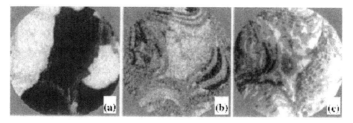

Fig. 6.8. SPLEEM images of the in-plane (**a**) and out-of-plane (**b**) magnetization components, taken with 1.2 eV electrons, and LEEM quantum size contrast image (**c**) taken at 3 eV of a five-monolayer-thick Co layer on W(110). Field of view 6 μm

are made visible by quantum size contrast in the LEEM image of Fig. 6.8c. The out-of-plane M component alters sign from monatomic terrace, to monatomic terrace so that a wrinkled magnetization results. The thickness dependence of the tilt angle of M was attributed to the influence of the second and fourth order interface anisotropy between Co and W. The sign change of the out-of-plane M component is due to the minimization of the dipolar energy, and its preferred occurrence at steps is caused by local anisotropies at steps at which there is a strong misfit perpendicular to the surface.

In addition, the evolution of the direction of M in island films with increasing thickness or temperature was studied [58]. At elevated temperature, but below the temperature at which alloying starts ($\approx$ 800 K) Co grows on W(110) in the Stranski-Krastanov mode with three-dimensional islands in a closely packed monolayer sea. The islands nucleate preferentially at W mesas, the remnants of previous alloying accidents, and grow preferentially in the floating layer direction W[001] (Fig. 6.9a). Independent of the aspect ratio of the islands M points in the easy direction (W[-110], Fig. 6.9b), indicating that the magnetoelastic anisotropy is in all cases larger than the shape anisotropy. For this reason, it is not surprising that the local magnetization is preserved when a thin continuous quasi-two-dimensional layer breaks up into many small three-dimensional islands upon annealing [58].

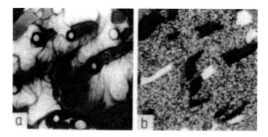

Fig. 6.9. LEEM image (left) and SPLEEM image (right) of a Co layer grown on W(110) at 790 K. The flat but three-dimensional Co crystals nucleate preferentially at W mesas and grow preferentially in the W[001] direction. Field of view 14 μm

One of the most interesting phenomena in quasi-two-dimensional layers are the standing waves in them that cause the quantum size oscillations discussed in Sect. 6.2 (Fig. 6.2). Standing waves in ferromagnetic films have become popular in connection with the interlayer coupling that will be discussed later. SPLEEM offers the possibility to study them with atomic depth.

The W(110) surface is atomically rather smooth and allows the Co layer to approach the bulk structure relatively fast. In contrast, the W(111) surface is atomically rough and imposes its lateral periodicity onto the growing Co film over a number of atomic layers, which depends upon the growth temperature [59]. Because of the large difference between the atomic diameters ($d_{Co} = 0.251$ nm, $d_W = 0.274$ nm), this pseudomorphic (ps) layer is strongly contracted perpendicular to the surface, so that three successive ps Co layers may be considered as a strongly corrugated Co(0001) layer with somewhat smaller packing density (17.30×10^{14} atoms/cm^2) than a Co(0001) layer in the bulk (18.37×10^{14} atoms/cm^2). On the basis of this consideration, one would expect no preferred M direction in the layer because the in-plane magnetocrystalline anisotropy of the Co(0001) plane is very small. The first experiments, however, showed a pronounced uniaxial anisotropy with large domains. This has been attributed to a slight miscut indicated by a small elongation of the LEED spots. On a second crystal that showed sharp LEED spots, small domains with several M orientations were observed whose size, shape, and distribution depended strongly upon the deposition conditions. This is compatible with negligible in-plane anisotropy and nucleation of non-interacting magnetic domains in different surface regions at zero external fields. On both substrates, the asymmetry is zero up to about ps 7 ML, then rises rapidly up to 8–9 ps ML and thereafter at a lower rate linearly up to the largest thickness studied (12 ps ML). Extrapolation to zero magnetic signal suggests that the first three ps ML, which correspond to less than one close-packed Co(0001) layer is nonmagnetic [32], at least at room temperature, similar to Co on W(110) [60]. This is not surprising in view of the strong hybridization of the $3d$ bands of the Co monolayer with the $5d$ bands of the W substrate, which causes a shift of the bands to lower energies, so that the minority spin states become more strongly occupied [61].

Co layers on W(100) have been studied only in a cursory manner because of their pronounced three-dimensional growth beyond 2–3 pseudomorphic monolayers. The Co crystals form long crystals with the (11–20) plane parallel to the substrate and the [0001] directions parallel to the W ⟨011⟩ directions. The first two monolayers were found to be nonmagnetic. From the crystals, weak magnetic contrast could be obtained only in films grown at room temperature. The long narrow crystals apparently have no (11–20) top faces, as judged by LEEM, so no specular beam and, therefore, no magnetic contrast could be obtained from them [59].

The SPLEEM study of Co films on Au(111) was motivated by SEMPA studies (see Chap. 7) of the same system in which the spin reorientation transition (SRT) with increasing coverage had initially been interpreted by a continuous M rotation within the domains [62], and later by a decay of the domain size and the coexistence of in-plane and out-of-plane domains in a narrow thickness range [63]. These studies were made in large thickness steps [62] or with wedges, in which the SRT took place

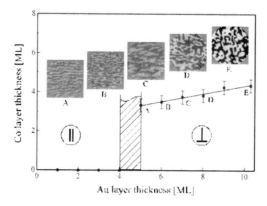

Fig. 6.10. Magnetization direction in thin Co layers on epitaxial Au layers on W(110) as a function of Au and Co thickness. To the left and above line AE the magnetization is predominantly in-plane. The inserts show the domain structure at the Au coverages corresponding to A–E and at Co film thicknesses increasing from 2.8 to 4.3 ML from A to E. Field of view 9–10 μm

within a narrow region [63–65]. The much faster image acquisition with SPLEEM allows a quasi-continuous study of the SRT during growth in much finer thickness steps and gives a more detailed picture of the SRT. In the SPLEEM study, (111)-oriented epitaxial layers on W(110) were used as substrates [66]. Previous work [67] had shown that the domain structure and the SRT depend on the thickness of the Au layer. In particular, at 2 ML Au, the domain structure was strongly influenced by the substrate steps. Therefore, the dependence of the domain structure on the thickness of the layer was studied in more detail [68]. M was found to be in-plane up to 4 ML Au. From about 5 ML Au upwards, it is out-of-plane up to a Au thickness-dependent Co thickness. This is indicated in Fig. 6.10, which also shows the out-of-plane domain structure in this region. Initially, when the anisotropic strain in the Au film still influences the domain structure, it is a striped phase, but with increasing Au thickness it develops into a more droplet-like phase, which is taken as an indication that the Au surface is nearly bulk-like at 10 ML Au.

On this surface, the SRT was studied in thickness increments as small as 0.05 ML [66]. The domain size was found to increase up to about 4 ML in agreement with [37] and with theoretical predictions that assume thin walls whose width is independent of thickness and small compared to the domain size [69]. The actual SRT takes place between 4.2 and 4.4 ML (Fig. 6.11). A detailed analysis of the correlation between the domain walls in the out-of-plane M images and of the domains in the in-plane M images [66] shows that the transition occurs via widening of the out-of-plane domain walls accompanied by a breakup of the out-of-plane domains. This scenario can be described to a first approximation by the Yafet-Gyorgy model [70]. Recently, Monte Carlo model calculations of the SRT also predicted wall broadening, but with a more complicated, twisted spin structure that develops into a vortex structure with decreasing perpendicular to dipolar anisotropy ratio [71]. Experiment shows neither a twisted structure during the transition nor a vortex structure above the transition.

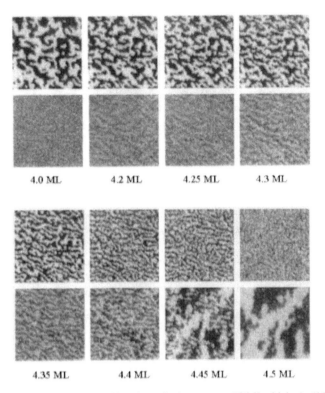

Fig. 6.11. Spin reorientation transition in a Co layer on a 10 ML thick Au(111) layer on W(110). Upper rows: out-of-plane, lower rows: in-plane magnetization images taken with 1.2 eV electrons. Coverage range 4.0–4.5 ML. Field of view 7 × 7 μm

The fact that above the transition the layer has uniaxial anisotropy similar to the layer grown directly on W(110) suggests that the twisted phase is suppressed by a still noticeable strain in the Au layer. Alternatively, the twisted phase could be a small-size or relaxation time effect in the Monte Carlo simulations. This is suggested by the fact that after long relaxation times only one vortex is left in the disk-shaped samples that is the lowest energy configuration for small samples. It is interesting to note that the evolution of the domain structure in the in situ SPLEEM [66] study agrees well with that of the ex situ SEMPA study of the annealed wedge that was covered with Au. Apparently during the slow growth used in the SPLEEM experiment, Au diffused onto the surface of the growing Co layer.

The magnetic properties and structure of Fe layers on Cu (100) have been the subject of many investigations, including a SEMPA study [72], because of their interesting dependence of structure and magnetism upon thickness, deposition conditions, and temperature, but the physical origin of this dependence is still not fully understood. SPLEEM is ideally suited to elucidate the connection between structure and magnetization. The system Fe/Cu(100) is a good example [34]. Fe films were grown in SPLEEM system B (see Sect. 6.2) at room temperature at deposition rates rang-

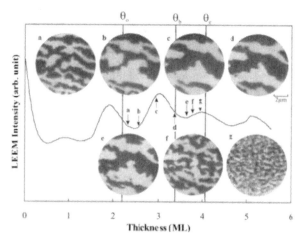

Fig. 6.12. LEEM intensity oscillations and out-of-plane magnetization images of a Fe layer on Cu(100) as a function of thickness taken with 1.8 eV electrons. Field of view 9 μm

ing from 0.07 to 0.49 ML/min. Figure 6.12 shows the LEEM intensity oscillations connected with layer-by-layer growth and out-of-plane SPLEEM images taken at the thicknesses marked by arrows. The lines marked by θ_a, θ_b, and θ_c indicate the onset of magnetism, the breakup of the domains, and the disappearance of magnetism, respectively. The asymmetry deduced from the SPLEEM images rises sharply from zero at 2 ML and becomes zero again at 4 ML. This is attributed to T_c being below room temperature in both cases. According to previous studies, the layer becomes antiferromagnetic above 4 ML with a ferromagnetic monolayer on top that has a low T_c. The Curie temperature and the decay of the magnetic domain structure during the approach of T_c could also be determined from the asymmetry. The dependence of the asymmetry upon thickness could be well fitted by the product of thickness and magnetization, taking into account the thickness dependence of the Curie temperature and assuming the two-dimensional Ising model for the temperature dependence of the magnetization, together with some broadening caused by nonuniform thickness. This supports the stipulation made in Sect. 6.2 that the asymmetry should be a measure of magnetization at least in very thin films.

6.4.2 Nonmagnetic Overlayers

Nonmagnetic overlayers change the surface anisotropy and, therefore, also the direction of M. A strong enhancement of the perpendicular anisotropy at 1–2 ML has been reported for a number of overlayers, including Au and Cu [73]. As the magnetic anisotropy arises mainly from spin-orbit coupling, it is not a priori clear why this enhancement should occur, and why it should be different in different studies of the same system. The structure of the overlayer seems to play an important role, and SPLEEM is a good technique to correlate it with the change of M. The influence of Au layers on Co layers on W(110) was studied as a function of Co layer thickness at

400 K, that of Cu layers on Co on W(110) as a function of temperature at constant Cu thickness. In both cases, no change in the magnetic domain size was observed, only a rotation of M toward the surface normal. In the case of Au overlayers, the perpendicular anisotropy increased with Au thickness up to 1.5–2 ML without a maximum for Co layers from 3–6 ML, and with a weak maximum at 2 Au ML on thicker films. This could be attributed to double layer island growth at this temperature. Cu overlayers, however, produced a clear maximum of the perpendicular anisotropy at coverages from 1 to nearly 2 ML, depending on growth temperature between room temperature and 430 K. The structure of the layer depended strongly upon temperature. At 365 K, the Cu islands were large enough so that the anisotropy could be measured locally. Initially, monolayer islands formed, on which at about 1.2 ML double layer islands grew. The enhanced perpendicular anisotropy appeared only in the monolayer regions and decreased in the 3 ML thick regions to the original value of the clean surface. Inasmuch as the layer structure depends strongly upon the growth conditions, the maximal enhancements vary with them. This explains the differences between different studies.

6.4.3 Sandwiches

It is well established that the interlayer coupling between two ferromagnetic layers through a non-ferromagnetic spacer layer depends strongly on the microstructure of the layers, in particular, of their interfaces. The coupling is usually collinear and oscillates with increasing spacer thickness between ferromagnetic (FM) and antiferromagnetic (AFM), but frequently non-collinear or 90° coupling is observed, too. This non-collinear coupling is due to a biquadratic coupling contribution to the interlayer exchange energy

$$E = J_1(1 - \cos\phi) + J_2(1 - \cos^2\phi), \tag{6.6}$$

in addition to the usual bilinear coupling that is responsible for the collinear coupling. Roughness causes spacer thickness fluctuations and magnetic dipoles. If some intermixing between ferromagnetic and spacer layers occurs, loose spins are created at the interface or in the spacer. All these phenomena can produce "extrinsic" biquadratic coupling [74–77], in addition to the "intrinsic" biquadratic coupling through atomically flat spacers. The intrinsic coupling is, however, too weak to explain the experimental observations. A deeper understanding of the influence of the roughness on the non-collinear coupling is, therefore, desirable. SPLEEM is well-suited for the correlation between roughness and magnetic structure.

Two systems have been studied: Co/Au/Co epitaxial sandwiches [78] and Co/Cu/Co epitaxial sandwiches [79], both on W(110). In all cases, the bottom Co layer was 7 ML thick, so it still had a small perpendicular M component (see Fig. 6.8). This allowed the simultaneous measurement of the coupling of perpendicular and in-plane magnetization. On top of the Co layer, Au layers were deposited in monolayer steps ranging from 3 to 8 ML, with some intermediate thicknesses of special interest. All layers except a few were deposited at room temperature to minimize roughness caused by three-dimensional crystal formation and intermixing. A few were deposited at about 400 K to check the influence of this roughness. The evolution of the domain structure in the top layer was followed in steps of one monolayer. The results will be illustrated now for Au spacers.

In the thickness range studied, the system Co/Au/Co with Au(111) orientation is known to have AFM coupling at about 5 ML and FM coupling below and above this thickness. This is confirmed by the SPLEEM studies that show, however, a more complicated coupling, as illustrated in Figs. 6.13 and 6.14 for FM and AFM coupling, respectively. In Fig. 6.13, the spacer layer is slightly more than 6 ML thick. The top row shows the domain structure images of the bottom layer with P parallel to the easy axis, hard axis, and perpendicular to the surface, respectively. In the out-of-plane M image, the contrast of the small domains of the perpendicular M component is small and no domains are recognizable in regions of high step density that were located before the deposition. Six Au ML attenuate the magnetic signal only slightly because of the large spin-independent inelastic mean free path in Au. However, already 1 Co ML causes a strong attenuation. At 2 and 3 Co ML, no in-plane signal is seen at all, while a strong perpendicular image contrast develops already at 2 ML, as expected for Co on Au (see Sect. 6.4.1). The second row in Fig. 6.13 shows the three images at 3 Co ML. At 4 ML Co, weak easy-axis contrast is already present, although the out-of-plane contrast is still strong, differing from the spin reorientation transition on Au without a Co underlayer. This is attributed to interlayer coupling. The influence of the substrate steps on the out-of-plane image that is still slightly seen at 2 Co ML has

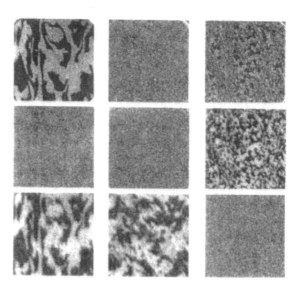

Fig. 6.13. In-plane easy axis (left), hard axis (center), and out-of-plane (right) SPLEEM images of a Co/Au/Co sandwich with ferromagnetic interlayer coupling taken with 1.2 eV electrons. The thickness of the top Co layer is from top to bottom 0, 3, and 7 ML. Room temperature deposition. Field of view 6 × 6 μm

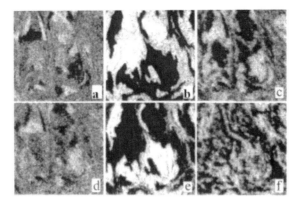

Fig. 6.14. SPLEEM images of a Co/Au/Co sandwich with antiferromagnetic coupling. Room temperature deposition. Energy and field of view as in Fig. 6.13. For explanation, see text

disappeared at 3 Co ML, so that not only FM-coupled domains are found above the domains in the bottom layer, but also domains, though smaller, in regions with high substrate step density. From 4 to 5 ML, the domain structure changes dramatically, as shown in the third row. The out-of-plane image has disappeared due to the SRT and has been replaced by a domain pattern in the hard-axis direction, while simultaneously the easy-axis image contrast has increased. This domain structure persists up to the highest Co thickness studied (8 ML), except for a slight coarsening of the hard-axis domain structure. If the gray levels are taken as a measure of the magnitude of the M components and the in-plane components are added vectorially, a complicated domain structure with the M directions inclined $+/-45°$ toward the easy axis in the bottom film arises, that is, the top layer shows non-collinear coupling with the bottom layer. Under the experimental conditions of this experiment, the layers are believed to grow as a three-level system, though with much smaller terraces than those shown in Fig. 6.8c, with the top face of the Au layer reproducing reasonably well the fluctuations of the top face of the bottom Co layer and with little intermixing, so that the non-collinear coupling most likely is of the dipolar type.

In the case of AFM coupling, the domain structure develops with thickness in the same manner as in the case of FM coupling. Figure 6.14 shows a few stages of this evolution for a spacer thickness of about 5 ML. Images (a) and (b) are out-of-plane and easy-axis in-plane images, respectively, of the bottom layer. The substrate region selected had a high density of atomic steps, so only a few domains can be recognized in the out-of-plane image. The out-of-plane image (c) is from a film with 3 Co ML on top of the Au layer. It shows AFM coupling in the regions in which the bottom layer has a perpendicular M component, but also perpendicular magnetization in other regions. Actually, M is not completely perpendicular because there is already a weak AFM-coupled in-plane component, as described above. At 4 Co ML, the out-of-plane M component has already significantly decreased (d) and the easy axis in-plane M component is nearly as large as shown in image (e) from a 6 ML thick top layer with perfect AFM coupling to the bottom layer. Simultaneously with the disappearance of the out-of-plane contrast, an in-plane hard-axis contrast develops, which is shown for the 6 ML thick top layer in image (f). Thus, the scenario for AFM coupling is exactly the same as for FM coupling. When the two in-plane gray levels are vectorially added pixel by pixel for the various spacer thicknesses, the resulting *M* rotation relative to the *M* direction in the bottom layer increases from $\phi = 15°$ at 4 Au ML to $\phi = 45°$ at 6 Au ML. With thicker spacers, the rotation angle increases even more, which indicates a biquadratic and bilinear coupling of equal magnitude. The interlayer coupling is quite sensitive to interfacial roughness. For example, in sandwiches grown at about 400 K, which have larger interface roughness, the domain pattern of the top layer shows no sign of in-plane interlayer coupling.

6.4.4 Other Topics

All the results discussed up to now have been obtained from samples prepared in situ. The question arises to what extent samples prepared ex situ can be studied with SPLEEM. In view of the small sampling depth of SPLEEM, contamination and other surface layers reduce or even eliminate magnetic contrast and have to be removed by sputtering or other feasible methods. That this can be done is illustrated by one of the first SPLEEM images taken (Fig. 6.15 [2]). It is from the (0001) surface of a bulk Co crystal that had been cleaned by simultaneous 1.5 keV Ar ion bombardment and annealing at 400 °C. Recent experiments demonstrate that this procedure is also applicable to thin films [80].

Important parameters in magnetic studies are external magnetic fields, temperature, and time. External fields parallel to the sample surface cannot be applied during imaging because they deflect the electron beam. Of course, with fields applied before

Fig. 6.15. SPLEEM image of the magnetic domain structure of a Co(0001) surface taken with 2 eV electrons. Field of view 12 μm

imaging, the remanent state of the sample is accessible. In fields perpendicular to the sample surface that are well aligned with the optical axis, imaging is possible. Only the focal length has to be changed and usually beam and/or sample tilt have to be corrected because perfect alignment is difficult. Without specially designed sample holders, fields are limited to a few hundred Gauss.

Cooling has been achieved up to 118 K in a SPLEEM instrument [80] with liquid nitrogen, but temperatures in the 10 K range should become accessible with liquid helium cooling. Temperatures in the 1 K range appear unrealistic because of the specimen holder/lens configuration, which does not allow effective thermal shielding. Time is an important parameter in connection with switching processes. The present image acquisition times in the 1 sec range can possibly be reduced to the 0.1 sec range with further instrument improvement, so that processes in this time domain range may become accessible to SPLEEM. The short switching times of technological interest are, however, outside the reach of SPLEEM.

6.5 Summary

SPLEEM is a non-scanning magnetic imaging technique that makes use of the spin dependence of the elastic scattering of slow electrons. As the spin dependence increases with decreasing energy, because of the energy dependence of the exchange potential, electrons in the 1 eV range are most useful. In this energy range, inelastic scattering and elastic attenuation in band gaps limit the sampling depth in ferromagnetic metals and alloys to a few monolayers. In nonmagnetic materials without band gaps, the sampling depth is usually considerably larger, as indicated by the "universal curve" of the inelastic mean free path.

The lateral resolution of SPLEEM is limited by the spherical and chromatic aberrations of the objective lens and is in practice presently not better than 10 nm. The vertical resolution is limited by the wavelength of the electrons and allows the imaging of monatomic steps and thickness variations via interference contrast. The dependence of the contrast on interference and diffraction is unique to SPLEEM and makes it an ideal method for the correlation between the magnetic structure, crystal structure, and morphology of the specimen. This strength is at the same time a weakness in the study of amorphous or polycrystalline specimens consisting of small crystals without preferred orientation. On these specimens, the intensity of the back-reflected electrons is distributed over a wide angular range, instead of being focused into a strong specular beam. Image acquisition times increase then from the 1 sec range, characteristic for SPLEEM of single crystalline or highly oriented surfaces, to the range needed in SEMPA.

The connection between the asymmetry obtained from two images with opposite polarization of the incident beam and the magnetization is not straightforward. SPLEEM averages over several monolayers and is at best a measure of the exponentially averaged mean value of the magnetization. In the 1 eV range, the situation is, however, much more favorable for a quantitative connection between asymmetry and magnetization than in the 10 eV range used in SPLEED, in particular, in the monolayer range. At present, a quantitative evaluation of SPLEEM images is possible for the ratio of the various components of the magnetization of a fixed layer thickness and a semi-quantitative comparison of these ratios as a function of layer thickness, temperature, or external field. This gives important data for the dependence of the magnetic domain structure on these parameters, from which the ratio of the various magnetic anisotropies may be deduced.

The incorporation of cooling and external fields in the existing systems and the development of commercial instruments, in which the weak points of the existing experimental systems are eliminated, should open up a wide field for SPLEEM in the future.

Acknowledgement. The development of SPLEEM was supported by the IBM Almaden Research Center and the Deutsche Forschungsgemeinschaft. Its application to thin film studies in the author's group is supported by NSF under grant number DMR-9818296.

References

1. R.J. Celotta, D.T. Pierce, G.-C. Wang, S.D. Bader, and G.P. Felcher, Phys. Rev. Lett. **43**, 728 (1979).
2. M.S. Altman, H. Pinkvos, J. Hurst, H. Poppa, G. Marx, and E. Bauer, Mat. Res. Soc. Symp. Proc. **232**, 125 (1991).
3. J.C. Slater, Phys. Rev. **81**, 385 (1951).
4. P. Hammerling, W.W. Shine, and B. Kivel, J Appl. Phys. **28**, 760 (1957).
5. B. Kivel, Phys. Rev. **116**, 926 (1959).
6. E. Bauer and H.N. Browne, in: Atomic Collision Processes, edit. by M. R. C. McDowell (North-Holland, Amsterdam, 1964), p. 16.
7. H.N. Browne and E. Bauer, unpublished work.
8. P.A.M. Dirac, Proc. Cambridge Phil. Soc. **26**, 376 (1930).
9. J.C. Slater, T.M. Wilson, and J.H. Wood, Phys. Rev. **179**, 28 (1969).
10. J. Kessler: Polarized Electrons (Springer, Berlin, 1985).
11. R. Feder, J. Phys. C: Solid State Phys. **14**, 2049 (1981).
12. R. Feder, Phys. Scripta T4, 47 (1983).
13. R. Feder and H. Pleyer, Surf. Sci. **117**, 285 (1982).
14. C.L. Fu and A.J. Freeman, J. Magn. Magn. Mater. **69**, L1 (1987).
15. G. Waller and U. Gradmann, Phys. Rev B **26**, 6330 (1982).
16. E. Tamura, R. Feder, G. Waller, and U. Gradmann, Phys. Stat. Sol. (b) **157**, 627 (1990).
17. E. Tamura and R. Feder, Solid State Comm. **44**, 1101 (1982).
18. A. Ormeci, B.M. Hall, and D.L. Mills, Phys. Rev. B **42**, 4524 (1990).
19. A. Ormeci, B.M. Hall, and D.L. Mills, Phys. Rev. B **44**, 12369 (1991).
20. K. Wagner, N. Weber, H.J. Elmers, and U. Gradmann, J. Magn. Magn. Mater. **167**, 21 (1997).
21. M. Plihal, D.L. Mills, H.J. Elmers, and U. Gradmann, Phys. Rev. B **51**, 8193 (1995).
22. R. Feder, S.F. Alvarado, E. Tamura, and E. Kisker, Surf. Sci **127**, 83 (1983).
23. G.A. Mulhollan, A.R. Koeymen, D.M. Lind, F.B. Dunning, E. Tamura, and R. Feder, Surf. Sci. **204**, 503 (1988).

24. J. Kirschner, Phys. Rev. B **30**, 415 (1984).
25. H.J. Elmers and J. Hauschild, Surf. Sci. **320**, 134 (1994).
26. T.L. Jones and D. Venus, Surf. Sci. **302**, 126 (1994).
27. H.J. Elmers, Internat. J. Modern Phys. B **9**, 3115 (1995).
28. H.J. Elmers, J. Hauschild, and U. Gradmann, J. Magn. Magn. Mater. 140–144, 671 (1995).
29. H.J. Elmers, J. Hauschild, and U. Gradmann, Phys. Rev. B **54**, 15224 (1996).
30. H.J. Elmers, J. Hauschild, G.H. Liu, and U. Gradmann, J. Appl. Phys. **79**, 4984 (1996).
31. K.L. Man, M.S. Altman, and H. Poppa, Surf. Sci. **480**, 163 (2001).
32. Th. Duden, R. Zdyb, M.S. Altman, and E. Bauer, Surf. Sci. **480**, 145 (2001).
33. E. Bauer, T. Duden, H. Pinkvos, H. Poppa, and K. Wurm, J. Magn. Magn. Mater. **156**, 1 (1996).
34. T. Scheunemann, R. Feder, J. Henk, E. Bauer, T. Duden, H. Pinkvos, H. Poppa, H. and K. Wurm, Solid State Commun. **104**, 787 (1997).
35. M.P. Gokhale, A. Ormeci, and D.L. Mills, Phys. Rev. B **46**, 8978 (1992).
36. M. Plihal and D.L. Mills, Phys. Rev. B **58**, 14407 (1998) and references therein.
37. J. Hong and D.L. Mills, Phys. Rev. B **62**, 5589 (2000).
38. D.L. Abraham and H. Hopster, Phys. Rev. Lett. **62**, 1157 (1989).
39. H.C. Siegmann, Surf. Sci. 307–309, 1076 (1994).
40. J.C. Groebli, D. Oberli, and F. Meier, Phys. Rev. B **52**, R 13095 (1995).
41. J. Hong and D.L. Mills, Phys. Rev. B **59**, 13840 (1999).
42. P.J. Feibelman and D.E. Eastman, Phys. Rev. B **10**, 4932 (1974).
43. E. Bauer, unpublished.
44. M.P. Gokhale and D.L. Mills, Phys. Rev. Lett. **66**, 2251 (1991).
45. D. Oberli, R. Burgermeister, S. Riesen, W. Weber, and H.C. Siegmann, Phys. Rev. Lett. **81**, 4228 (1998).
46. D.L. Abraham and H. Hopster, Phys. Rev. Lett. **59**, 2333 (1987).
47. C. Li, A.J. Freeman, and C.L. Fu, J. Magn. Magn. Mater. **94**, 134 (1991).
48. E. Bauer, Surf. Rev. Lett. **5**, 1275 (1998).
49. E. Bauer, Rep. Prog. Phys. **57**, 895 (1994).
50. T. Nakanishi, S. Okumi, K. Togawa, C. Takahashi, C. Suzuki, F. Furuta, T. Ida, K. Wada, T. Omori, Y. Kurihara, M. Tawada, M. Yoshioka, H. Horinaka, K. Wada, T. Matsuyama, T. Baba, and M. Mizuta, In CP421, *Polarized Gas Targets and Polarized Beams*, edit. by R.J. Holt and M.A. Miller (American Institute of Physics, 1998), p. 300.
51. D.T. Pierce, In: *Experimental Methods in the Physical Sciences*, Vol. 29A, edit. by F.B. Dunning and R.G. Hulet (Academic Press), p. 1.
52. T. Duden and E. Bauer, Rev. Sci. Instrum. **66**, 2861 (1995).
53. E. Bauer, In: *Handbook of Microscopy, Methods II*, edit. by S. Amelinckx, D. van Dyck, J. van Landuyt, and G. van Tendeloo (VCH, Weinheim, 1997) p. 751.
54. K. Grzelakowski, T. Duden, E. Bauer, H. Poppa, and S. Chiang, IEEE Trans. Magnetics **30**, 4500 (1994).
55. K. Grzelakowski and E. Bauer, Rev. Sci. Instrum. **67**, 742 (1996).
56. H. Pinkvos, H. Poppa, E. Bauer, and J. Hurst, Ultramicroscopy **47**, 339 (1992).
57. T. Duden and E. Bauer, Phys. Rev. Lett. **77**, 2308 (1996).
58. H. Pinkvos, H. Poppa, E. Bauer, and G.-M. Kim, In: *Magnetism and Structure in Systems of Reduced Dimension*, edit. by R. F. C. Farrow, B. Dieny, M. Donath, A. Fert, and B.D. Hermsmeier (Plenum Press, New York, 1993), p. 25.
59. K.L. Man, R. Zdyb, Y.J. Feng, C.T. Chan, M.S. Altman, and E. Bauer, to be published.
60. H. Fritsche, J. Kohlhepp, and U. Gradmann, Phys. Rev. B **51**, 15933 (1995).
61. H. Knoppe and E. Bauer, Phys. Rev. B **48**, 1794 (1993).

62. R. Allenspach, M. Stampanoni, and A. Bischof, Phys. Rev. Lett. **65**, 3344 (1990).
63. M. Speckmann, H.P. Oepen, and H. Ibach, Phys. Rev. Lett. **75**, 2035 (1995).
64. H.P. Oepen, Y.T. Millev, and J. Kirschner, J. Appl.Phys. **81**, 5044 (1997).
65. H.P. Oepen, M. Speckmann, Y. Millev, and J. Kirschner, Phys. Rev. B **55**, 2752 (1997).
66. T. Duden and E. Bauer, MRS Symp. Proc. **475**, 283 (1997).
67. M.S. Altman, H. Pinkvos, and E. Bauer, J. Magn. Soc. Jpn **19**, 129 (1995).
68. T. Duden, Ph.D. thesis, TU Clausthal (1996).
69. B. Kaplan and A. Gehring, J. Magn. Magn. Mater. **128**, 111 (1993).
70. Y. Yafet and E.M. Gyorgy, Phys. Rev. B **38**, 9145 (1988).
71. E.Y. Vedmedenko, H.P. Oepen, A. Ghazali, J.-C.S. Lévy, and J. Kirschner, Phys. Rev. Lett. **84**, 5884 (2000).
72. R. Allenspach and A. Bischof, Phys. Rev. Lett. **69**, 3385 (1992).
73. T. Duden and E. Bauer, Phys. Rev. B **59**, 468 (1999).
74. J.C. Slonczewski, Phys. Rev. Lett. **67**, 3172 (1991).
75. J.C. Slonczewski, J. Appl. Phys. **73**, 5957 (1993).
76. S. Demokritov, E. Tsymbal, P. Gruenberg, W. Zinn, and I.K. Schuller, Phys. Rev. B **49**, 720 (1994).
77. R. Arias and D.L. Mills, Phys. Rev. B **59**, 11871 (1999).
78. T. Duden and E. Bauer, Phys. Rev. B **59**, 474 (1999).
79. T. Duden and E. Bauer, J. Magn. Magn. Mater. **191**, 301 (1999).
80. E.D. Tober, G. Witte, and H. Poppa, J. Vac. Sci. Technol. A **18**, 1845 (2000).

7

SEMPA Studies of Thin Films, Structures, and Exchange Coupled Layers

H.P. Oepen and H. Hopster

Scanning electron microscopy with polarization analysis (SEMPA) has developed into a powerful technique to study domains in ultrathin films. In this chapter, we discuss from a very general point of view the instrumental aspects of the method. Examples of thin film investigations are given that demonstrate unique features of SEMPA. New solutions around apparent limitations of the technique are presented at the end, i.e., analyzing samples with contaminated surfaces and imaging in external fields.

7.1 Introduction

In 1982, triggered by the investigations of the energy dependence of the spin-polarization of secondary electrons (SE), the idea emerged to use this effect in a microscope to image magnetic structures [1,2]. The combination of a conventional Scanning Electron Microscope (SEM) and a spin-polarization analyzer promised the potential of investigating magnetic microstructures with high spatial resolution. In 1984, the first **spin-SEM** was realized by Koike and coworkers [3], followed less than one year later by a microscope built at NIST [4]. The latter group introduced the abbreviation **SEMPA**, which stands for **S**canning **E**lectron **M**icroscope with **P**olarization **A**nalysis. We will use both acronyms interchangeably. The next instruments followed in Europe [5,6]. A sketch of SEMPA is given in Fig. 7.1. Due to the low depth of information, ultra-clean surfaces are essential, requiring ultrahigh vacuum (UHV) conditions. Since conventional SEMs usually do not operate in UHV, appropriate UHV-compatible columns (or guns) are rare and expensive. Hooked on is the detector system that consists of an electron optic and a spin-polarization analyzer. The optic has to focus the secondaries into the polarization analyzer. The most important feature of the electron optic is the acceptance angle for SE. It is extremely important that the optic collects electrons emitted in the full 2π solid angle.

Several microscopes have been realized [7–14], and a few more have been proposed. Basically, all the systems look very similar. Some have attachments for surface

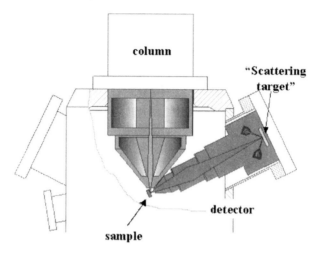

Fig. 7.1. Sketch of SEMPA. The electron microscope column gives a narrow primary electron beam that creates secondary electrons at the sample. The secondary electrons are focused onto a scattering target, and scattered intensities are measured. The scattering targets and energies differ in the various SEMPA depending on the polarization analyzer used. The whole system works under ultrahigh vacuum conditions

preparation or to grow films in situ. Most of the microscopes use a Mott spin-polarization analyzer based on the scattering of electrons off a gold film at energies from 20 – 100 KeV [15–17]. Two groups use different kinds of low energy spin-polarization analyzers. In the NIST system, the low energy diffuse scattering from Au is used to perform spin analysis [18,19]. The microscopes at MPI in Halle [14] and Hamburg [5] are equipped with the LEED spin-polarization analyzer (Low Energy Electron Diffraction) [20,21].

In the beginning, semi-infinite samples were studied, focusing on devices [22,23], or surface properties of bulk systems like the microstructure of domain walls [24–26]. The unique properties of the technique have been discussed by the various groups [5, 6, 22, 23, 27]. The biggest advantage of SEMPA is that the orientation of magnetization can be measured directly via the spin-polarization of the secondaries, since the spin-polarization vector is anti-parallel to the magnetization [28]. The polarization orientation is achievable since most of the analyzers measure two perpendicular polarization components simultaneously. It is a question of geometry then to obtain full vector information. In a more sophisticated setup, one can use two perpendicular spin analyzers yielding access to all three components, though not truly simultaneously since one has to switch between the two detectors. In two SEMPA systems, this has been realized [5,29], while another approach is to use a spin-rotator [30] within the optics [13,30]. Another advantage of the spin-SEM is the high surface sensitivity that allows imaging of domain structure in ultrathin films, i.e., films with thicknesses of a few monolayers [31–33]. The spatial resolution has been improved to 20 nm by establishing second-generation microscopes [34]. Recently, a resolution of about

5 nm has been reported for the third generation of spin-SEM [35]. This is competitive with the resolution of Lorentz microscopy (see Chap. 4 in this book).

Some of the setups have been described in detail, discussing the specifications of every component of the microscope [5, 23, 29, 36]. In the following, we would like to concentrate on the essential features that can be deduced from general considerations about the physical processes involved. We will stick to the situation of the microscope operating in the counting mode, as in most of the existing SEMPAs the signal is very low. The reasons for this are the low intensity of SEM primary beams and the extremely low efficiency of spin-analyzers. The statistical error in the measurement of polarization [15] δP is

$$\delta P = 1/\sqrt{FJt}, \tag{7.1}$$

with F the figure of merit and J the electrons per second entering the detector and t the measuring time per pixel. The figure of merit describes the overall efficiency of a detector. It is

$$F = S^2 \eta, \tag{7.2}$$

with S being the Sherman function, which describes the polarization sensitivity, i.e., how well the detector separates spin-up and spin-down electrons, and η the reflectivity, i.e., the number of electrons detected divided by the total number of electrons entering the detector [5, 22, 23]. From this, the relative error in polarization detection $\delta P/P$ is

$$\frac{\delta P}{P} = \frac{1}{P\sqrt{FJt}}, \tag{7.3}$$

which determines the relative precision. This quantity sets the ultimate limit of what can be resolved as the smallest change in polarization and thus in magnetic structure. The inverse quantity is more common in SEM, i.e., the contrast that can be achieved [5, 22, 23]. The formula is important for spin-SEM, and it is the fundamental expression for the following discussion. The performance of the entire system is determined by its three components: gun as the excitation source, secondary electron emission processes, and the spin detection system. We will discuss these in the next sections.

7.2 Instrumentation

7.2.1 Basics: Secondary Electron Emission

We will use the intensity distribution and the spin polarization of the secondaries to work out some general features for a SEMPA system (always concentrating on the case of Fe). The intensity distribution of the secondary electrons is well known from scanning electron microscopy [37]. The energy dependence is described by an analytic function:

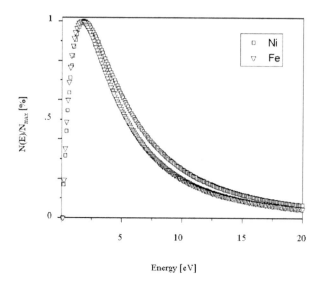

Fig. 7.2. Intensity distribution of secondary electrons. The graph is normalized to the value of the maximum of the distribution. The work function for Fe (Ni) are 4.5 eV (5.15 eV), taken from [123]

$$N(E) \propto E/(E+\Phi)^4 , \qquad (7.4)$$

with E the energy of the secondaries and Φ the work function of the material. $N(E)$ is plotted for Ni and Fe in Fig. 7.2. Energy resolved spin polarizations have been measured for Fe, Co, Ni, and also alloys [38–40]. In all cases there is a strong spin polarization enhancement at very low energies. In Ni [40] this feature is very sharp, while in Fe it is much broader [41]. As $P(E)$ depends on the excitation energy [42], we chose measurements for excitation energies well above 1000 eV. Figure 7.3 shows a fit according to

$$P_n(E) = P_1 + P_2 \exp(-E/E_H) , \qquad (7.5)$$

with the parameters $P_{1,2}$ and E_H given in the figure caption.

It is most fortunate that high intensity and high polarization are both found at very low SE energies. Thus, it is clear that a SEMPA system has to analyze the very low energy secondaries. On the other hand, the energy distribution is quite broad, which can pose experimental problems with electron optics. The question arises: What are the optimum performance conditions. The relevant quantity for counting statistics is $P^2 J$.

With the model curves for $P(E)$ and $J(E)$ (i.e., $N(E)$ from Fig. 7.2), we can calculate what can be achieved. We model the energy acceptance by a window over energy interval of width D centered about a mean pass energy E_D. For the sake of simplicity we assume normal emission and full acceptance of the electrons emitted into $\Omega = 2\pi$ and a transmission of 100% within the energy window and zero outside.

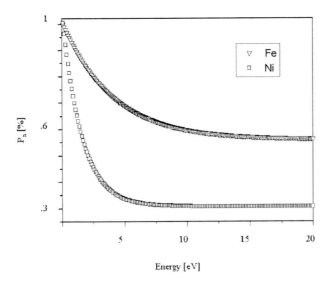

Fig. 7.3. Normalized secondary electron polarization. The *curves* are obtained by fitting experimental *curves* from [40,41]. For $P_n = P/P_0$ ($P_0 =$ is the polarization at zero energy) and the energy dependence $P_n(E) = P_1 + P_2 \exp(-E/E_H)$, we obtain for Fe (Ni) the following fitting parameters. $P_1 = 0.56$ (0.3075), $P_2 = 0.44$ (0.6984), and $E_H = 4$ eV (1.56 eV)

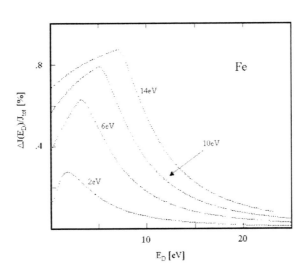

Fig. 7.4. Secondary electron intensity ΔI for given energy window D versus pass energy E_D. The different energy intervals D are given as parameters in the plot. The intensity is normalized to the total secondary electron intensity, i.e., the intensity between $0-50$ eV

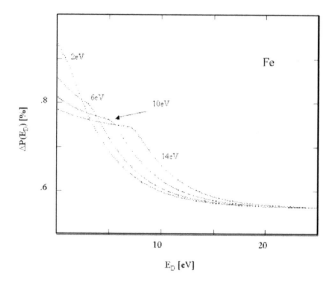

Fig. 7.5. Secondary electron polarization ΔP for given energy window D versus pass energy E_D. The different D are given as parameters in the plots. The graphs have been attained utilizing the P_n distributions from Fig. 7.3

The results of this averaging is shown in Fig. 7.4 and Fig. 7.5 for the intensity and polarization, respectively. With these results, $\Delta P^2 \Delta J$ can be calculated, which is shown in Fig. 7.6 as a function of pass energy, with the energy width as a parameter. Figures 7.4 and 7.6 show very similar overall shape. From the experimental point of view this is an extremely important feature. It means for any SEMPA tuning with given energy spread the system should be assembled in such a way as to find the intensity maximum, which is automatically the maximum of the quantity $P^2 J$. It is not necessary to measure polarization in finding optimal performance. This fact makes a real-time tuning feasible.

From the above considerations about the transmitted intensity and polarization one can derive important conclusions for designing a new microscope and its operation characteristics:

1. Figure 7.6 directly indicates that it is not desirable to select an energy spread that is very small, as this costs a lot of performance. The microscope becomes very delicate to handle, as $\Delta P(E)$ is extremely sensitive to any change of E_D. This will contribute to apparatus asymmetries. A sophisticated energy spectrometer is necessary to prevent such problems [7, 10].
2. On the other hand, it is also not necessary to take ΔE_D too large. The gain in the relevant quantity $\Delta P^2 \Delta J$ is very low above $\Delta E_D = 10\,\text{eV}$. From Fig. 7.6 the increase is about 5% going from $\Delta E_D = 10\,\text{eV}$ to $\Delta E_D = 14\,\text{eV}$, while it is 12.7% for changing from 6 to 10 eV. A problem of the large energy spread is that it becomes difficult to attain the full emission angle for higher electron

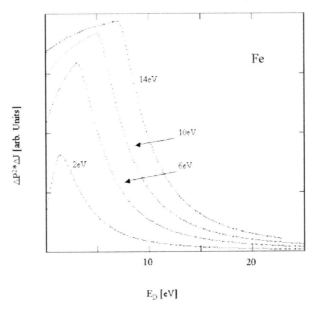

Fig. 7.6. The product of ΔP^2 and ΔJ versus E_D for Fe. D is given as a parameter in the plot.

energies. This effect will reduce the nominal gain, which is given in the above calculations for the ideal system. The optical problems of focusing the electrons into the spin-analyzer will also rise considerably due to chromatic aberration of the electron optics.

3. These considerations reveal that it is favorable to select an energy spread of $\Delta E_D = 6-10\,\text{eV}$, which has to be handled by the attached detector system. Some of the SEMPA systems have been characterized by energy spreads that lie within this span [5, 29].

7.2.2 Spin-Polarization Analyzer

A major point in SEMPA design is the energy width of the detector system. What is the energy spread the spin-polarization analyzer can tolerate for desired specifications, i.e., figure of merit F and/or sensitivity S. Or, vice versa, what degradation of performance of the polarization analyzer does the desired energy width cause. The latter consideration immediately excludes spin-polarization analyzers that need very sharp energy distributions, like analyzers utilizing scattering at very low energies [43–46]. The compromise that has to be made with such analyzers will be considerably worse than utilizing detectors that accept a very broad energy distribution without loss of performance, like the Mott-polarimeter [13, 15] or the Low Energy Diffuse Scattering spin analyzer (LEDS) [19, 22]. With such analyzers, however, the energy width is defined by another element of the detector system. This can be the focusing optics, due to its chromatic aberration or an energy analyzer that is incorporated into the system prior to the polarization analysis [5, 7, 10]. In the latter case, a stable situation is achieved. In the former case, however, the detector system will show more or less strong dependence on the imaging condition of the optics and most often on the lateral position of the point of electron emission (due to the scanning of the primary beam) [29]. The handling of the spin-SEM can become very delicate, as any change of the electron trajectories can change ΔE_D, which causes ΔP to vary. A position-dependent apparatus asymmetry is the consequence. Particularly, when the detector sensitivity is not high, this can cause severe problems. Very sophisticated beam corrections (steering) are necessary to minimize such unwanted effects. An excellent solution to this problem with a perfect correction has been incorporated into the NIST microscope [29]. An alternative is to use a polarization analyzer that works as its own energy filter with desired energy spread. This can be achieved, for example, by an appropriate geometrical layout of the LEED-detector [20, 47] utilizing the low energy diffraction of electrons to perform as the energy dispersive element. If an active energy analyzer is incorporated, the optics should provide 100% transmission at least over the same energy window.

Now we can put the results of the two previous sections together and discuss the consequences with respect to the spin analyzers and electron sources. With the values ΔP and ΔJ averaged over the energy window, the relative error of polarization detection is

$$\frac{\delta P}{P} = \frac{1}{\Delta P \sqrt{F \Delta J t}} . \tag{7.6}$$

Common to all spin-polarization analyzers is a very low efficiency of about $F = 10^{-4}$, which is surprisingly similar for all kinds of analyzers used in spin-SEM [5, 16, 17, 29, 34, 36]. Many attempts have been made to overcome this problem. Some analyzers with slightly higher efficiency have been reported [43–46]. Besides the problems due to the small energy spread allowed, the other issue in SEMPA is polarization vector analysis. While the analyzers currently used in SEMPA give access to two polarization components simultaneously, the new analyzer designs are sensitive to only one component at a time. Putting $F = 10^{-4}$ into the formula we obtain

$$\frac{\delta P}{P} = \frac{100}{\Delta P \sqrt{\Delta J t}} . \tag{7.7}$$

Next, we will make a best-case approximation in the sense that we assume the highest ΔP that can be observed. This is found for clean Fe surfaces. For the ideal detector tuned to the optimum ΔE_D range one can expect for Fe with the polarization of $P_0 = 50\%$ at zero energy (see Fig. 7.5) [1]

---

[1] For Ni it is worse. Due to the strong reduction of the polarization when averaging over the energy window and the low value of P_0, the polarization becomes $\Delta P = 0.085$. For comparison, the ratio of the polarization values

$$\frac{\Delta P_{Ni}}{\Delta P_{Fe}} = \frac{1}{4.7}$$

is most important.

$$\Delta P = 0.8 \times P_0 = 0.4 \tag{7.8}$$

and thus

$$\frac{\delta P}{P} = \frac{250}{\sqrt{\Delta J t}} \, . \tag{7.9}$$

If we want to have a relative accuracy of 10% (which means in SEM terms a contrast of 10) we end up with

$$\Delta J t = 6.25 \times 10^6 \tag{7.10}$$

When imaging magnetic microstructures it is highly desirable to keep the dwell time as short as possible. Favorable dwell times are in the range of milliseconds, which immediately puts ΔJ, the electrons per second entering the detector, into the range of 10^{10} counts/s. As the reflectivity η is in the range of 10^{-3} for most of the analyzers, it puts the limits on the counting facilities. Obviously, in high-end systems counting must be fast and the equipment should at least handle count rates up to 10 MHz without problems. Limits are pushed even higher, and problems can arise if η is higher, which is true for the LEDS-detector [18, 19]. Switching to an analog operation mode can become necessary [29]. A low reflectivity means that the polarization sensitivity is higher and vice versa for analyzers with similar efficiency ($F = 10^{-4}$). A higher sensitivity, S, minimizes problems with apparatus asymmetry, the lower reflectivity fits better the working range for pulse counting when using the best columns available at the moment. Hence, it is advisable to select the analyzer that has the higher spin sensitivity (S) and the lower reflectivity to prevent severe difficulties.

7.2.3 Electron Column

SE-detectors used in conventional SEM typically have an efficiency close to 100%. A high probe current is usually not a design criterion of highest priority for SE-columns. Consequently, only very few columns are commercially available (and usually very expensive) that have a sufficiently high primary intensity and high spatial resolution for imaging with SEMPA.

For a given detector system and sample, ΔP is fixed. With the efficiency of the spin-polarization analyzer at a value around 10^{-4}, it is the product of ΔJ and the dwell time that have to be optimized to obtain the desired accuracy. If dwell times are kept short, we will obtain the specification for the column. The number ΔJ given in the last equation can be expressed by the rate of primary electrons and the secondary electron yield Y. For reasonable values of ΔE_D, 60 to 80% of the true secondaries are used in the spin analyzer. For the further estimation, we take 70%. The secondary electron yield depends strongly on the primary electron energy. It varies monotonically from 0.2 for primary energies of 20 keV [29] to 1 for 3 keV [42]. For the following, we take 0.2, as most of the SEM columns are designed to be used and show best performance (spatial resolution, probe current) at highest energies.

This means we get only $0.7 \times 0.2 = 0.14$ secondary electrons per incoming primary electron. Thus, the primary current required is correspondingly larger by a factor of about seven. Converting from number of electrons to charge ($I = Je$) gives

$$I_P t = 7.15 \times 10^{-12} C \qquad (7.11)$$

to achieve the previously discussed magnetization contrast. The measuring time is limited by experimental constraints, i.e., mechanical stability, vacuum condition, and image size. A dwell time of 10^{-3} seconds gives a time per image of $T = 65$ s (260 s) for an image size of 256×256 pixel (512×512). To achieve the 10% accuracy for Fe, a probe current of about 7 nA (for $Y = 1$, it is $I_P = 1.5$ nA) is required. These numbers are quite high for SEM columns, particularly when the spatial resolution should be high. For dwell times of 10^{-2} sec, the current has to be in the range of 0.7 nA, which means that the time to take an image increases to about 10 min (45 min). This measuring time puts some stronger constraints on the vacuum condition and mechanical stability. If the goal is to resolve fine structures of the magnetic microstructure, the accuracy has to be increased. For 1% uncertainty in spin-polarization measurement, the primary current would have to be raised by a factor of 100. Probe currents of this range are far out of reach when high spatial resolution is required. Dwell times have to be raised and the number of pixels has to be reduced. Line scans are the best approach under such circumstances, which were actually employed in the investigation of domain walls [25, 26].

High probe current is at variance with high spatial resolution [48]. Particularly for thermionic guns, a current of nA is already considered high with resolution better than 0.1 µm. The best choice under these prerequisites seems to be the field emission (FE) column. Due to the high brightness of the electron source (called tip), focusing with highest resolution for probe currents in the desired range becomes feasible. The field emission guns offer the possibility of spatial resolution below 10 nm with probe currents in the range of nA [34]. This property of field emission is the reason why most of the spin-SEMs are equipped with such sources [3, 5, 6, 11].

FE guns, however, are not easy to work with. The problems with cold field emission are short-term fluctuations and long-term drift. While the fluctuations do not pose a problem in SEMPA, due to the normalization procedure to obtain the quantity "polarization", the latter problem has to be considered important. Due to contamination of the tip, the emission current drops dramatically around two orders of magnitude on the time scale considered here, i.e., an acquisition time per image in the range of minutes. Hence, the above-mentioned advantage of high resolution and high probe current has to be modified in the sense that it is only true for a very short time after preparation of a clean tip. To bypass that problem, hot field emission is performed, delivering slightly lower resolution but higher stability as the contamination of the tip is prevented [14, 34].

Another advantage of field emitter systems is the fact that low energy operation is possible with still reasonably good spatial resolution. This is favorable, as the secondary electron yield increases considerably when the primary energy is reduced. This allows another optimization of the achievable secondary electron rate by tuning the column. A peak in the total secondary electron rate will show up as a compromise of increasing SE yield and decreasing column performance, i.e., probe current, at lower primary energies.

Field emission guns are, however, generally limited to a maximum current that they can deliver. This is different with thermionic emitters. If the resolution is not of primary concern, thermionic guns will be the best choice. They are much easier to handle and give access to faster data acquisition in combination with analog signal processing [29]. The latter condition is not achievable with field emission. The resolution for LaB_6 emitters (thermionic gun) is some 10 nm in the desired current range (nA) [48].

7.2.4 Polarization Vector Analysis

As discussed, SEMPA can measure the magnetization vector. For in-plane magnetization P_x, P_y is measured for every pixel. Instead of displaying images, one can display information as scatter plots [7]. A scatter plot of the vector in the $P_x P_y$ plane is the frequency distribution of the P vector. Any information about the lateral position is dismissed. An example of a scatter plot is given in Fig. 7.7b, derived from the domain image given in Fig. 7.7a, which shows a vortex structure in a small disc (only pixels from inside the disc have been used in the scatter plot). The resulting scatter plot exhibits a ring revealing the uniform distribution of the magnetization vector in all in-plane directions. If the magnetic microstructure exhibits well defined domains, the plot will show a clustering of polarization vectors around the domain magnetization directions [7]. Hence, such plots directly display the symmetry and

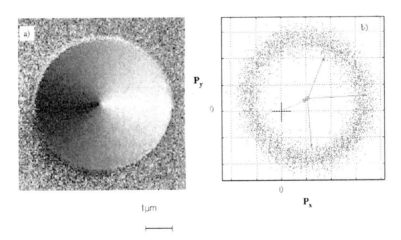

Fig. 7.7. Domain image of a thin film disc and scatter plot of measured polarization vectors. The domain image (**a**) exhibits a vortex structure in a small soft magnetic disc. The scatter diagram (**b**) is created from the two polarization values measured at every spot inside the disc. The cross in the scatter diagram represents the origin, while the *thicker arrow* to the center of the ring gives the vector of apparatus asymmetry. Every dot in the scatter plot represents the polarization doublet obtained at one pixel of the image of the domain structure. Courtesy G. Steierl

the magnetization direction of the domains (i.e., easy axes of magnetization). The center of the ring (Fig. 7.7b) is obviously shifted with respect to the origin. The vector assigned to the displacement of the center of the ring from the origin is the apparatus asymmetry within the plane of polarization detection. A large vector of apparatus asymmetries indicates that the detector system is not well aligned or technical problems of mechanical (alignment) or electrical (detection efficiency, counting electronics, saturation effects) origin exist.

The scatter plot further allows a direct transformation of the domain structure into domain images with full vector information, i.e., the spatial distribution of magnetization vector orientation. Two different situations, can in general occur. First, the polarization is oriented completely parallel to the plane of the detector. In this situation, the complete vector information is contained in the image. The scatter plot will show a symmetry with points equally spaced from the center of a ring like in Fig. 7.7b. The distance between the center of clusters and the asymmetry-corrected origin or the radius of the ring is the absolute value of polarization. The scatter plot yields a very easy way of coding the domain image. A color wheel can be assigned to the in-plane angle directly [29, 49]. The second situation appears when the planes of the polarization detection and magnetization orientation do not coincide. In this situation, the tilting angle can be determined and corrected. If, however, domains exist with magnetization pointing out of the plane of detection, the scatter plot will show clustering at points with different distances to the origin. Scatter points that are farthest away represent in-plane domains and give the absolute value of the polarization. The scatter points that are closer to the origin (corrected for apparatus asymmetry) can be used to determine the tilt angle of the polarization vector. Using a color wheel proposed by Hubert and coworkers [50], the transformation can yield domain images that reflect information about all three components.

7.3 Case Studies

Very different techniques are used to investigate the magnetic microstructures in ferromagnets. All of these techniques have their strengths and drawbacks that make them best suited for the investigation of different aspects of magnetism. In the first section, we will give some examples that show the strengths of spin-SEM, concentrating on ultrathin film systems. The second section will discuss apparent drawbacks of the technique and possible ways around them.

7.3.1 Ultrathin Films

One advantage of SEMPA is its high surface sensitivity, which gives the signal, i.e., polarization, as an average over only a few atomic layers. This fact makes it feasible to investigate systems with thicknesses in the nanometer and sub-nanometer range, i.e., of a few atomic layers only. The technique works in reflection, which puts no limitation on the dimensions of the support of the ultrathin ferromagnet. In particular, that means spin-SEM allows the investigation of ideal systems like single crystal ferromagnets or, utilizing its surface sensitivity, the best thin films one can fabricate by growing them epitaxially on single crystal surfaces. Those well characterized ultrathin films allow for a better and easier interpretation, as the magnetic properties can be correlated with structure and morphology. Another very special feature of ultrathin films is that the magnetization orientation is identical throughout the thickness of the film as long as the thickness is smaller than the characteristic magnetic length λ. The characteristic length is either $\lambda_K = \sqrt{\frac{A}{K}}$ or $\lambda_{ms} = \sqrt{\frac{A}{2\pi M_S^2}}$, with A the exchange stiffness, K the anisotropy constant, and M_S the saturation magnetization, depending on the energy that determines the microstructure, i.e., either the anisotropy or the dipolar energy [50]. The unique situation is that the magnetic microstructure observed at the surface is that of the complete ferromagnet. This makes the interpretation and the setup of the magnetic energy balance straightforward, without the need for assumptions one is forced to make for bulk ferromagnets. A direct correlation between magnetic microstructure and various energy contributions become feasible.

We will first discuss the influence of surface and interface contributions to the magnetic anisotropy of films with perpendicular magnetization. The second section deals with films magnetized in the film plane. In such systems, magneto-static energies due to wall structures are dominant and determine the microstructure. The last subsection is devoted to the exchange coupling across very thin spacer layers.

7.3.2 Films with Perpendicular Magnetization

In films with magnetization perpendicular to the film plane, a competition between two magnetic energies is found. One is the surface/interface anisotropy, which in distinct systems prefers the vertical magnetization orientation. The other is the magneto-static energy. Due to magnetic charges appearing at the surface of a vertically magnetized film, a field is created inside the magnet that is oriented opposite to the sample magnetization. The magnetization is destabilized by its own field, and the system is in a high energy state. In an infinite film, no field is created when the magnetization is lying in the plane. Hence, the in-plane magnetized state has a lower energy with respect to magneto-statics.

While the interface and surface contribution is constant with film thickness, i.e., gives a constant energy per film area, the magneto-static contribution depends on thickness. The latter can be easily visualized as the number of magnetic moments in the internal field (which is constant, as the charges at the interface do not change) increases linearly with thickness. Hence, the related energy per area E^S can be given in first-order anisotropy approximation as

$$E^S = K_1^S \sin^2 \Theta - t \left(2\pi M_S^2 - K_1^V\right) \sin^2 \Theta . \tag{7.12}$$

using the convention that θ is the angle to the normal and a positive value for K favors vertical magnetization. M_S is the saturation magnetization and t the film thickness. A possible volume anisotropy contribution is included (K^V) in the formula. In case the surface and interface favors a vertical magnetization, the first expression wins for ultrathin films. The second part becomes dominant with increasing thickness.

As the magneto-static energy is usually larger than K^V, the magnetization turns into the film plane, a so-called spin reorientation. This unique behavior in ultrathin films, which gives an easily accessible control parameter, caught the attention of many physicists as soon as perpendicular magnetized films were discovered (for an overview, see [6, 51–53], see also Chaps. 1 and 6 in this book).

The Co/Au(111) system [54–56] was the first in which the spin reorientation was studied by means of spatially resolved techniques [33, 57]. In the following, we will discuss the results in the framework of the physics of spin reorientation.

The growth of Co on Au(111) was found by no means to be perfect [58]. The cobalt nucleates as double-layer islands at the elbows of the herringbone reconstruction of the Au(111) surface. The islands grow when more material is deposited and finally coalesce. Even at a coverage of seven atomic layers, or monolayers (ML), the film has a granular structure and is rough. The film can be made considerably smoother by heating [55, 59]. This smoothing has a very strong effect on the domain structure [59]. One remarkable feature of the domain structure after heating is shown in Fig. 7.8. A sequence of zooms into one particular thickness regime of a wedge-shaped film is displayed. The images show the vertical component of magnetization only. The Co wedge was grown through a mask that yields a Co stripe on the Au(111) single crystal surface. An overview of the domain structure is given in Fig. 7.8a. In the low thickness regime, domains pointing in or out of the film are coded as black and white. At higher thicknesses the contrast diminishes, e.g., the contrast is the same as on the gold substrate. In that part, the magnetization is in the film plane. On the left-hand side, where the film is very thin, domains exist that are similar to those obtained in non-annealed films. This indicates that the annealing cannot wipe out the roughness in the low coverage regime and pinning of domains [6] is still dominant. Interesting, however, is the reappearance of small domains at higher thicknesses just before the magnetization flips into the film plane. Here, the domains become even smaller than in the regime of low thicknesses, which is demonstrated by the zooms into that regime (Fig. 7.8b,c). The domain size seems to collapse upon thickness increase. The origin of this behavior is purely magnetic. The gain of magneto-static energy lets the film break up into small domains. The gain in total energy, however, is achieved at the expense of domain wall energy.

The energy balance of a thin ferromagnet with stripe domains was theoretically worked out a long time ago [60], while an analytical approximation for the semi-convergent series given in [60] was published only a few years ago [61, 62]. With the analytical solution for the magneto-static energy, the domain size D can be given as [61]

$$D = xt \exp\left(\frac{\sigma_w}{4 t M_S^2}\right), \qquad (7.13)$$

with t as thickness and σ_W as domain wall energy. The same formula is obtained for stripe and checkerboard patterns [61]. The geometry factor x accounts for the two different geometries [61]. The analytical solution allows one to fit the thickness-dependent domain size of the SEMPA images. The influence of the competing energies is included in the ansatz for the Bloch wall energy. An effective anisotropy

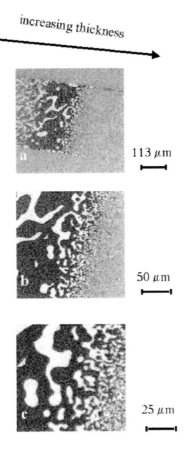

Fig. 7.8. Domain structure in a Co wedge on Au(111). The thickness increases from left to right. The pictures give the vertical component of magnetization. Black/white means that the magnetization points into/out of the plane of drawing. Gray indicates that no vertical component exists. The magnetization is either orientated in the film plane or zero. (**a**) shows a survey displaying the Co stripe and parts of the Au(111) surface. At the low thickness side, domains exist that are determined by film morphology. (**b**) and (**c**) are zooms into the range where the vertical magnetization is fading away. A collapse of the domain size is found

constant has to be introduced. The Bloch wall energy σ_W is thus

$$\sigma_W = 4\sqrt{A\, K_{\rm eff}}\,, \tag{7.14}$$

with A the exchange stiffness and the effective anisotropy $K_{\rm eff}$ taken from above, i.e.,

$$K_{\rm eff} = \frac{K_1^S}{t} + K_1^V - 2\pi M_S^2\,. \tag{7.15}$$

Taking the literature value for the first order volume (or bulk) anisotropy constant of hcp-Co, the experimental results can be fitted using the first order surface anisotropy as the only fit parameter. The result is $K_1^S = 0.83\,\text{erg/cm}^2$ (checkerboard), which stands for the sum of the two interface anisotropies involved, i.e., the surface and the Au-Co interface [59]. The value fits quite nicely into the span of published interface anisotropies for Co/Au(111) that have been obtained by other methods [52].

With the value for K_1^S we can calculate the magnetic energies at the thickness where the smallest domains appear in Fig. 7.8 (domain size ~ 500 nm, $t \sim 4.75$ ML). With the formula given in [61] we obtain for that domain size a gain in magneto-static energy of 1.6% of the total magneto-static energy, while for that thickness the expense of energy due to domain wall creation is 0.64% of the total magneto-static energy. As the gain in magneto-static energy is highest where the domains are smallest, it will be smaller in the lower thickness regime. We thus may conclude that the concept of an effective anisotropy, which takes as a correction $2\pi M_S^2$ for the magneto-static energy, is a very good approximation. The error margin of this approximation is below 1.6% for all thicknesses considered in the fit. This estimation further demonstrates that the speculation that magneto-static interactions throughout the wedge might change the physics [6] is unjustified on the much larger scales of the typical slopes of the wedges.

The effective anisotropy at the thickness of collapsing vertically magnetized domains is ~ 10% of the total magneto-static energy. With that effective anisotropy one can calculate the domain wall width. The resulting wall width is in the range of 1/10 of the domain size, i.e., 30–50 nm. This ratio of domain wall width to domain size is within the range of the validity of the approximation [61, 62]. This gives another justification for the fit.

The non-zero effective anisotropy poses the question at what thickness the effective anisotropy is actually vanishing. A straightforward calculation using the above value for the surface/interface anisotropy reveals that this should happen at a thickness of 5 ML, i.e., the film thickness at which the first order anisotropy is canceled by the magneto-static energy. In the experiments, this thickness turned out to be very special. At this thickness the first in-plane magnetized domains appear [63]. Moreover, it was found that a magnetic field applied perpendicular to the film drives the film into a single domain state (except of the very thin film region) with a borderline at about 5 ML [63]. This means that upon applying a field, a state of vertical magnetization is created that spans a larger thickness range than the multi-domain vertical state. The latter represents the equilibrium or, more precisely, a state with lower energy, as the estimation of the involved energies has revealed. The former state, however, gives the total thickness range of possible vertical magnetization. Hence, as even the external fields cannot push the vertical magnetization to higher thicknesses, the upper borderline (at 5 ML) must represent the line of absolute stability above which the in-plane magnetization becomes the only stable state. In other words, we may conclude that the experiments with fields confirm the calculated borderline (or thickness) of first order anisotropy cancellation. It must be emphasized that the calculations are based on the outcome of the magnetic microstructure analysis of the state of equilibrium, while the single borderline appears after magnetizing procedures.

The influence of the magnetic field and the meaning of the appearing borderline can be utilized to study the influence of the interface properties on covering the

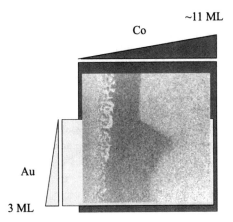

Fig. 7.9. Domain structure in Co on Au(111) partially covered by Au. The vertically magnetized domains are displayed. A double-wedge structure has been fabricated with Co thickness increasing from left to right. The thickness of the Au capping layer increases from top to bottom starting in the upper half of the image. A magnetic field has been applied vertically prior to imaging. The degrading image quality in the lower part is due to the increasing Au coverage that reduces the polarization of the secondaries

surface by nonmagnetic material. Figure 7.9 shows a double-wedge experiment. On a Co wedge a Au wedge was grown. The gradients in thickness are perpendicular to each other in the two wedges. The film was magnetized in a vertical field. At low thickness, domains persist due to the roughness in the film. In the thicker range, the film is driven into the single domain state (dark gray) by the field. A sharp borderline appears where the magnetization turns into the film plane (light gray). The borderline is shifted to higher Co thicknesses as soon as the Co is covered by Au, which means that the interface anisotropy is becoming stronger. In the sub-monolayer Au coverage, a steep increase is found that indicates how sensitively the surface and interface anisotropy depends on coverage (see also Chap. 6). Similar behavior has also been found due to carbon contamination [64]. The extreme sensitivity of film anisotropy on the condition of the interface and surface is most likely responsible for the large span of reorientation thicknesses or interface anisotropies published for the different systems. The behavior of the perpendicular magnetic anisotropy on coverage has been extensively studied [65–67]. The significance of the SEMPA image, however, is that it represents a snapshot of the whole span of coverage with infinitely small increments of changes, while conditions of growth and preparation are identical.

All experiments found an offset between the thickness where the perpendicular domains disappear and the thickness where in-plane domains emerge. The field experiments reveal that between these two borderlines, or lines of discontinuities, the films can exhibit two different states of magnetization, i.e., meta-stability. Meta-stability will only exist if different states of magnetization have a local minimum in their angle-dependent free energy at the same value of the driving parameter, here the thickness [63]. Such a coexistence of states was proposed for ferromagnets with uniaxial anisotropy in which the first order anisotropy becomes weak and the second order anisotropy contribution comes into play. A phase diagram was suggested describing the ranges of stability in the anisotropy space [68]. Under the condition that the second order anisotropy depends on thickness in the same way as the first order, a thickness variation is represented by a straight line in the phase diagram in anisotropy space [69]. This fact allows the direct assignment of the two lines of discontinuity appearing in the domain images to the two borderlines of the state of coexisting phases. The corresponding thicknesses can in turn be used to calculate the second order surface anisotropy with high precision [63].

The remaining question from the SEMPA investigation is about the origin of the lines of discontinuity. In the experiment, these lines are visible due to a reduction in signal in the range of coexistence. A reduced remanence was also reported in early investigations of the reorientation in low-temperature grown Fe/Cu(001) [70, 71]. The non-spatial resolving experiment was resumed by Allenspach performing spin-SEM investigations [72]. It was clearly demonstrated that the range in thickness and temperature with reduction of remanence was not as large as that found in the first studies. This was attributed to the fact that domains were not completely erased in the previous studies [6]. Still, the very careful investigation revealed a reduction in signal in a very small thickness and temperature range. This result is actually equivalent to the finding in the Co/Au(111). Allenspach found small in-plane domains that were assumed to be responsible for the reduction. Very small in-plane domains were found also in spin-polarized LEEM experiments on Co/Au(111) (see also Chap. 6) [73]. Hence, we might expect the same to happen in Co/Au(111), reducing the signal as in the case of Fe/Cu001) [6]. In that particular range of anisotropy values there is apparently a possibility for domains with both orientation of magnetization (vertical and in-plane) to coexist. The latter scenario has been theoretically confirmed and worked out by Monte-Carlo simulations [74]. The simulation reveals that the size of the different domains depends on the depth of the minima for the different magnetization orientations [74].

The above examples demonstrate how the magnetic microstructure can be used to extract magnetic quantities by a careful examination of the domain structure. In experiments on thickness-dependent properties, the slope of the wedges allows for extremely high thickness resolution. The variation of the slope of the wedge makes it feasible to separate thickness-dependent properties, morphology-determined behavior, and micromagnetic behavior. The examples given here are considered ideal systems as the behavior, in particular the spin reorientation, is solely driven by magnetic properties. The findings prove that assumption true. More often, the spin reorientation is a secondary effect driven by changes in structure or electronic properties. One of the most widely studied thin film ferromagnets, room-temperature-grown Fe/Cu(001) [75], belongs to this class of systems. Here, the spin reorientation is accompanied by a structural change [76]. Most likely the structural change drives the change of magnetization orientation. Those systems cannot be treated solely on magnetic grounds. In such films, however, the magnetic properties are an extremely sensitive probe for resolving the transition of structural or electronic origin. Hence, the magnetic investigation can be utilized to study e.g. structural phase transitions with extremely high accuracy. Such spatially resolved experiments have not been performed yet.

7.3.3 Films with In-Plane Magnetization

In this paragraph, we deal with ultrathin films with magnetic anisotropy that favors in-plane magnetization or systems with very small magnetic anisotropy. From the micromagnetic point of view, such films seem to be by far not as interesting as systems with perpendicular magnetization [6]. This is due to the fact that for infinitely large films, i.e., with lateral dimensions much larger than the thickness, the state of lowest energy is the single domain state since the magneto-static energy of a mono-domain with in-plane magnetization is almost zero [77, 78]. So there is no driving force for the creation of domains, as the domain wall energy, which has to be expended, is not balanced by a gain in anisotropy energy. Actually, it was found in the ultrathin films that the as-grown state is single domain [79]. The magneto-static energy, however, is exactly zero only in the case of infinitely extended films. With finite size the magnetic poles created at the edge of the film are responsible for a local field that can generate small domains just at the edges, called edge domains [31, 79]. Edge domains were also observed in thick films, i.e., in films with thickness up to 1 μm, some time ago [80]. In spite of the fact that the mono-domain state is the state of lowest energy, domains could be created in ultrathin films and turn out to be stable [32]. The domain pattern is characterized by a very irregular structure. In spite of a reasonably strong fourfold magnetic anisotropy of the Co/Cu(001) films [81], the domain structure revealed no remnants of this symmetry in contrast to common experience. The same was also found in uniaxial in-plane magnetized films of Co/Cu(1 1 13) [82]. In brief, the reason for that special feature of the domain structure are the domain walls, which are Néel walls [83]. A Néel wall prevents surface charges while volume charges are created, because in the wall the magnetization rotates within the film plane. The walls have a narrow core and long-ranging tails that are created by the volume charges on both sides of the wall. This has been found for films with very small magnetic anisotropy [84]. The tails have been described theoretically by a logarithmic dependence [85, 86]. Qualitatively, the same wall profiles have been observed in the ultrathin films in spite of the magnetic anisotropy [87]. Due to the wall tailing, a strong interaction of walls acts in the ultrathin film that creates the irregular domain structure. While the domain structure is affected, the interaction also changes the wall profile, which causes difficulties in determining the unperturbed profile, i.e., particularly the tail.

Although the domain structure is not the state of lowest energy the domain structure is stable. It is not clear which mechanism stabilizes the domains. It is not the influence of imperfections that have been shown to be of minor or even no importance [6,79]. The understanding of the domain structures, however, would yield more information about magnetic or structural properties. Simulations are difficult due to computation limitations, as large sample sizes have to be considered. Yet, the understanding of the main features of the microstructure will become an important

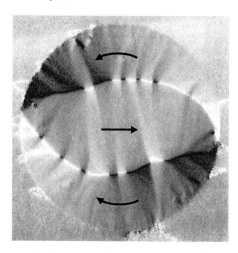

Fig. 7.10. SEMPA image of a Co disc (polycrystalline) with a diameter of 10 μm. The image gives the angle distribution of magnetization within the film plane. The magnetization orientation has been calculated from the measurements of two orthogonal magnetization components. The resulting orientation is coded in gray. For the sake of simplicity the *arrows* are drawn that reveal the flow of the magnetic flux. The domain walls are cross-tie walls which have to be expected for the thickness of the structures

issue as such films are considered for the use in future storage or (hybrid) electronic devices.

Complex structures of domain walls are known for thicker films of soft magnetic materials, i.e., cross-tie walls (see also Chap. 11) [88]. Cross-tie walls are created in certain thickness ranges when the wall structure switches from Néel- to Bloch-like walls. Although such walls have large lateral extensions, they are better understood [89–91]. They consist of legs with Néel-type character and Bloch-lines that transform to Bloch-walls upon thickness increase. Cross-tie walls can be seen in the SEMPA image shown in Fig. 7.10. The thin film structure (thickness = 40 nm) is made from polycrystalline Co, which has a low magnetic anisotropy. Due to its finite dimensions, the magneto-static energy overcomes the anisotropy energy and determines the domain structure counterbalanced by the exchange energy. The minimization of the magneto-static energy lets the system prevent any pole at the structure edges, which creates a so-called flux closure structure. Following the *arrows* in the image it becomes obvious how the magnetization is closed in itself. The cross-tie walls that are formed extend in some cases throughout large parts of the structure. A detail of the wall is shown in Fig. 7.11 along with a sketch of the cross-tie. Two different Bloch lines appear in a regular sequence. One is the center of a swirl, while the second type appears at points where the Néel-type legs merge into the wall center. The spatial resolution for magnetic structure in this image is less than 30 nm [92]. Nonetheless the structure of the Bloch lines is not resolved. The Bloch lines are singularities where, for symmetry reasons, the magnetization is pointing out of the

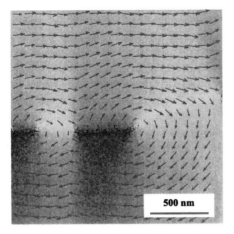 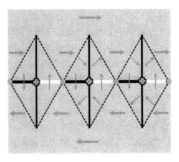

Fig. 7.11. Detail from Fig. 7.10 showing the fine structure of the cross-tie. The *arrows* given in the domain image on the left-hand side are the results of the SEMPA investigation. A sketch of the cross-tie wall is given on the right-hand side [50]

film plane [50]. The fine-structure has not yet been experimentally resolved (see also Chap. 11).

7.3.4 Exchange Coupled Films

The discovery of antiferromagnetic coupling in Fe/Cr/Fe structures and of the associated giant magnetoresistance (GMR) effects was a milestone and triggered an explosive growth in the field of ultrathin magnetic films [93, 94]. RKKY-type oscillatory exchange coupling was soon found in many systems with nonmagnetic (paramagnetic) spacer layers [95]. In a very short time these systems found their way into magnetic devices. Here, we will concentrate on the unique contributions made by SEMPA. In a series of studies by the NIST group, various spacer layer materials were investigated: Cr [96–98], Ag [99, 100], Au [101, 102], and Mn [103, 104]. The use of wedged interlayers, Fe whiskers as substrates, and the ability of SEMPA to directly measure the magnetization direction with high spatial resolution revealed many of the details and intricacies of the exchange coupling mechanisms. The Fe whiskers provide exceptionally smooth surfaces compared with ordinary single crystal substrates. The classical example of Fe/Cr/Fe has contributed most to our knowledge. The epitaxy of this system is of very good quality and, therefore, shows the various coupling effects most clearly. Furthermore, the SDW structure of Cr adds another dimension to the problem.

The surface sensitivity of SEMPA allows the use of ultrathin top Fe layers, which are typically 2 nm thick. On the wedged samples, the oscillatory coupling is directly visible in the SEMPA images. Short and long periods are observed in the Fe/Cr/Fe system, depending on growth conditions. Only samples with an elevated growth temperature show the short period (two atomic layer) oscillations, while for the low-temperature-grown samples, the short period oscillations are absent due to interface roughness. This study led to detailed investigations of the Fe/Cr growth. Combined SEMPA and STM showed the importance of interface alloying [97,98].

SEMPA studies on Fe/Cr/Fe, in which the polarization vector was measured, revealed the presence of perpendicular (biquadratic) coupling. This was found primarily at interlayer thicknesses where the bilinear (collinear) coupling was small. In this case, biquadratic coupling can even become dominant. Similar effects were seen in other systems and perpendicular coupling is now a rather common phenomenon associated with interface roughness. In the most general case, the magnetic coupling can result in non-collinear alignment, i.e., the magnetization directions are at intermediate angles. This has been studied in detail in the Fe/Mn/Fe system [103, 104]. A disadvantage of SEMPA is that coupling strengths cannot be measured directly, since SEMPA cannot be applied in a magnetic field (but see later). The NIST group combined SEMPA studies with MOKE [102].

Antiferromagnetic (AF) exchange coupling is, of course, not restricted to AF interlayer materials (like Cr or Mn). However, AF order is crucial in a different context. The phenomenon is now known as exchange bias and has been known for a long time [105]. The effect consists of a shift of the hysteresis loop of a ferromagnetic (FM) layer in a field-cooled AF/FM film structure. The AF layer pins the FM magnetization. It has been clear from the beginning that this must be due to the interface magnetic structure, but details have not been worked out. There has been a growing interest in this field. Exchange biasing is widely used in layered magnetic devices. For a recent review of exchange biasing see, e.g., [106]. Until recently, there was no experimental probe available of AF order at surfaces or interfaces (see also Chaps. 2 and 9). Since SEMPA requires a net spin polarization, it is not sensitive to AF order. There is, however, one situation where SEMPA can be applied to AF systems. When the AF order consists of layer-by-layer alternating magnetic moments there will be a net spin-polarization signal due to the surface sensitivity. This was first demonstrated for growth of Cr on Fe(100) in spin polarized electron scattering experiments and also for Mn [107, 108]. This case does not even require spatially resolved measurements as long as the surface layer has a macroscopic net magnetic moment. The NIST group showed the oscillating magnetization of Cr surface, grown as a wedge on Fe whiskers [109]. In principle, one could use this to extract surface moments [110]. However, there are many parameters that enter into this model. So it seems rather difficult to extract reliable quantitative information [111].

Two recent SEMPA studies deal with the magnetic structure of FM layers exchange coupled to AF materials, namely ultrathin, Fe films on Cr(100) [12] and NiO(100) [112]. Temperature dependent SEMPA images of 2 nm Fe on Cr(100) are shown in Fig. 7.12. At elevated temperature (slightly above RT), the Fe film can be saturated showing a single domain state in remanence. This is shown in the top panel of Fig. 7.12. When the temperature is lowered, a spin reorientation occurs. The Fe moments show a tendency to turn toward the perpendicular (still in-plane) direction.

However, this is not a uniform rotation, but strong variations of the local turn angle on the 10 micron scale exist (center panels of Fig. 7.12). The turning is attributed to biquadratic coupling due to frustration because of atomic steps [113–115]. The

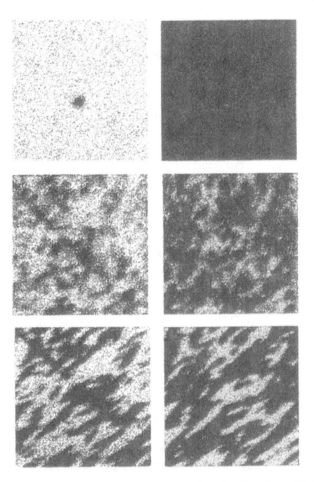

Fig. 7.12. Temperature-dependent images of 2 nm Fe on Cr(100). *Top:* above RT; single domain state after magnetization. *Center:* temperature lowered to about 100 K. *Bottom:* warmed to slightly above room temperature. The *left* and *right columns* show images of the two in-plane magnetization directions, respectively. The image size is about 70 μm

spatially varying magnetization angle is due to varying step densities on this length scale [116].

One has to remember that terrace widths are on the tens of nanometer scale, while the magnetization varies on the micron scale, thus about a hundred times larger. Thus, one is dealing with average step densities. Upon warming to above the Cr ordering temperature, the driving force on the Fe magnetization disappears and the Fe film forms conventional domains with the magnetization pointing in the easy directions. This is shown in the lower panel of Fig. 7.12.

A SEMPA image of 0.9 nm Fe on NiO(100) is shown in Fig. 7.13 [112]. Comparing the Fe magnetization directions with the expected NiO(100) surface spin

Fig. 7.13. Magnetic domains in 0.9 nm Fe on Ni(100). Image size is 70 μm (after [112])

orientation, it was concluded that a near-perpendicular coupling exists. Of course, it would be highly interesting in the future to determine both the directions of the AF moments (e.g., by linear dichroism measurements) and the FM moments at the same time.

7.3.5 Decoration Technique

The high surface sensitivity of SEMPA is often cited as a major problem that limits the application of the technique to ideal systems. "Ideal" is meant in the sense that the surface of the ferromagnet has to be prepared, i.e., cleaned in-situ before imaging, or, in case of thin films, that the films have to be fabricated in or transferred under UHV conditions into the microscope. In particular, investigations of systems and devices developed for commercial applications having undergone several steps of preparation seem to be inaccessible with SEMPA. This problem has been recognized and discussed since the first realization of the spin-SEM. In the very early stages of the technique, an elegant idea was put forward to overcome that problem. It was suggested to create a fresh and clean surface by depositing a very thin layer of iron on top of the sample. If the Fe layer is kept thin compared to the system under investigation, the film will not alter the magnetic properties while mirroring the domain structure. This decoration technique was successfully proven [117]. It was speculated that the cloning of the surface microstructure is due to the exchange coupling of the adlayer and the ferromagnet, though this has not yet been proven. The consequences of this idea for SEMPA investigations are manifold. First, the trick can be used to enhance contrast at least up to the maximum value, i.e., that of Fe [27, 118]. Secondly, it widens the field of application of SEMPA to the whole class of ferromagnetic materials. While nonitinerant ferromagnets can be investigated [119], it also gives access to insulating ferromagnetic materials. Finally, the decoration technique makes the quality of the surface less important and the investigation of nearly all kinds of samples becomes feasible. An example of the latter situation is given in Fig. 7.14. The thin film structures are fabricated by lithographical methods. After lithography, the surface of the sample is strongly contaminated. Moreover, as can be seen from the topographical image, remnants of the structuring process are randomly spread over the surface. In principle,

7 SEMPA Studies of Thin Films, Structures, and Exchange Coupled Layers 161

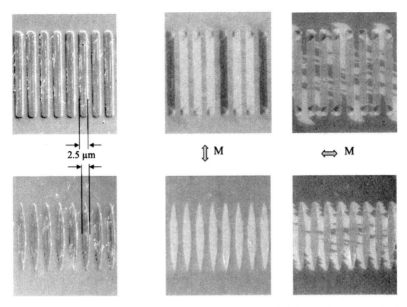

Fig. 7.14. Permalloy thin film structures. The structures were fabricated by means of e-beam lithography. The topography is displayed on the left-hand side. The images at the center and on the right-hand side show the magnetic microstructure obtained in two perpendicular in-plane directions of magnetization. All images in one line were taken simultaneously. Prior to imaging with SEMPA, the sample was covered by an iron film of about 10 monolayers in thickness. The component of magnetization along the horizontal direction (right-hand side) exhibits domains that are induced by the stray fields of the structures

that makes the study in spin-SEM impossible, since the secondary electrons created in the nonmagnetic contamination layer are not spin polarized. Even general SEM becomes difficult due to charging. The topography image in Fig. 7.14 (left-hand side) reveals that some areas are still charging up although the surface has been covered by roughly 10 monolayers of Fe. The right-hand side shows the magnetic microstructure obtained by two perpendicular in-plane components of magnetization [92]. Both images were taken simultaneously. Obviously, a magnetic microstructure is attained. The bars seem to prefer the orientation of magnetization along the long axis. They have been magnetized in a field parallel to the long axis of the bars. While the bars with pointed ends are all magnetized in the same direction, the ones with flat ends are not. A closer look reveals that at the ends of the rectangular bars closure or vortex structures appear that were also found in the so-called acicular structures by Lorentz-Microscopy [120, 121]. Hence, from the equivalence one can deduce that the domain structure of Fig. 7.14 is that of the underlying ferromagnet, which is determined by its morphology although the structures are covered by Fe. In other words, the magnetic microstructure in the Fe layer reflects the domain pattern of the soft magnetic bars.

The thin ferromagnetic layer imparts even more information about the system. The image achieved with the second component of magnetization reveals a magnetic structure in the thin film in the region where there is no magnetic structure beneath. Parts of this structure, particularly above and below the bars, reflect the stray field distribution caused by the ferromagnetic structures. During deposition, the Fe atoms are aligned in the stray fields of the structure and form domains around the structures. The Fe film (at least outside the structures) seems to exhibit a uniaxial anisotropy, which follows from the fact that any domain structure is missing in the vertical, in-plane component. Only the horizontal, in-plane component (probably the easy axis) shows domains. Particularly, the left-right asymmetry at the bottom and top of the oppositely magnetized flat-ended bars are created by stray fields. Without understanding all the details, which would require a sophisticated simulation, one can immediately deduce that the structure at the flat end of the bars does not completely terminate the magnetic flux. Magnetic poles still exist that cause the stray field and its characteristic traces in the film adjacent to the structures. A similar situation explains the domain structure in the film close to the bars with pointed ends. The stray fields are similar for all structures that cause the black/white contrast to appear in a regular left/right sequence.

The two effects determining the domain structure in the film are (a) the reflection of the domain structure of underlying magnetic material and (b) the effect of the stray fields. In this investigation, the film magnetic properties can also be identified, i.e., the films behave uniaxially with an easy axis perpendicular to the orientation of the structures. That fact poses the question for the film properties on top of the magnetic structures. Either it is the same and we have to postulate a coupling of the film to the structure that is stronger than the anisotropy, or the structural and magnetic properties are different on the substrate and the magnetic structure. The question has not been solved, since the sample surface is not very well characterized. Assuming a similar strongly contaminated surface on the whole sample, one might interpret the finding as an indication for exchange coupling between the two ferromagnets.

7.3.6 Imaging in Magnetic Fields

It is commonly stated as a disadvantage of SEMPA that it is incompatible with external fields. In fact, that statement is not generally applicable, since the response to a magnetic field does not necessarily yield more information than what can be extracted from the magnetic microstructure. The only problem is to understand the microstructure and make appropriate calculations, which is often a formidable task. The modeling then reveals the magnetic properties involved. Hence, the imaging in an in-situ field is not always of general importance. That situation has changed recently as nanomagnets become a big issue in research and development. Particularly, the application of nanomagnets in new devices puts the spatially resolved investigations of magnetic reversal into focus. In that research field, the spatial resolution is necessary to attain the response of the low-dimensional structure to the magnetic field. This trend changes the situation and the demand for the combination of SEMPA, with its power of spatial resolution, with external fields is reasonably founded.

What are the problems when using a magnetic field in SEMPA? First, electron beams are deflected in magnetic fields, and complicated steering and corrections are necessary. The second problem is spin precession, which destroys, or at least changes, the information that is used in spin-SEM. Both effects can be reduced to acceptable limits when the impact of the field on the electrons is kept to a low level either by small fields or small field extensions. The former situation was realized using fields up to 800–1200 A/m (10–15 Oe) [36, 118]. While in vertical fields the problems are very low because the field is along the direction of electron motion [36], the use of in-plane fields is more delicate [118].

The second approach, i.e., utilizing very localized magnetic fields, has been realized recently [122]. For the SEMPA investigation, a small-scale yoke has been designed that can be positioned as close as 10 μm to the film surface. The yoke has a gap of ~ 100 μm, and the pole pieces have a thickness of 100 μm. Magnetic fields up to ~ 15 000 A/m (~ 185 Oe) can be applied, which could be determined by measuring the deflection of the primary beam running through the field [122]. The field direction is mainly parallel to the sample surface. The primary beam, as well as the secondary electron beam move through the gap. This setup was successfully used to study the switching of a permalloy microstructure. A small sequence of the field-dependent domain structure is shown in Fig. 7.15. The magnetic field direction is given by the arrow, while the field strength is given at the corresponding domain image. Wall movement is found in low fields, while in high fields, domains are

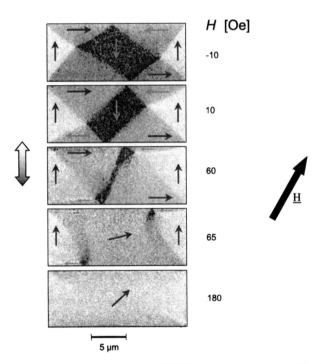

Fig. 7.15. Permalloy thin film structure. The SEMPA images are taken while the fields are applied. The field direction is given by the *arrow*, while the strength (in Oe) is indicated at the individual domain pictures

irreversibly annihilated and finally the magnetization rotates. Very similar behavior is known from literature and textbooks [50, 120].

The examples given in this paragraph demonstrate the potential of SEMPA even in applications where it was commonly believed that SEMPA could not work successfully. The few examples prove the applicability of SEMPA investigations in the new field of applied research, i.e., the device development based on nanostructures called "Spintronics."

7.4 Conclusions

Spin-SEM provides a powerful tool for the investigation of magnetic domain structures in ultrathin films. The very unique feature is that the magnetization orientation is measured with high spatial resolution. This has been successfully applied in studying the internal structure of domain walls, spin-reorientation in ultrathin films, and exchange coupling in films. Further developments are still under progress, pushing the limits of spatial resolution into the range of a few nanometers. A big step forward would be achieved if spin-polarization analyzers with higher sensitivity were available. Preparation techniques have been developed that allow the investigation of nearly all kinds of samples with SEMPA. The first in-field imaging has been successfully demonstrated.

Acknowledgement. HPO would like to thank G. Steierl and J. Kirschner for allowing their images to be included prior to publication. Support from the "Bundesministerium für Bildung und Forschung" (No. 13N7331/4) is gratefully acknowledged. HH gratefully acknowledges research support from the NSF. The SEMPA project was also supported by the Sony Research Center.

References

1. R.J. Celotta and D.T. Pierce in: Microbeam Analysis-1982, K.F.J. Heinrich (ed.), San Francisco Press, p. 469.
2. J. Kirschner, Scanning Electron Microsc. **III**, 1179 (1984).
3. K. Koike and K. Hayakawa, Jpn. J. Appl. Phys. **23**, L187 (1984).
4. J. Unguris, G. Hembree, R.J. Cellota, and D.T. Pierce, J. Microscopy **139**, RP1 (1985).
5. H.P. Oepen and J. Kirschner, Scanning Micros. **5**, 1 (1991).
6. R. Allenspach, J. Magn. Magn. Mater. **129**, 160 (1994).
7. T. VanZandt, R. Browning, C.R. Helms, H. Poppa, and M. Landolt, Rev. Sci. Instrum. **60**, 3430 (1989).
8. E.A. Seddon, Y.B. Xu, D. Greig, J.R.M. Wardell, D. Rubio-Temprano, and M. Hardiman, J. Appl. Phys. **81**, 4063 (1997).
9. Y. Iwasaki, M. Takiguchi, and K. Bessho, J. Appl. Phys. **81**, 5021 (1997).
10. Y. Lee, A.R. Koymen, and M.J. Haji-Sheikh, Appl. Phys. Lett. **72**, 851 (1998).
11. C. Stamm, F. Marty, A. Vaterlaus, V. Weich, S. Egger, U. Maier, U. Ramsperger, H. Fuhrmann, and D. Pescia, Science **282**, 449 (1998).

12. H. Hopster, Phys. Rev. Lett. **83**, 1227 (1999).
13. J. Barnes, L. Mei, B.M. Lairson, and F.B. Dunning, Rev. Sci. Instrum. **70**, 246 (1999).
14. G. Steierl, H.P. Oepen, and J. Kirschner, to be published.
15. J. Kessler: "Polarized Electrons", 2nd edn., Springer, Berlin, 1985.
16. T.J. Gay and F.B. Dunning, Rev. Sci. Instrum. **63**, 1635 (1992).
17. G.C. Burnett, T.J. Monroe, and F.B. Dunning, Rev. Sci. Instrum. **65**, 1893 (1994).
18. J. Unguris, D.T. Pierce, and R.J. Cellota, Rev. Sci. Instrum. **57**, 1314 (1986)
19. M.R. Scheinfein, D.T. Pierce, J. Unguris, J.J. McClelland, R.J. Celotta, and M.H. Kelley, Rev. Sci. Instrum. **60**, 1 (1989).
20. J. Kirschner and R. Feder, Phys. Rev. Lett. **42**, 1008 (1979).
21. J. Kirschner: "Polarized Electrons at Surfaces", Springer, Berlin, 1985.
22. G.G. Hembree, J. Unguris, R.J. Celotta, and D.T. Pierce, Scanning Microsc. Intern. Suppl. **1**, 229 (1987).
23. K. Koike, H. Matsuyama, and K. Hayakawa, Scanning Microsc. Intern. Suppl. **1**, 241 (1987).
24. K. Koike, H. Matsuyama, K. Hayakawa, K. Mitsuoka, S. Narishige, Y. Sugita, K. Shiiki, and C. Saka, Appl. Phys. Let. **49**, 980 (1986).
25. H.P. Oepen and J. Kirschner, Phys. Rev. Lett. **62**, 819 (1989).
26. M.R. Scheinfein, J. Unguris, R.J. Celotta, and D.T. Pierce, Phys. Rev. Lett. **63**, 668 (1989).
27. J. Unguris, to be published in: "Magnetic Imaging and Its Application to Materials", see http://physics.nist.gov/Divisions/Div841/Gp3/epg_files/pub.html.
28. J. Kirschner and S. Suga, Solid State Commun. **64**, 997 (1987).
29. M.R. Scheinfein, J. Unguris, M.H. Kelley, and R.J. Celotta, Rev. Sci. Instrum. **61**, 2501 (1990).
30. T. Kohashi, H. Matsuyama, and K. Koike, Rev. Sci. Instrum. **66**, 5537 (1995).
31. J.L. Robins, R.J. Celotta, J. Unguris, D.T. Pierce, B.T. Jonker, and G.A. Prinz, Appl. Phys. Lett. **52**, 1918 (1988).
32. H.P. Oepen, M. Benning, H. Ibach, C.M. Schneider, and J. Kirschner, J. Magn. Magn. Mater. **86**, L137 (1990).
33. R. Allenspach, M. Stampanoni, and A. Bischof, Phys. Rev. Lett. **65**, 3344 (1990).
34. H. Matsuyama and K. Koike, J. Electron. Microsc. **43**, 157 (1994).
35. T. Kohashi and K. Koike, Jpn. J. Appl. Phys. **40**, L1264 (2001).
36. R. Allenspach, IBM J. Res. Develop. **44**, 553 (2000).
37. See Chapter 4 in [48].
38. E. Kisker, W. Gudat, and K. Schröder, Solid State Commun. **44**, 591 (1982).
39. J. Unguris, D.T. Pierce, A. Galejs, and R.J. Cellota, Phys. Rev. Lett. **49**, 72 (1982).
40. H. Hopster, R. Raue, E. Kisker, G. Güntherodt, and M. Campagna, Phys. Rev. Lett. **50**, 70 (1983).
41. M.S. Hammond, G. Fahsold, and J. Kirschner, Phys. Rev. B **45**, 6131 (1992).
42. J. Kirschner in "Surface and interface characterization by electron optical methods", A. Howie & U. Valdré (eds), Plenum Publishing Corporation, 267.
43. H.C. Siegmann, D.T. Pierce, and R.J. Celotta, Phys. Rev. Lett. **46**, 452 (1981).
44. D. Tillmann, R. Thiel, and E. Kisker, Z. Phys. B **77**, 1 (1989).
45. R. Bertacco, M. Merano, and F. Ciccacci, Appl. Phys. Lett. **72**, 2050 (1998).
46. R. Bertacco, D. Onofrio, and F. Ciccacci, Rev. Sci. Instrum. **70**, 3572 (1999).
47. H.P. Oepen, unpublished.
48. L. Reimer: "Scanning Electron Microscopy: Physics of Image Formation and Microanalysis", Springer Series in Optical Sciences Vol. 45, Berlin, 1985.

49. J. Unguris, M.R. Scheinfein, D.T. Pierce, and R.J. Cellota, Appl. Phys. Lett. **55**, 2553 (1989).
50. A. Hubert and R. Schäfer: "Magnetic Domains: The Analysis of Magnetic Microstructures", Springer, Berlin, 1998.
51. U. Gradmann in "Handbook of Magnetic Materials" Vol. 7, ed. K.H.J. Buschow, North Holland, Amsterdam, 1993.
52. J.A.C. Bland and B. Heinrich (eds.): "Ultrathin Magnetic Structures" Vol. 1 & 2, Springer, Berlin, 1994.
53. M. Farle, Rep. Prog. Phys. **61**, 755 (1998).
54. C. Chappert, D. Renard, P. Beauvillain, J.P. Renard, and J. Seiden, J. Magn. Magn. Mater. **54-57**, 795 (1986).
55. F.J.A. den Broeder, D. Kuiper, A.P. van de Mosselaer, and W. Hoving, Phys. Rev. Lett. **60**, 2769 (1988).
56. C.H. Lee, H. He, F. Lamelas, W. Vavra, C. Uher, and R. Clarke, Phys. Rev. Lett. **62**, 653 (1989).
57. J. Pommier, P. Meyer, G. Pémissard, J. Ferré, P. Bruno, and D. Renard, Phys. Rev. Lett. **65**, 2054 (1990).
58. B. Voigtländer, G. Meyer, and N.M. Amer, Phys. Rev. B **44**, 10354 (1991).
59. M. Speckmann, H.P. Oepen, and H. Ibach, Phys. Rev. Lett. **75**, 2035 (1995).
60. Z. Málek and V. Kamberský, Czech. J. Phys. **21**, 416 (1958).
61. B. Kaplan and G.A. Gehring, J. Magn. Magn. Mater. **128**, 111 (1993).
62. Y.T. Millev and J. Phys. Cond. Matt. **8**, 3671 (1996).
63. H.P. Oepen, M. Speckmann, Y.T. Millev, and J. Kirschner, Phys. Rev. B **55**, 2752 (1997).
64. M. Dreyer, M. Kleiber, A. Wadas, and R. Wiesendanger, Phys. Rev. B **59**, 4273 (1999).
65. P. Beauvillain, A. Bounouh, C. Chappert, R. Mégy, S. Ould-Mahfoud, J.P. Renard, P. Veillet, D. Weller, and J. Corno, J. Appl. Phys. **76**, 6078 (1994).
66. J. Kohlhepp and U. Gradmann, J. Magn. Magn. Mater. **139**, 347 (1995).
67. B.N. Engel, M.H. Wiedmann, R.A. Van Leeuwen, and C.M. Falco, Phys. Rev. B **48**, 9894 (1993).
68. H.B.G. Casimir, J. Smit, U. Enz, J.F. Fast, H.P.J. Wijn, E.W. Gorter, A.J.W. Duyvesteyn, J.D. Fast, and J.J. de Jong, J. Phys. Radium **20**, 360 (1959).
69. Y.T. Millev and J. Kirschner, Phys. Rev. B **54**, 4137 (1996).
70. D.P. Pappas, K.-P. Kämper, and H. Hopster, Phys. Rev. Lett. **64**, 3179 (1990).
71. D.P. Pappas, C.R. Brundle, and H. Hopster, Phys. Rev. B **45**, 8169 (1992).
72. R. Allenspach and A. Bischof, Phys. Rev. Lett. **69**, 3385 (1992).
73. T. Duden and E. Bauer, Mater. Res. Soc. Symp. Proc. **475**, 273 (1997).
74. E.Y. Vedmedenko, H.P. Oepen, and J. Kirschner, Phys. Rev. B submitted.
75. C. Liu, E.R. Moog, and S.D. Bader, Phys. Rev. Lett. **60**, 2422 (1988).
76. J. Thomassen, F. May, B. Feldmann, M. Wuttig, and H. Ibach, Phys. Rev. Lett. **69**, 3831 (1992).
77. C. Kittel, Phys. Rev. **70**, 965 (1946).
78. R. Carey & E.D. Isaac: "Magnetic Domains and Techniques for Their Observation", The English Universities Press Limited, London, 1966.
79. H.P. Oepen, J. Magn. Magn. Mater. **93**, 116 (1991).
80. E. Feldtkeller in Proceedings of the International Symposium held in Clausthal-Göttingen, eds.: R. Niedermayer and H. Mayer, Van de Hoeck & Ruprecht, Göttingen, 1966.
81. P. Krams, F. Lauks, R.L. Stamps, B. Hillebrands, and G. Güntherodt, Phys. Rev. Lett. **69**, 3674 (1992).

82. A. Berger, U. Linke, and H.P. Oepen, Phys. Rev. Lett. **68**, 839 (1992).
83. L. Néel, Compt. Rend. Acad. Sci. Paris, **241**, 533 (1955).
84. E. Feldtkeller, Z. Angew. Phys. **15**, 206 (1963).
85. R. Kirchner and W. Döring, J. Appl. Phys. **39**, 855 (1968).
86. H. Riedel and A. Seeger, Phys. Stat. Sol. (b) **46**, 377 (1971).
87. A. Berger and H.P. Oepen, Phys. Rev. B **45**, 12596 (1992).
88. E.E. Huber Jr., D.O. Smith, and J.B. Goodenough, J. Appl. Phys. **29**, 294 (1958).
89. R.M. Moon, J. Appl. Phys. **30**, 82S (1959).
90. S. Methfessel, S. Middelhook, and H. Thomas, J. Appl. Phys. **31**, 302S (1960).
91. S. Middelhook, J. Appl. Phys. **34**, 1054 (1963).
92. G. Steierl, W. Lutzke, H.P. Oepen, S. Tegen, C.M. Schneider, and J. Kirschner, Magnetoelektronik, Beiträge zum BMBF-Statusseminar Magnetoelektronik, Dresden 2000, VDI-Technologiezentrum, Düsseldorf, 341 (2000).
93. P. Grünberg, R. Schreiber, Y. Pang, M.B. Brodsky, and H. Sowers, Phys. Rev. Lett. **57**, 2442 (1986).
94. M.N. Baibich, J.M. Broto, A. Fert, F. Nguyen Van Dau, F. Petroff, P. Etienne, G. Creuzet, A. Friederich, and J. Chazelas, Phys. Rev. Lett **61**, 2472 (1988).
95. S.S.P. Parkin, N. More, and K.P. Roche, Phys. Rev. Lett. **64**, 2304 (1990).
96. J. Unguris, R.J. Celotta, and D.T. Pierce, Phys. Rev. Lett. **67**, 140 (1991).
97. D.T. Pierce, J.A. Stroscio, J. Unguris, and R.J. Celotta, Phys Rev. B **49**, 14564 (1994).
98. D.T. Pierce, J. Unguris, R.J. Celotta, and M.D. Stiles, J. Magn. Magn. Mater. **200**, 290 (1999).
99. J. Unguris, R.J. Celotta, D.T. Pierce, and J.A. Stroscio, J. Appl. Phys. **73**, 5984 (1993).
100. J. Unguris, R.J. Celotta, and D.T. Pierce, J. Magn. Magn. Mater. **127**, 205 (1993).
101. J. Unguris, R.J. Celotta, and D.T. Pierce, J. Appl. Phys. **75**, 6437 (1994).
102. J. Unguris, R.J. Celotta, and D.T. Pierce, Phys. Rev. Lett. **79**, 2734 (1997).
103. D.A. Tulchinsky, J. Unguris, and R.J. Celotta, J. Magn. Magn. Mater. **212**, 91 (2000).
104. D.T. Pierce, A.D. Davis, J.A. Stroscio, D.A. Tulchinsky, J. Unguris, and R.J. Celotta, J. Magn. Magn. Mater. **222**, 13 (2000).
105. W.H. Meiklejohn and C.P. Bean, Phys. Rev. **102**, 1413 (1956).
106. J. Nouges and I.K. Schuller, J. Magn. Magn. Mater. **192**, 203 (1999).
107. T.G. Walker, A. Pang, H. Hopster, and S.F. Alvarado, Phys. Rev. Lett. **69**, 1121 (1992).
108. T.G. Walker and H. Hopster, Phys. Rev. B **48**, 3563 (1993).
109. J. Unguris, R.J. Cellota and D.T. Pierce, Phys. Rev. Lett. **69**, 1125 (1992).
110. P. Fuchs, V.N. Petrov, K. Totland, and M. Landolt, Phys. Rev. B **54**, 9304 (1996).
111. D.T. Pierce, R.J. Celotta, and J. Unguris, J. Appl. Phys. **73**, 6201 (1993).
112. H. Matsuyama, C. Haginoya, and K. Koike, Phys. Rev. Lett. **85**, 646 (2000).
113. J.C. Sloncewski, Phys Rev. Lett. **67**, 3172 (1991).
114. J.C. Slonczewski, J. Magn. Magn. Mater. **150**, 13 (1995).
115. N.C. Koon, Phys. Rev. Lett. **78**, 4865 (1997).
116. H. Hopster, J. Appl. Phys. **87**, 5475 (2000).
117. T. VanZandt, R. Browning, and M. Landolt, J. Appl. Phys. **69**, 1564 (1991).
118. Y. Iwasaki, K. Bessho,. J. Kondis, H. Ohmori, and H. Hopster, Applied Surface Science **113/114**, 155 (1997).
119. M. Haag and R. Allenspach, Geophysical Research Letters **20**, 1943 (1993).
120. K.J. Kirk, J.N. Chapman, and C.D.W. Wilkinson, Appl. Phys. Lett. **71**, 539 (1997).
121. T. Schrefl, J. Fiedler, K.J. Kirk, and J. Chapman, J. Magn. Magn. Mater. **175**, 193 (1997).
122. G. Steirerl, G. Lin, D. Jorgov, and J. Kirschner, Rev. Sci. Instrum. **73**, 4264 (2002).
123. D.E. Eastman, Phys. Rev. B **2**, 1 (1979)

8

Spin-SEM of Storage Media

K. Koike

The study of both the magnetization distribution of recorded and non-recorded media is needed if magnetic recording densities are to be increased. There are various magnetic recording media depending on the type of recording drive such as hard disc drive, tape drive, magneto-optical disc drive, etc. The magnetization of the recorded bits of each medium has its own direction, i.e., parallel, oblique, or perpendicular to the medium plane. Thus, spin-SEM is advantageous to the study of these recording media, since it can detect the three components of the magnetization vector with a spatial resolution of several nanometers. In this chapter, I review how the spin-SEM contributes to increasing the recording density of various recording media.

8.1 Introduction

Drawing pictures and writing characters on media such as rocks and paper was long the main method for storing information. At the end of the 19th century, Edison invented the first method for storing voices and other sounds. In this method, sounds were recorded in a spiral groove on a metal cylindrical medium and retrieved using a metal stylus combined with a membrane, which collected or emitted sound waves. These mechanical media, however, were not rewritable.

At the beginning of the 20th century, Poulsen invented a rewritable medium. He used a magnetic wire and an electromagnet to record sounds magnetically. This approach has evolved into various methods to meet the needs of various applications. During this evolution, various methods have come and gone. Today, in what we call the "information age," the commonly used magnetic recording systems are hard disc drives (HDDs), floppy disc drives (FDDs), magneto-optic (MO) disc drives, and tape drives.

The need for larger and larger storage capacities and smaller and smaller storage systems has led to a number of studies on ways to increase recording density, and the size of recorded magnetic bits is now in the sub-micrometer to nanometer range. These studies of recording systems have thus required the high-resolution observation of recorded bits. In this chapter, I describe the role spin-polarized scanning electron microscopy (SEM) [1] has played in the increase of recording density.

8.2 HDD Recording Media

Because of their high read and write speeds, short random access time, and large capacity, HDDs are used in almost every kind of computer, from personal ones to mainframes, as nonvolatile internal and/or external storage devices. Their market share is the largest among the various kinds of storage devices and will increase further as new applications, such as full-time recording of multichannel television programs, are found for them.

The recording media of current HDDs are magnetic thin polycrystal films on aluminum discs. Figure 8.1 shows how the normalized output signal and media noise power depend on the linear recording density for two different recording materials: $Co_{86}Cr_{10}Ta_4$ and $Co_{80}Cr_{16}Ta_4$ [2]. The noise power of the latter is about twice that of the former over the range shown. The noise power of both is characterized by three regions: (1) The power initially increases linearly with the recording density up to about 90 kFCI (flux change per inch), (2) it increases supralinearly from about 90 to 170 kFCI, and (3) it stays almost constant above about 170 kFCI.

Figure 8.2 shows representative bit images at linear densities of 50, 100, and 140 kFCI, or at nominal bit lengths of 0.5, 0.25, and 0.18 μm [2]. It shows that the bit boundaries have random zigzag shapes, that the average amplitudes of the zigzags are larger the higher the density, and that the average wavelength of the zigzags is larger for $Co_{86}Cr_{10}Ta_4$ than for $Co_{80}Cr_{16}Ta_4$. The major media noises in longitudinal recording come from these random zigzag bit boundaries and are called "transition noise" [3].

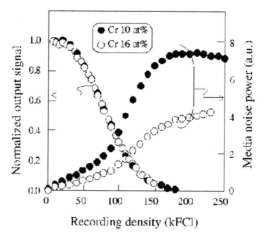

Fig. 8.1. Normalized output signals and media noise powers of $Co_{86}Cr_{10}Ta_4$ and $Co_{80}Cr_{16}Ta_4$

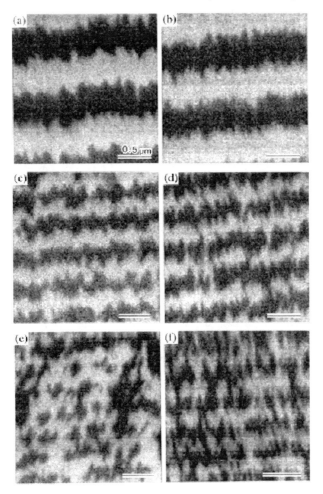

Fig. 8.2. Bit images of magnetization component along track (vertical) direction. Images (**a**), (**c**), and (**e**) are from $Co_{86}Cr_{10}Ta_4$; images (**b**), (**d**), and (**f**) are from $Co_{80}Cr_{16}Ta_4$. Linear densities are 50, 100, and 140 kFCI from top to bottom

The origins of the random bit boundaries are threefold. One is the random size and shape of the grains. The materials used in these media, nonmagnetic Cr and Ta, segregate at the grain boundaries, reducing the exchange couplings among the grains. Thus, the boundaries, or domain walls, run along the boundaries outside the grains because they have less energy than those running inside the grains, and the bit boundaries have zigzag shapes reflecting the geometry of the grains. Another origin is magnetostatic interaction among the magnetic charges appearing at bit boundaries. The zigzag bit boundaries have lower charge densities, i.e., lower magnetostatic energies, than those of the straight ones. The zigzag bit boundaries due to these two origins do not change their geometrical properties if the bits are longer than some critical length and the noise/bit is constant. Therefore, the noise per unit length increases linearly with the recording density. The noise due to these two origins corresponds to the recording density region (1). The other origin of the random bit boundaries is magnetostatic interaction between the magnetic field produced by the magnetic charge at the bit boundaries and the magnetization switched during recording [4]. When the bit length becomes smaller, the magnetization is switched not only by the magnetic field from the recording head, but also by the random field from the boundaries of the adjacent bits recorded immediately before, and the zigzag amplitude increases. Since this magnetostatic interaction increases as the bit length is reduced, noise increases more rapidly with the recording density than it did in region (1). The noise due to this origin corresponds to recording density region (2). When the nominal recording density is high enough so that the bit length is less than the grain size, recording corresponds to ac-erasing, and the noise changes very little with the nominal recording density. The noise thus corresponds to region (3).

The difference in noise between $Co_{86}Cr_{10}Ta_4$ and $Co_{80}Cr_{16}Ta_4$ can be explained as follows. Bertram et al. showed that the noise power of a recorded medium is proportional to the cross-track correlation width at the transition center or to the average wavelength of the random zigzag bit boundaries [5,6]. Since the average wavelength of the zigzag boundaries of $Co_{86}Cr_{10}Ta_4$ with larger noise is about twice that of $Co_{80}Cr_{16}Ta_4$, $Co_{86}Cr_{10}Ta_4$ should have about twice as much noise, as illustrated in Fig. 8.1. The origin of the difference in wavelengths between the two media is the difference in the exchange coupling among the grains. Since $Co_{80}Cr_{10}Ta_4$ contains nonmagnetic elements, which segregate at grain boundaries, its exchange interaction is stronger than that of $Co_{86}Cr_{16}Ta_4$. This means that if the wavelength of $Co_{80}Cr_{10}Ta_4$ is shorter, the domain walls are longer and the total wall energy is higher. Therefore, the wavelength is longer at the cost of magnetostatic energy due to magnetic charging at the walls.

These experimental results indicate that the transition noise can be reduced and the area density can be increased by reducing the size of the grains, which are magnetically separated from each other by segregated nonmagnetic elements. This is the approach currently being followed by industry.

8.3 Obliquely Evaporated Recording Media

Perpendicular recording was proposed by Iwasaki and Nakamura to reduce the self-demagnetizing effect of longitudinal recording and thus increase the recording density [7]. However, saturation recording is more difficult with perpendicular recording than with longitudinal recording because of spacing loss. A compromise method, shown in Fig. 8.3, is to record on media with an inclined easy axes of magnetization, i.e., not to record in-plane or perpendicular, but to record inclined by some angle toward the track direction. Doing this enables longitudinal saturation recording with a leading-side magnetic field from a ring head. The inclined easy axis is obtained by oblique evaporation, and the media are called "obliquely evaporated recording media." This recording method has been commercialized and is used for high-band

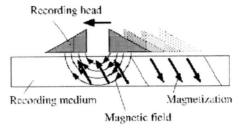

Fig. 8.3. Illustration of recording on obliquely evaporated medium with a ring head

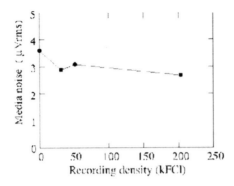

Fig. 8.4. Media noise on an obliquely evaporated Co-CoO tape

8-mm videocassette recorders. It is also proposed for use in large-capacity storage systems for high-definition television sets and mainframe computers.

Figure 8.4 shows the change in media noise with the linear recording density for a Co-CoO film obliquely evaporated on polyamide tape [8]. The film thickness was 200 nm, and the easy axis of magnetization inclined 30° from the film surface normal. The noise generally decreased with an increase in the recording density, a different relationship than that of longitudinal recording media (Fig. 8.1).

Figure 8.5 shows bit images at linear densities of (a) 10 and 300, and (b) 50 and 100 kFCI, or nominal bit lengths of 2.5 and 0.085, and 0.5 and 0.25 μm. The easy axis was tilted to the left, and the recording head moved from right to left over the medium. A 10°-azimuthal angle of the bits was used to reduce crosstalk during playback. For this recording media, even the bits of 300 kFCI look well recorded at this magnification.

Figure 8.6 shows higher magnification bit images at linear densities of (a) 10, (b) 50, and (c) 300 kFCI [8]. The characteristic features of the bits shown in (a) and (b) are that island-like domains about 80 nm on average are scattered randomly among the bits, band-like domains about 130 nm in width appear at the bit boundaries, and the bit boundaries fluctuate with almost the same period and amplitude as does the island-like domains. The island-like domains and bit-boundary fluctuations cause the noise. At 300 kFCI, the bits are formed only by band-like domains (c).

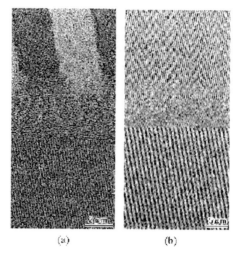

Fig. 8.5. Recorded bit images. Linear densities are (**a**) 10 (*upper*), 300 (*lower*) and (**b**) 100 (*upper*), 50 kFCI (*lower*)

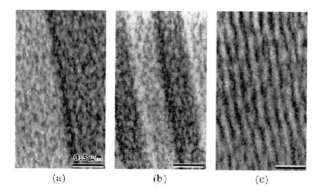

Fig. 8.6. Recorded bit images. Linear densities are (**a**) 10, (**b**) 50, and (**c**) 300 kFCI

The island-like domains are thought to be reversed domains that reduce the magnetostatic energy due to the magnetic charge at the film surface. We could confirm this if we could observe the crosssectional domain structure of the film. In the case of tape media, however, it is difficult to obtain a clear cross section and to observe the domain structure by SEM because a charge accumulates on the insulating tape. We thus made another obliquely evaporated Co-CoO medium with the same physical and magnetic properties on a conductive Si substrate. The cross section of this sample is easily obtained by cracking the substrate.

Figure 8.7 shows (a) top and (b) cross-sectional views of the domain structure of this medium in remanence state after a magnetic field was applied perpendicularly to the medium plane [8]. Figure (a) shows a domain structure similar to that shown in Figs. 8.6 (a) and (b), although the average island-like domain size of about 130 nm

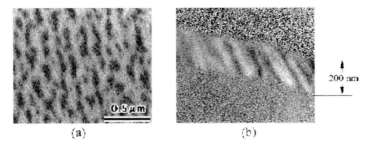

Fig. 8.7. (a) Top and (b) cross-sectional views of domain structure of Co-CoO obliquely evaporated recording medium in remanence state

was larger than that of Figs. 8.5 (a) and (b). Figure 8.7 (b) shows that domains obliquely penetrated the medium of the film, suggesting that the island-like domains shown in Fig. 8.7 (a) are reversed domains. The results illustrated in Fig. 8.7 suggest that the island-like domains shown in Figs. 8.6 (a) and (b) are also reversed domains.

The difference in domain size between Figs. 8.6 and 8.7 is probably due to the final direction of the magnetic field, which determines the magnetization direction of the medium. We found that when the angle between the easy axis of magnetization and the applied magnetic field is smaller, the size of the island-like domains in remanence state is larger [9]. In the case of Fig. 8.7, the angle was 30°, whereas in the case of Fig. 8.6, it was larger because the trailing-edge side of the final magnetic field was applied to the medium, as illustrated in Fig. 8.3.

From the domain observations described so far, we constructed a schematic model of the magnetization distribution of the recorded bits [8]. As shown in Fig. 8.8, band-like domains with almost no reversed domains appear only at the leading-edge side of the bits. This can be explained as follows [10]. In the recording process, the leading-edge-side magnetic field from the ring head saturates the magnetization; this is followed by demagnetization caused by the demagnetization field of the film and by the trailing-edge-side magnetic field. The demagnetization produces island-like reversed domains among the bits. When the magnetic field from the head is reversed, the trailing-edge-side field from the head and the field from the bit boundary region of the new bit suppress the formation of island-like domains near the bit-boundary region

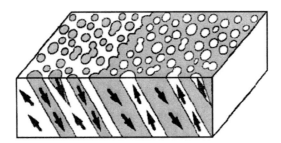

Fig. 8.8. Schematic model of magnetization distribution

of the neighboring bits written immediately before. Since the latter field is effective only near a bit boundary, band-like domains are formed only on the leading-edge-side of the bit-boundary region. Transition noise due to fluctuations of bit boundaries should thus increase with the linear recording density. On the other hand, noise due to reversed domains should decrease with an increase in the density, since the area with reversed domains decreases in size with an increase in the density because of the existence of band-like domains with a density-independent width. The total noise is the sum of these two kinds of noise. That the total noise decreases with the density indicates that the contribution of the reversed domains to the total noise is larger than that of the fluctuation in bit boundaries. To reduce the noise, the number of reversed domains and the amount of fluctuation in the bit boundaries must be reduced. This could be done by increasing the crystal anisotropy and reducing the size of the grains.

8.4 Magneto-Optical Recording Media

Magneto-optical (MO) recording media have perpendicular magneto-anisotropy with a relatively low Curie temperature of around 200 °C [11]. The read and write methods for the media are completely different from that of HDD and obliquely evaporated media; a data bit with up (down) magnetization is formed after the media is heated with a focused laser beam to or near to the Curie temperature while a magnetic field is applied with an up (down) direction. MO recording media have a large storage capacity (e.g., 2.3 GB for a 3.5 in disk) and are used for personal-computer peripheral storage devices, mini-discs for music storage devices, and libraries for various kinds of mass data storage devices.

Figure 8.9 shows the media noise (Nd) in the DC erased state and the recording noise (Nm) for TbFeCo recording films grown on substrates with four different surface roughnesses, Ra. Here, Nm is the difference between Nd and the total noise

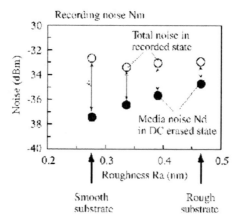

Fig. 8.9. Media noise Nd in DC erased state and recording noise Nm from TbFeCo recording films grown on substrates with four different surface roughnesses

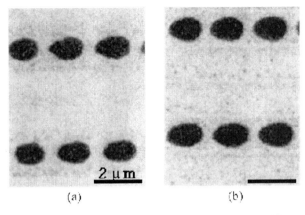

Fig. 8.10. Images of bits recorded on films with (**a**) smooth (Ra = 0.28 nm) and (**b**) rough (Ra = 0.47 nm) substrates

superimposed on a read-back signal from the recorded media. As shown in the figure, Nd increased and Nm decreased with an increase in Ra. The increase in Nd is thought to be due to the fluctuation of the magnetic easy axis, since the easy axis tends to be perpendicular to the substrate surface.

Figure 8.10 (a) and (b) show images of bits recorded on film on a smooth (Ra = 0.28 nm) or rough (Ra = 0.47 nm) substrate [12]. The bits shown in (a) had irregular distorted shapes, whereas those shown in (b) had regular, smooth, oval shapes. These results indicate that the origin of the larger noise for the smooth substrate is the distorted bit shapes.

The next question is why? Figure 8.11 shows bit images we obtained to answer this question; (a) and (b) show bits recorded with an applied magnetic field of zero, or with polarity opposite to that for normal recording, for the smooth and rough substrates, respectively [12]. This recording was made possible by the magnetic field from the region around the bits of interest. For the smooth substrate, the recorded bits had different sizes with irregular shapes when the magnetic field was zero. They split into smaller domains when the field was −50 Oe and almost completely disappeared when the field reached −100 Oe. For the rough substrate, the sizes and shapes of the bits were almost the same for all the applied fields, although the reversed domains were smaller the larger the negative magnetic field, as can be seen more clearly in the magnified images shown in (c).

These results can be interpreted by the different numbers of domain-wall-pinning sites for the films on smooth and rough substrates [13], as schematically shown in Fig. 8.12, where the illustrations correspond to parts of the domain images in Figs. 8.10 and 8.11. The film on the smooth substrate probably had fewer pinning sites, so the bit shapes, which are determined by the domain walls connecting the sparse pinning sites, became irregular (Fig. 8.12 (a)). On the other hand, the film on the rough substrate probably had more pinning sites because of spatial variation in the magnetic parameters of the materials [14,15], so that the bit shapes became smooth (Fig. 8.12

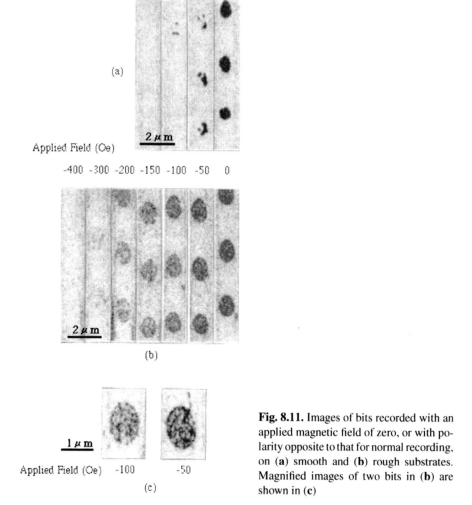

Fig. 8.11. Images of bits recorded with an applied magnetic field of zero, or with polarity opposite to that for normal recording, on (**a**) smooth and (**b**) rough substrates. Magnified images of two bits in (**b**) are shown in (**c**)

(b)). Unfortunately, in the present case, the decrease in Nm with the increase in Ra is canceled out by the increase in Nd, meaning that the total noise is independent of Ra. Consequently, the best way to decrease total noise is to grow a smooth recording film with lots of pinning sites.

8.5 Concluding Remarks

In this chapter, I explained how spin-SEM has contributed to the study of magnetic recording media. Spin-SEM has excellent characteristics. It has high spatial resolution, it can detect the magnetic direction, and it can obtain magnetic images even for rough or not well-defined surfaces. These characteristics are particularly important

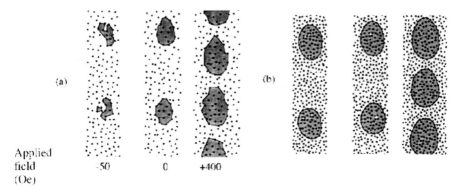

Fig. 8.12. Schematics illustrating formation of bit shapes shown in Figs. 8.10 and 8.11. (**a**) Film on smooth substrate had fewer pinning sites, and the bit shapes (determined by the domain wall connecting the sparse pinning sites) became irregular. (**b**) Film on rough substrate had more pinning sites, and the bit became smooth

for the study of commercially available recording media. Since spin-SEM was first developed, its performance has been continuously improved so that its spatial resolution has now reached 5 nm [16]. This resolution is high enough for studying magnetic recording media with a density of more than 1 Tbit/in^2. Further contributions of this technique to the study of magnetic recording media is expected.

Acknowledgement. I am grateful to H. Matsuyama, T. Kohashi, F. Tomiyama, Y. Shiroishi, A. Ishikawa, H. Aoi, T. Takayama, C. Haginoya, H. Miyamoto, J. Ushiyama, and H. Awano for their help with various parts of this work and to M. Futamoto for his useful comments and suggestions. This work was done in the Advanced Research Laboratory, Hitachi, Ltd.

References

1. K. Koike and K. Hayakawa: Jpn. J. Appl. Phys., **23**, L187 (1984).
2. H. Matsuyama, K. Koike, F. Tomiyama, Y. Shiroishi, A. Ishikawa, and H. Aoi: IEEE. Trans. Magn., **MAG-30**, 1327 (1994).
3. R. A. Baugh, E. S. Murdock, and B. R. Natarajan: IEEE Trans. Magn. **MAG-19**, 1722 (1983).
4. J.-G. Zhu: IEEE Trans. Magn., **MAG-29**, 195 (1993).
5. H. N. Bertram and R. Arias: J. Appl. Phys. **71**, 3439 (1992).
6. H. N. Bertram, I. A. Beardsley, and X. Che: J. Appl. Phys. **73**, 5545 (1993).
7. S. Iwasaki and Y. Nakamura: IEEE Trans. Magn., **MAG-13**, 1272 (1977).
8. H. Matsuyama, T. Kohashi, K. Koike, and T. Takayama: J. Mag. Soc. Jpn., **20**, 795 (1996).
9. T. Kohashi, H. Matsuyama, C. Haginoya, K. Koike, and T. Takayama: J. Appl. Phys., **81**, 7915 (1997).
10. I. Tagawa, Y. Shimizu, and Y. Nakamura: J. Mag. Soc. Jpn., **15**, 827 (1991).
11. Y. Mimura and N. Imamura: Appl. Phys. Lett., **28**, 746 (1976).
12. T. Kohashi, H. Matsuyama, C. Haginoya, K. Koike, H. Miyamoto, J. Ushiyama, and H. Awano: J. Mag. Soc. Jpn., **20**, 303 (1996).

13. T. Satoh, Y. Takatsuka, H. Yokoyama, S. Tatsukawa, T. Mori, and T. Yorozu: IEEE Trans. Magn., **MAG-27**, 5115 (1991).
14. R. Giles and M. Mansuripur: Comput. Phys., **MAR/APR**, 204 (1991).
15. M. Mansuripur, R. Giles, and G. Patterson: J. Mag. Soc. Jpn., **15**, 17 (1991).
16. T. Kohashi and K. Koike: Jpn. J. Appl. Phys., **40**, L1264 (2001).

9

High Resolution Magnetic Imaging by Local Tunneling Magnetoresistance

W. Wulfhekel

An introduction to spin-polarized scanning tunneling microscopy with a soft magnetic tip is given. After illustrating the fundamental physical effect of tunneling magnetoresistance and giving a short historical background, it is shown how magnetic and topographic information can be separated using a modulation technique of the tip magnetization. Important for the functionality of the method is to avoid magnetostriction in the tip during reversal of its magnetization. It is shown that this is theoretically and experimentally possible with an appropriate tip material of very low magnetostriction. The closure domain structure of Co(0001) is studied and ultrasharp 20° domain walls of only 1.1 nm width are found. This narrow width is explained on the basis of a micromagnetic model, and a lateral resolution of the technique better than 1 nm is shown. The limits of the technique due to the stray field of the magnetic tip are illustrated. In the case that the stray field of the tip influences the sample under investigation, the local magnetic susceptibility can be measured. Furthermore, we focus on the contrast mechanism and give evidence that the tunneling magnetoresistance depends on the barrier height in agreement with Slonczewski's model. Finally, the possibility of magnetic imaging through a non-magnetic overlayer is discussed.

9.1 Introduction

Since the invention of scanning tunneling microscopy (STM) in 1981 by Binning and Rohrer [1], the technique has developed into an invaluably powerful surface analysis tool due to its real space imaging capabilities with atomic resolution [2]. Working in the field of magnetic imaging, one may ask the simple question: Is it possible to develop a technique similar to STM to image magnetic domains with high resolution? The invention of spin-polarized scanning tunneling microscopy (SP-STM) is the direct answer to this question. In an SP-STM not only is the electron charge used to map the surface topography, but also the electron spin is utilized to image the spin structure of the sample, which is directly related to the sample magnetization. The principle of operation of an SP-STM is based on a fundamental property of ferromagnets. Due to the spin-sensitive exchange interaction between localized electrons

(Heisenberg model) or electrons in a delocalized electron gas (Stoner model), the electronic density of states splits up into different minority and majority densities (Fig. 9.1a). This is in contrast to paramagnetic substances, where the distributions of spin-up and spin-down electrons are identical. It was Jullière [3] who discovered in 1975 the consequences of this imbalance of majority and minority electrons, i.e., the spin polarization, on tunneling between two ferromagnets. In his fundamental experiment, two magnetic films, Fe and Co, were isolated by a thin Ge film to form a tunnel junction. The two magnetic films had the same easy axis of magnetization but different coercive fields. This permitted the alignment of their magnetization parallel or antiparallel as a function of an applied magnetic field. Jullière found that the tunneling conductance G depends on the relative orientation of the magnetization of the two layers. For parallel orientation, the conductance G was 14% higher than for antiparallel orientation. He explained his finding with the spin polarization of the tunneling electrons. Under the assumption of a small bias voltage across the junction and in the absence of spin-flip scattering during the tunneling process, the electrons in the ferromagnets near the Fermi energy determine the tunneling conductance of the junction. For a parallel orientation, the majority/minority electrons of the first electrode tunnel into the majority/minority states in the second electrode, respectively, as depicted in Fig. 9.1a. In the simple case that the transmission through the barrier material itself shows no spin dependence, the conductance $G_{\uparrow\uparrow}$ is proportional to the density N of initial and final states and hence is proportional to the product of the initial and final majority and minority densities:

$$G_{\uparrow\uparrow} \propto N_\uparrow(1)N_\uparrow(2) + N_\downarrow(1)N_\downarrow(2) \tag{9.1}$$

For antiparallel orientation of the magnetization (see Fig. 9.1b), majority/minority electrons tunnel into minority/majority states and the conductance $G_{\uparrow\downarrow}$ is proportional to:

$$G_{\uparrow\downarrow} \propto N_\uparrow(1)N_\downarrow(2) + N_\downarrow(1)N_\uparrow(2) \ . \tag{9.2}$$

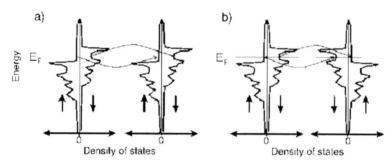

Fig. 9.1. Schematic drawing of spin-conserved tunneling between two ferromagnetic materials represented by their density of states. In (**a**) the two ferromagnets are magnetized parallel such that majority electrons from one electrode tunnel into minority states of the other electrode, while in (**b**) the ferromagnets are magnetized antiparallel such that majority electrons from one electrode tunnel into minority states of the other

With the spin polarization $P_i = (N_\uparrow(i) - N_\downarrow(i))/(N_\uparrow(i) + N_\downarrow(i))$ of the electrons of electrode i, the relative variation of the conductance is given by:

$$\Delta G/G_{\uparrow\uparrow} = 2P_1 P_2/(1 + P_1 P_2) \,. \tag{9.3}$$

This variation of the conductance and consequently of the resistance is called the tunneling magnetoresistance (TMR). More than a decade later, Slonczewski treated the problem of spin-polarized tunneling rigorously in the free electron model [4]. Neglecting higher order spin effects like spin accumulation, he calculated the dependence of the conductance on the angle θ between the magnetization of the two layers:

$$G = G_0(1 + P'_1 P'_2 \cos\theta) \,. \tag{9.4}$$

Where P'_i is the effective spin-polarization of the combination of ferromagnet i and the barrier, while G_0 is the mean conductance containing no parameters of the magnetization direction of the layers. The $\cos\theta$ dependence is strict, since it originates from the quantum mechanical rotation behavior of the spin $1/2$ tunneling electrons, i.e., it reflects the electron spin. Later, Slonczewski's prediction for the angular dependence of the TMR effect was also experimentally confirmed [5].

During the last decade, many attempts have been made to use the TMR effect in an STM to obtain spin sensitivity. Two different approaches have been of major importance: First, the use of ferromagnetic tips that lead to a spin-polarized tunneling according to Jullière's model discussed above, and, secondly, the use of GaAs tips with spin-polarized carriers that are created by optical pumping with circularly polarized light [6]. Early attempts in the beginning of the 1990s to use ferromagnetic tips and utilize the TMR effect were of limited success. The experiments by Johnson and Clarke [7], who used bulk Ni tips to image the magnetic structure of surfaces in air, were dominated by spurious effects like magnetostriction and mechanical vibrations of the tip or sample. Almost at the same time, Wiesendanger et al. [8] reported spin-polarized vacuum tunneling at room temperature between a ferromagnetic CrO_2 tip and the topological antiferromagnetic Cr(001) surface [9]. Using a tungsten tip, topographic constant current line scans revealed atomic steps on Cr(001) of the expected step height of 0.14 nm, while using a ferromagnetic CrO_2 tip, alternating step heights of 0.16 and 0.12 nm were observed. This was attributed to the TMR effect between the ferromagnetic tip and the ferromagnetically ordered Cr atoms on the terraces. When the spin polarization of the tip and the Cr terrace atoms are parallel, the tunneling current is enhanced due to the TMR effect (see Eq. 9.4) and in the constant current mode of the STM, the tip is retracted by a small amount (0.02 nm). On the adjacent atomic terrace on Cr, the spin polarization of the terrace atoms is opposite due to the topological antiferromagnetic order of Cr(001) [9]. Therefore, on this terrace, the TMR effect leads to a reduction of the current and the STM tip approaches. This mechanism results in alternating step heights seen with a spin polarized tip. However, no separation of topography and spin information could be obtained in this approach, and reference measurements had to be acquired with nonmagnetic tips.

In the mid-1990s, a more promising approach for magnetic imaging using optically pumped GaAs tips in combination with a lock-in technique to separate topographic and magnetic information was established [10–12]. By using circularly polarized light, spin polarized carriers are excited into the conduction band of the tip and then tunnel into the sample. The spin polarization of the electrons can be selected by the helicity of the light [6]. This is the key to separating spin information from topographic information. By modulating the helicity of the light, and by this the spin polarization of the carriers, modulations in the tunneling current are induced due to spin-dependent tunneling. The modulations were detected with a lock-in amplifier. The signal is used to construct magnetic images, as shown in Fig. 9.2. Hence, the modulation of spin polarization enables one to separate spin information (Fig. 9.2a) from topographic information (Fig. 9.2b), although only one physical parameter, i.e., the tunneling current, is measured. The spin information is contained in the AC part of the tunneling current at the frequency of the optical modulation, while the topographic information is contained in the DC component. This optical modulation technique, however, suffers from a rather low contrast. Further, an unintended additional magneto-optical contrast of limited lateral resolution is present due to the interaction of the light and the sample [13]. Only a few studies on domain patterns have been published using this technique, and no experiments have been presented

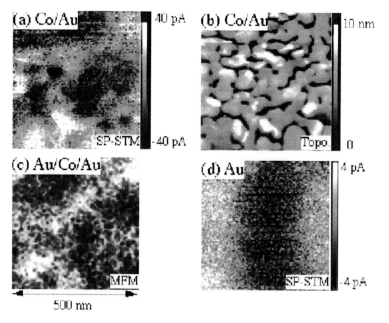

Fig. 9.2. (a) Magnetic SP-STM image and (b) topographic image of the same location of a Co film on Au. The polarization image shows magnetic domains that are similar to those obtained by MFM on a Au-covered sample depicted in (c). Magnetic SP-STM image of an Au film showing a non-vanishing contrast. All images are 0.5×0.5 μm. Figure taken from [14]

that rigorously prove the magnetic origin of the observed domains. Moreover, non-magnetic films are reported to show in some cases a considerable signal (see Fig. 9.2c) similar to the domains in magnetic films [14], raising questions about the reliability of this method.

Recently, different groups revived the first approach, the use of ferromagnetic tips. In these new approaches, spin and topographic information could be separated by two methods [15, 16]. Bode et al. used spin-polarized scanning tunneling spectroscopy to obtain spin information. This approach is described in detail in Chap. 10. The second method is addressed in the following section.

9.2 Experimental Setup

In this section, we focus on the use of ferromagnetic tips in combination with a modulation technique of the spin polarization. Analogous to the concept of optically pumped GaAs tips, a modulation of the spin polarization is used to separate spin information from topographic information in the tunneling current. In this approach to SP-STM, a soft magnetic tip is chosen as the STM tip. The longitudinal magnetization of the tip is switched periodically with the frequency f by the magnetic field induced by a small coil wound around the tip, as depicted in Fig. 9.3. The whole volume of the tip is ferromagnetic such that the reversal of the whole tip is driven by the field of the coil at the backside of the tip, and the apex is switched between the two energetically favored longitudinal magnetized states, as will be discussed in detail below. In this way, the spin polarization of the electrons at the tip apex is periodically reversed. Magnetic contrast is separated from the topographic information by phase-sensitive detection of the Fourier component of frequency f via a lock-in amplifier. Due to the local tunneling magnetoresistance effect between the magnetic tip and the surface of the specimen, the tunneling current shows an AC component related to the spin polarization of the sample. If the switching frequency f of the tip is chosen well above the cut-off frequency of the feedback loop of the STM, the variations in the tunneling current are not compensated by the current feedback loop and may be detected in the tunneling current with the lock-in amplifier. The DC component in the tunneling current is used to map the topography simultaneously with the spin structure. Important for the functionality of the method is that mechanical vibrations are avoided during the switching process of the tip and that the tip–sample distance is kept constant. Primarily, this is necessary to prevent the tip from crashing into the sample surface, since it is positioned only a few Å in front of the sample during STM operation. Secondly, one has to avoid changes in the tunneling current due to distance changes in order not to cover the small modulations of the tunneling current caused by the tunneling magnetoresistance. Due to the exponential dependence of the tunneling current on the gap width, the tolerable changes of the distance are only on the order of 0.05 Å. For larger mechanical vibrations, the variations in the tunneling current are larger than those caused by the TMR effect, which under favorable conditions is in the range of several 10% [17]. To achieve this low level of mechanical vibrations during the switching of the tip magnetization,

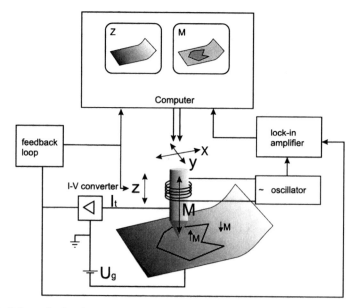

Fig. 9.3. Schematic drawing of the experimental setup. The magnetic tip of the STM can be scanned in x and y directions over the surface, while the z component is regulated with a constant current feedback loop such that the tip follows the topography of the sample. During scanning, the tip is periodically switched by the magnetic field of a coil wound around the tip. The resulting variations of the tunneling current are detected after preamplification with a lock-in amplifier to construct the magnetic image of the surface

special care has to be taken in the choice of the tip material. For optimal performance, one needs low coercive fields of the material to minimize magnetic dipolar forces between the tip and the exciting coil. Furthermore, a vanishing magnetostriction of the tip material prevents changes in the tip length during switching. Magnetization losses should be low to avoid energy dissipation and thus periodic heating and thermal expansion of the tip. Best results were obtained with an amorphous metallic glass of the CoFeSiB family with high Co concentration [18]. The material offers extremely low coercivities in the range of 50 μT with very high initial magnetic susceptibility, negligible magnetostriction ($< 4 \times 10^{-8}$) [19] and a low saturation magnetization of 0.5 T combined with low magnetization losses at frequencies up to 100 kHz.

The magnetic tips were electrochemically etched from specially designed thin CoFeSiB wires of 130 μm diameter. As etching agent, a dilute mixture of HCl and HF was used that was suspended by surface tension as a thin liquid membrane in a Pt ring during etching. The pH value was tuned such that the formation of silica from the Si in the amorphous wire was prevented. Using low etching currents on the order of 250 μA, sharp and pointed tips were created, as can be seen in Fig. 9.4a. The cone angle of the tip is typically between 8 and 15°, and the radius of curvature can be as low as 20 nm. These tips were then fixed with conducting glue to a nonmagnetic tip

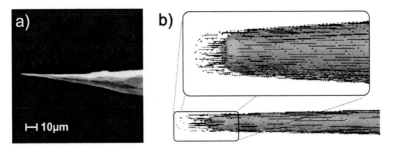

Fig. 9.4. (a) Scanning electron microscopy image of an etched CoFeSiB tip, (b) micromagnetic simulation of the tip magnetization. The inset shows the apex of the tip in higher magnification. The total length of the tip is 500 nm

shaft, around which the magnetic coil was wound. The coil was mechanically fixed to the shaft by insulating glue to avoid vibrations. The coil, which is used to switch the longitudinal magnetization of the tip, is sufficiently light, such that it can be scanned together with the tip during imaging of the surfaces.

9.3 Magnetic Switching and Magnetostriction of the Tip

To understand which magnetic information can be obtained with the soft magnetic tips, the magnetic configuration of the tip and its switching behavior were investigated. Micromagnetic calculations of the end of the tip were carried out to find the stable domain configuration. The simulations have been performed with a micro magnetic finite element algorithm based on direct energy minimization. The shape of the tip is approximated by a cone of an aperture angle of 12°, capped at its end with a hemisphere of 30 nm diameter. Since the whole magnetic tip is too large to be modeled in the framework of numerical micromagnetism, we analyzed only the last 500 nm of the end of the tip. This length, however, is well above the single domain particle diameter such that the end of the tip is free to form domains in the volume that is included in the simulations. As the stable configuration, we found the single domain state with a homogeneous magnetization pointing along the axis of the tip (see Fig. 9.4b) in agreement with what one would expect for an elongated object. The apex of the tip is free from vortices. On the outer surface of the cone-shaped tip, the magnetization points along the axis of the tip and hence does not lie in the surface of the tip. This can be explained by the limited saturation magnetization of the tip material. If the magnetization of the tip followed the contours of the cone, all the flux of the tip would be concentrated at the tip apex, where the flux would exceed the saturation of the material. However, the magnetization of the material is limited by the saturation magnetization, and the flux leaks out of the tip. This leaking keeps the magnetization exactly along the tip axis on the outer surface of the cone-shaped tip. Due to symmetry, the configuration of magnetization opposite than that depicted in Fig. 9.4b has the same energy. There are two stable configurations, and, therefore, the end of the tip shows a bistable behavior. The switching of the magnetization of the tip is a more complex process. As has been shown using Kerr microscopy, the wires show a multidomain structure on the millimeter scale and lack a single, large Barckhausen jump [20]. Nevertheless, due to their extreme magnetic softness, they exhibit a high magnetic susceptibility. As a consequence, when a magnetic field is applied with the small coil at the backside of the tip, the flux created is dragged into the needle-shaped tip. Applying a field below saturation results in a movement of the internal 180° domain walls such that the flux induced by the coil at the backside is fully kept inside the tip. Magnetostatically, it is unfavorable for the flux to leak out at the side of the tip and instead it is guided to the apex. It is then only the direction of the flux that determines which of the two single domain configurations of the end of the tip is the more stable one. If pinning of domain walls does not hinder switching of the end of the tip, it is efficiently switched between the two states just by the collected flux from the backside of the tip. We confirmed this switching behavior of the tip by micromagnetic simulations that revealed no energetic barrier for domain wall motion in the vicinity of the tip apex and showed complete switching between two states of opposite longitudinal magnetization. In this way, sensitivity for the perpendicular component of the sample magnetization is achieved with our SP-STM. The simulations of the switching process also give an estimate of the expected magnetostriction of the tip during the process of switching. As switching of the tip proceeds by domain wall formation and movement and not by coherent rotation of the entire magnetization of the tip, magnetostriction is active in the magnetic domain walls only. Thereby, the length of the tip is changed by the magnetostriction in the wall, when it exits (or enters) the apex of the tip. The width of an 180° domain wall at the end of the tip is around 20 nm, as the micromagnetic calculations show. Together with the low magnetostriction constant of the material of $< 4 \times 10^{-8}$ this results in an undetectable distance change on the order of 10^{-5} Å. Theoretically, vibration due to magnetostriction can safely be neglected.

We checked experimentally for magnetostriction and other mechanical vibrations of the tip by performing test measurements of the SP-STM setup on a nonmagnetic Cu(001) sample. Figure 9.5a displays the topography of a Cu(001) crystal as obtained with a CoFeSiB tip while applying an alternating field of about 1 mT at 20 kHz. Terraces separated by atomic steps are clearly visible. Obviously, vibrations due to magnetostriction or other effects are small enough to get stable STM images. Note that the weak vibrations visible as ripples in the topography are not related to the switching of the tip, but are due to insufficient damping of vibrations of the building. In the signal obtained from the lock-in amplifier, however, one observes a weak contrast at the step edges (see Fig. 9.5b). This cross talk from the topography is on the order of 0.3% of the tunneling current and is due to small mechanical vibrations of the tip caused, e.g., by eddy currents acting on the tip in the alternating field of the coil. These vibrations can be avoided when the exciting field is reduced by one order of magnitude (see Fig. 9.5c and d). The lock-in signal using an exciting field of 100 μT is zero and does not show any crosstalk from the topography while the magnetization of the end of the tip is still switched, as will be discussed below. Hence, vibrations due to magnetostriction can be excluded down to the sensitivity

9 High Resolution Magnetic Imaging by Local Tunneling Magnetoresistance

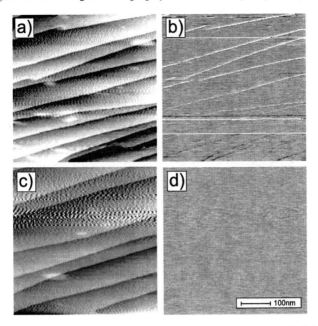

Fig. 9.5. (**a**), (**c**) STM scans of the topography and (**b**), (**d**) the spin signal of the same areas of Cu(001). During scanning an alternating magnetic field of 20 kHz was created by the coil around the tip. (**a**), (**b**) The field was set at 1.1 mT and (**c**), (**d**) 100 μT. For the higher field, mechanical vibrations of the tip are observed causing a cross-talk from the topography into the spin signal. (**b**), (**d**) Both spin images are normalized to a black and white contrast corresponding to 0.3% of the tunneling current

of the lock-in detection of $< 0.1\%$ of the tunneling current. Taking the well-known exponential dependence of the tunneling current on distance [21], one can estimate the vibrations in the narrow frequency bands around the modulation frequency f and its second harmonic. The lock-in signals correspond to distance changes between tip and sample of less than 5×10^{-4} Å, i.e., mechanical vibrations of the tip due to magnetostriction or other forces can experimentally be neglected. Using CoFeSiB, it is feasible to measure the tunneling magnetoresistance locally between the tip and the sample without unwanted mechanical vibrations.

9.4 Magnetic Imaging of Ferromagnets

Since the magnetic contrast of the SP-STM is based on the tunneling magnetoresistance effect, i.e., an interface effect, the instrument is mostly sensitive to the topmost atomic layer of the sample. As a consequence, atomically clean sample surfaces are required. The same holds for the apex of the tip. Therefore, the SP-STM experiments have to be performed in ultrahigh vacuum. After transferring new tips to the STM, the magnetic tips have to be cleaned in-situ by sputtering with 1 keV

Ar$^+$ ions to remove the native oxide at the apex. Samples were cleaned by cycles of argon sputtering (1 keV) and annealing until no traces of contamination could be found in Auger electron spectra. After sample and tip preparation, tunneling images of the topography, as well as the magnetization were recorded simultaneously at room temperature. After the initial tests for vibrations on a nonmagnetic substrate, we focused on imaging ferromagnetic surfaces. As a first example, a polished but polycrystalline Ni disk is imaged. On large scans (several μm^2) of the Ni surface, strong magnetic contrasts can be found in the spin signal, as displayed in Fig. 9.6a. The image of the spin signal shows two regions, i.e., domains with different intensities, separated by a fine, bright line, i.e., a domain wall. The observed domains in the spin signal are not related to the topography, as can be seen by comparing the topography of Fig. 9.6b with the spin signal of the very same area (white box in Fig. 9.6a). This excludes the possibility that the observed domains are caused by a crosstalk from the topography. In agreement with the theoretically predicted bistable behavior of the tip, the domains in the spin signal disappear abruptly when the size of the exciting field is lowered below 40 μT and reappear for fields above 50 μT. Upon further increase of the field, the contrast in the domain images does not rise further. The width of the domain walls observed on the polycrystalline Ni disk is between 100 and 150 nm and hence in qualitative agreement with calculated wall widths of 85–200 nm, depending on the wall type and the crystal orientation of Ni [22]. This gives a first hint of the good lateral magnetic resolution. The domains in the spin signal, however, are changing on the time scale of hours during repeated scanning, pointing at an influence of the magnetic tip on the observed domains. This might be attributed to the magnetically soft nature of polycrystalline Ni.

To learn more about SP-STM, its capabilities and limitations, a better defined surface than polycrystalline Ni was chosen for further studies. The (0001) surface of

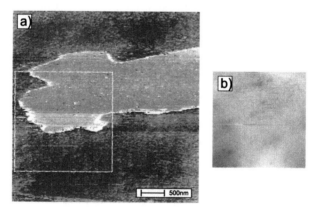

Fig. 9.6. (a) SP-STM scans of the spin structure of a polished polycrystalline Ni surface. To show that the observed spin contrast is not related to the morphology, the topography (b) and the spin signal (white box in (a)) of the very same area were recorded. The black-white contrast in the spin signal is 0.5% of the tunneling current. The peak-to-peak roughness of the topography corresponding to full black-white contrast is 3 nm

hcp Cobalt was investigated. Co is magnetically much harder than Ni and displays a strong uniaxial magetocrystalline anisotropy with an easy direction along the c-axis, i.e., perpendicular to the selected (0001) surface. Due to the minimization of the stray field energy, the single-domain state, however, is unstable and splits up into a Lifshitz closure domain pattern. Since for Co the magnetic anisotropy and the dipolar energy are of the same order of magnitude [22], no simple closure domain structure occurs but a complex, dendritic structure is observed [23], where the magnetization of most areas on the surface is strongly rotated away from the surface normal. Figure 9.7a shows an MFM image of the dendritic closure domain structure at the surface of Co(0001) taken in air with a tip magnetized perpendicular to the sample plane, i.e., the perpendicular magnetization component of the sample is imaged. Typical for this surface is the ramified pattern with domains that successively branch into finer structures. The magnetization flows out of and into the branches and lies almost in the surface plane between the branches. The magnetization does not have a fixed out-of-plane component, but varies continuously. As scanning electron microscopy with polarization analysis (SEMPA, see Chap. 7) measurements have shown, the perpendicular component varies between almost fully perpendicular orientation in the center of the branches to an in-plane orientation in the gray regions between the branches [23]. The refining branches of the domain pattern are an excellent object to test the experimental resolution of SP-STM. Figure 9.7b shows an SP-STM image of the typical branching structure at the same scale taken in ultrahigh vacuum from the same crystal as the MFM measurements. The domain structures observed with MFM and SP-STM are similar, although the images were not recorded on the same spot of the surface. At this magnification, the resolution limit of MFM on the order of several 10 nm to 100 nm (see Chaps. 11,12) becomes obvious. The branch structure seems to be blurred in comparison with the images taken with SP-STM at the same magnification, and the ends of the branches seem rounded, while STM shows pointed ends of the branches.

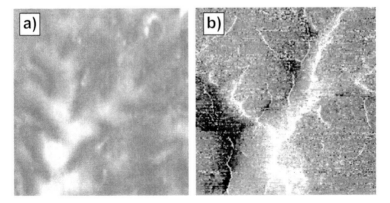

Fig. 9.7. (a) MFM and (b) SP-STM image of the closure domain pattern of Co. Both images are on the same scale of $4 \times 4\,\mu m^2$ [29]

SP-STM offers the possibility of further zooming into the closure domain structure of Co(0001) up to much higher magnifications than MFM. When focusing on the ends of the fractal branches, sharp features in the otherwise smooth contrast can be observed. The contrast across these sharp features resembles domain walls (see Fig. 9.8a). By applying a magnetic field and observing the movement of the magnetic domain wall with respect to the sample topography, it was confirmed that the contrast is indeed of magnetic origin. The contrast across the domain walls is much smaller than the full contrast observed on larger scales in the closure pattern. This can be realized when looking at the measured perpendicular magnetization component across the wall in comparison with that of the large-scale domain pattern. Figure 9.8b displays a line scan between two points of maximal contrast, i.e., between two points that are located in the center of the branches. As can be seen, the perpendicular component varies continuously along several micrometers between the two extrema. Across the domain wall at the end of a dendritic branch, the contrast changes abruptly on a length scale of about 2 nm, while the value of the change is only 20% of the maximal contrast centered just in the middle between the extrema (see Fig. 9.8c). Taking the maximal observed contrasts as perpendicular up and down magnetization, one can estimate a maximum angle of rotation across the wall of only 20° centered around the in-plane direction. To estimate the wall width w, we fit the line profile $m_z(x)$ with the standard wall profile for uniaxial systems [22]

$$m_z = \tanh(2x/w) , \tag{9.5}$$

resulting in the width of 1.1 ± 0.3 nm. At first sight, this ultra-narrow width seems to be unphysical and to contradict common knowledge about domain walls. The wall shown in Fig. 9.8c is one order of magnitude narrower than the Bloch wall in bulk Co of about 11 nm [22] and a factor of five narrower than the magnetic exchange length of Co. This is surprising, since the walls observed on the surface originate from bulk domains of the crystal. To exclude instrumental reasons for the

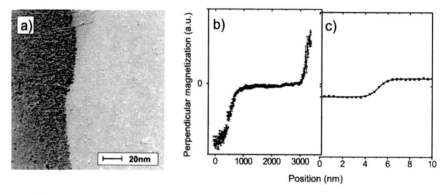

Fig. 9.8. (a) Detail of a sharp domain wall at an end of a branch in high magnification, (b) line scan through points of maximal perpendicular magnetization component, and (c) line scan across the ultrasharp domain wall at an end of a dendritic branch, including the fitted wall profile of 1.1 nm width

observation of such sharp walls, we take the following consideration. One mechanism that could cause sharper walls would be a nonlinear response of the instrument to the perpendicular component of the magnetization, e.g., a response like a step function. Theoretically, we can exclude such a nonlinear response, since the magneto-tunnel effect is a linear effect with the magnetization projection along the tip axis (see Eq. 9.4). Experimentally, we can exclude a step-like response function since we observe sharp and smooth contrasts in the same image, while a step-shaped response function would result in entirely sharp contrasts for all structures, including even the line scan of Fig. 9.8b. An alternative scenario could be to pick up the domain wall with the magnetic tip and drag it along during scanning until it snaps back due to the tension of the wall. In that case, a sharp transition would be observed at the point of snapping back. This is a common artifact in scanning probe techniques. To test for this mechanism, the wall was recorded while scanning from the right to the left and from the left to the right (see Fig. 9.9). If the wall was dragged along and snapped back, an opposite displacement of the wall for scanning in the two directions should have been seen. However, the domain wall appeared at exactly the same position for both directions, ruling out any significant dragging. Hence, the observed ultrasharp domain walls are real and need a physical explanation.

To understand the origin of the specific type of 20° wall and to calculate its expected width, we again focus on the closure domain pattern of Co(0001). In Co(0001) the magnetocrystalline anisotropy favors a magnetization along the surface normal. To reduce the stray field energy, domains of opposite magnetization along the normal separated by 180° domain walls are formed in the bulk of the crystal (see Fig. 9.10). This magnetization configuration reduces the overall stray field, but still produces a large number of surface charges, since the flux is not kept inside the crystal. As Hubert et al. [24] suggested, the system can reduce the amount of surface charges by a partial flux closure with tilted surface domains. In these surface domains, the magnetization rotates away from the magnetocrystalline easy direction, which cost anisotropy energy. However, the system saves dipolar energy due to the partial flux closure. For Co, the ratio between magnetocrystalline and dipolar energy is 0.4.

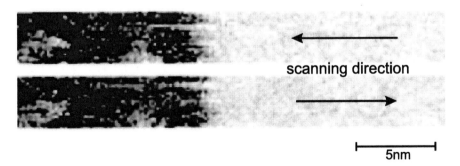

Fig. 9.9. Detailed SP-STM images of an ultra sharp domain wall scanning from the right to the left (*top*) and from the left to the right (*bottom*) excluding dragging of the domain wall during the scanning process

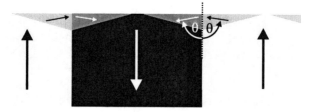

Fig. 9.10. Schematic cross section of the closure domain pattern of Co with tilted surface domains. Due to the competition of magnetocrystalline and shape anisotropy, the flux closure is incomplete at the surface and closure domains with tilted magnetization are present

From this value, one can easily calculate the angle θ, by which the magnetization is tilted from the surface normal by minimizing the free energy of the surface closure domain configuration. As depicted in Fig. 9.11a, a clear minimum of the energy at a large angle of $\theta = 80°$ is found, i.e., the flux closure is obtained by almost in-plane magnetized surface domains. Hence, under the condition that there are well-defined domains, one expects to find 20° domain walls on the surface in agreement with our SP-STM observations. Next, the expected domain wall width of such a 20° domain wall is calculated in a one-dimensional model by minimizing the sum of the magnetic exchange energy and anisotropy energy in the wall [25]. Energetic contributions of the dipolar energy are neglected in our calculation, since they only give a small correction to the wall energy. Figure 9.11b shows the wall energy as a function of the wall width w. As parameters for the calculations, the exchange constant of Co $A = 1.5 \times 10^{-11}$ J/m and the first order magnetic anisotropy of Co $K_u = 5 \times 10^5$ J/m^3 [22] were taken. The minimum of the energy is found at a wall width of only 1.5 nm. This is in good agreement with the experimentally observed wall width of 1.1 ± 0.3 nm. The small deviation of the theoretical wall widths might be due to neglecting dipolar fields or higher order anisotropies and surface anisotropies of Co(0001). This explains the occurrence and nature of the ultrasharp domain walls. There is the possibility, that a domain wall is much narrower than the magnetic exchange length $\sqrt{A/K_u}$ without violation of micromagnetic rules, but only if the angle of rotation across the wall is small. A similar wall width can be estimated by a rule of thumb argument. A 180° domain wall has a width of about 11 nm in bulk Co [22]. A 20° domain wall should have a width of a fraction of 20/180 of this. The finding of sharp domain walls on the surface of Co(0001) also gives some experimental evidence for the theoretical predictions of Hubert and Rave that sharp wall-like transitions can be formed at the surface of a closure domain pattern [26], especially when higher order in-plane or out-of-plane anisotropy terms are present, as is the case for Co(0001). Why the sharp walls are only observed close to the ends of the dendritic branches of the closure domain pattern remains an open question. Possibly only at these special points is the magnetic flux compensated in such a way that the total anisotropy term becomes stationary [26] and well-defined domains may form. For a more detailed discussion of the domain walls observed on Co(0001), see [27].

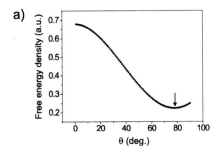 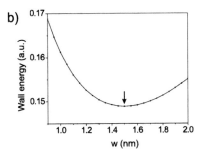

Fig. 9.11. (a) Free energy density of the tilted closure domain configuration as a function of the tilting angle θ [24] and (b) energy density as a function of width w of a 20° domain wall

9.5 Magnetic Susceptibility

With tips etched to a sharp and pointed shape, high lateral magnetic resolution of 1 nm or better can be achieved as demonstrated above with the example of a 1-nm-wide magnetic domain wall. These sharp tips are the limiting case only. In some cases, etching does not result in perfectly sharp tips as deduced from optical inspection under a microscope, or a sharp end of a tip is destroyed by a tip crash during scanning or tip approach. In these cases, the lateral resolution is worse. Besides this, there are two additional effects. After a crash, the microscopic shape of the tip is changed and by this its micromagnetic shape factor. This can result in a magnetization direction that does not lie along the tip axis anymore, and an unspecified direction of sensitivity is obtained in the magnetic contrast. The changed direction of sensitivity shows up in SP-STM images on Co as a change of the observed domain pattern from the dendritic pattern of the perpendicular magnetization to a pattern of well-defined in-plane domains similar to those in-plane domains observed on Co(0001) with SEMPA [23]. The second effect is that a dull tip produces a higher and extended magnetic stray field. In contrast to sharp tips, which produce a rather localized and limited stray field such that the domain walls of hard magnetic materials are not affected and are resolved with high resolution, the magnetic stray field of dull tips may influence the magnetic objects under investigation. The domain walls can be moved by the magnetic tips and are smeared out during imaging, or even small domains can be destroyed. In the previous section, we explained how to avoid this magnetostatic influence of the tip on the sample by using pointed tips. For dull tips, this influence is strong and magnetic imaging of undisturbed structures becomes practically impossible. The primary unwanted influence can, however, be used to measure an additional property of a magnetic sample, the local magnetic susceptibility, with high lateral resolution. As an example for a relatively dull tip, a tip cut from a piece of CoFeSiB was used. The tip was dull from optical inspection and not pointed. Nevertheless, the topographic resolution obtained with cut tips is still reasonable, as can be seen in Fig. 9.12a. Magnetic images obtained with this kind of tip often show domain walls which are smeared out over a range of up to 1 μm, as depicted in Fig. 9.12b. Note that there is a slight cross talk from the topography in the magnetic image due to

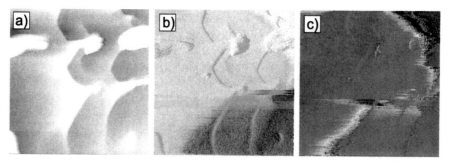

Fig. 9.12. (a) Topographic STM image, (b) magnetic image ($1f$-component) and (c) image of the local magnetic susceptibility ($2f$-component) of the same area on Co(0001). The switching frequency f was 41 kHz. The magnetic tip in this experiment was cut and not etched and was rather dull. All images are 8×8 μm

a somewhat higher roughness of the sample. The smearing out of the walls is due to a periodic domain wall movement induced by the alternating field of the tip. The walls rapidly oscillate with the switching frequency f such that the resolution is limited to about 1000 nm, while the topographic resolution is still good. The sample magnetization, however, cannot instantaneously follow the stray field of the tip, and a phase difference between the magnetization of the tip and the sample exists. This phase-shifted variation of the sample magnetization with frequency f induces a double frequency modulation of the tunneling current, as it is the product of the sample and tip magnetization. The higher harmonic in the tunneling current is related to the local magnetic susceptibility of the sample and can be detected simultaneously with a second lock-in amplifier with the spin signal [28]. This mechanism may be used to obtain domain wall contrast, as shown in Fig. 9.12c. From the observed width of the susceptibility signal around the wall and the switching frequency f, a local domain wall speed of around 10 cm/sec can be estimated. Hence, not only can static measurements of the sample magnetization be carried out with SP-STM, but also dynamic studies while, at the same time, recording magnetization and topography. This technique, in combination with higher switching frequencies, might even allow local studies of the switching behavior of individual magnetic nanostructures. Note that for the sharp tips described in the previous section, no measurable susceptibility signal was detected in the domain walls, showing that the magnetostatic interaction in that case can be suppressed efficiently.

9.6 The Contrast Mechanism

The difference in the lateral resolution of the two scanning probe techniques, MFM and SP-STM, is based on the physical phenomena underlying their contrast mechanisms. In MFM, the force between the magnetic volume of the end of the tip and the sample is used. In order to minimize the cross talk from the topography in the magnetic image, the tip is positioned several tens of nanometers above the sample during magnetic imaging. Due to the size of the effective magnetic volume of the tip, as well as the distance between the magnetic center of that volume and the sample surface, magnetic forces are averaged over a large region, which limits the lateral resolution. As discussed in Chaps. 11 and 12, the upper resolution limit is on the same order as the distance between the magnetic center of the tip and the sample surface, i.e., it is typically several tens of nanometers. In SP-STM, the spin-polarized tunneling current between the surface of the sample and the apex of the tip is used to obtain magnetic information. Here, the resolution is only limited by the sharpness of the tip, which can be atomic under favorable circumstances. Above, we demonstrated a resolution of at least 1 nm, which is impossible to obtain with an MFM using bulk magnetic CoFeSiB tips, since the magnetic volume of the tip is extremely large. As a price for the better resolution of SP-STM, however, the contrast mechanism requires clean samples and tips and imaging has to be carried out under ultrahigh vacuum conditions. The exposure of the clean sample surface and tip to just 10 Langmuir of molecular oxygen leads to a practically complete vanishing of magnetic contrast showing the high surface sensitivity of the contrast mechanism. For a more detailed comparison of the contrast mechanisms of SP-STM and MFM, see [29].

Besides high resolution magnetic imaging, SP-STM can also be used to learn more about the fundamental physics of its contrast mechanism, i.e., the tunneling magnetoresistance effect across a vacuum barrier. The technique allows continuous variations of parameters that are inaccessible in planar junctions with oxide barriers. Here, we focus on the dependence of the TMR on the barrier resistance. In SP-STM, the resistance can be continuously varied by choosing the feedback parameters, i.e., the tunneling voltage and the current, for the topographic stabilization of the tip. Recording the observed magnetic contrast across a domain wall of Co(0001) as a function of feedback current and fixed voltage, the TMR as a function of resistance is measured. As depicted in Fig. 9.13, one notices a more or less constant TMR above resistances of 10^7 Ω and a continuous drop below 10^7 Ω. First of all, this

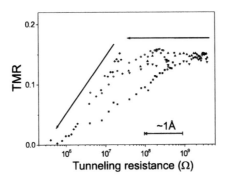

Fig. 9.13. Size of tunneling magnetoresistance (TMR) as a function of resistance of the tunneling gap. The arrows are guides to the eye only. Different symbols represent measurements with different tips and different bias voltages between 20 and 1000 mV. The strongest drop is observed for bias voltages around 200 mV

dependence gives a guideline – what feedback parameters to use in SP-STM to get an optimal magnetic contrast. One should avoid resistances below 10^7 Ω. Secondly, one may learn more about the TMR effect itself. When increasing the tunneling current and, by this, approaching the tip toward the surface (one order of magnitude in the tunneling resistance corresponds roughly to an approach of 1 Å), the TMR effect starts to decrease below a critical gap width. This is in contrast to Julliére's model, which predicts that the TMR effect is only a function of the spin polarization of the two ferromagnets and not a function of the gap width. The same holds for the more elaborate model of Slonczewski (see Eq. 9.4), where only the average conductance G_0 depends on the gap width. However, Slonczewski predicted that the TMR in the free electron approximation depends on the barrier height [4], as the effective polarization of the electrodes P'_i not only contains information of the ferromagnet, but also information of the barrier. P'_i is given by

$$P'_i = (k_\uparrow - k_\downarrow)/(k_\uparrow + k_\downarrow) * (\kappa^2 - k_\uparrow k_\downarrow)/(\kappa^2 + k_\uparrow k_\downarrow) \tag{9.6}$$

where $k_\uparrow$ and $k_\downarrow$ are the momenta of the majority/minority electrons at the Fermi energy and κ the imaginary momentum of the electrons in the barrier, which is directly related to the barrier height Φ. The first factor of equation 9.6 reflects the

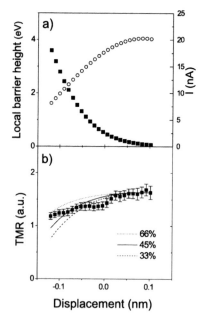

Fig. 9.14. (a) Filled squares show the tunneling current as a function of displacement, where negative displacements correspond to an approach of the tip. The open circles are the barrier height of the gap calculated from the tunneling current. (b) Filled squares represent the measured TMR as a function of displacement, while the lines are the TMRs calculated with the measured barrier heights and the spin polarizations as indicated. All TMR values have been normalized to the TMR at large distance (0.1 nm displacement)

spin polarization of the free electron gas at the Fermi energy, while the second factor is a correction factor that is related to the ferromagnet/insulator matching. Hence, one can expect that experimentally the TMR effect depends on the barrier height. To measure the barrier height of the vacuum barrier between tip and sample as a function of the gap width S, the tunneling current I as a function of the displacement was recorded, as depicted in Fig. 9.14a. The barrier height Φ can be determined from

$$\Phi(\text{eV}) = 0.952(d\ln(I)/dS)^2 \tag{9.7}$$

for small bias voltages [30]. Indeed one observes a drop of Φ when coming closer than the corresponding gap resistance of $\sim 5 \times 10^7$ Ω (see Fig. 9.14a). Simultaneous to the tunneling current, the TMR effect was recorded as shown in Fig. 9.14b. It displays the characteristic decrease below 10^7 Ω. Using Slonczewski's model with a spin splitting of 1 eV for Co [31] and a spin polarization of 45% [17], the distance dependence of the effective polarization P_i' and the TMR can be calculated from the experimentally observed barrier height. The result for the TMR is plotted as a solid line in Fig. 9.14b, showing good agreement with the measured TMR as a function of the gap width. Therefore, the drop of the TMR at small gap width can be attributed to the reduction of the barrier height. For comparison, we also calculated the expected drop of the TMR for 33% and 65% spin polarization. Qualitatively, the same drop is observed. The reduction of the barrier height at resistances below 10^7 Ω has been observed also for other surfaces [21,30] and can be attributed to the overlap of electron densities of tip and sample at short distances [30]. Due to the overlap, electrons do not have to overcome the whole work function to go from one electrode to the other, but feel only a fraction of it. Since this effect is rather fundamental, one should expect that the TMR decreases for other materials, as well when the tip is approached, giving a general rule for optimal magnetic imaging conditions. For a more detailed discussion of this effect, see [32].

Finally, we discuss the possibility of observing magnetic domains through a thin overlayer of a nonmagnetic material, e.g., Au(111). For the case of planar tunneling junctions, the use of a nonmagnetic layer inserted between the insulator barrier and one of the ferromagnetic electrodes has been studied intensively [33–36]. Conflicting results for the dependence of the TMR on the thickness of the nonmagnetic spacer layer have been reported. The TMR was found to decay very rapidly with the overlayer thickness [34, 35] such that only a couple of atomic layers led to the complete vanishing of spin polarization, while others find a much slower decay [33]. It has been shown that the decay is different when the layer is inserted between the bottom electrode and the insulator or above the insulator, hinting at an influence of different growth morphologies of the spacer layer [36]. Several groups have also addressed this problem from the theoretical side [37–40], showing a rather slow decay of the spin polarization of the tunneling current with overlayer thickness. For magnetic imaging with SP-STM, nonmagnetic overlayers of noble metals may be used to protect the ferromagnetic sample from oxidation in air. If imaging is possible through the overlayer, operation under ambient conditions could be possible with an inert tip. To test the theoretical predictions for the idealized structures and achieve magnetic imaging through a protective layer, we deposited a thin Au film on top of the Co(0001)

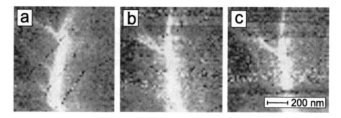

Fig. 9.15. SP-STM images of nearly the same area of a Co(0001) crystal at different Au overlayer thicknesses. (**a**) clean Co(0001), (**b**) 1.3 ML Au, (**c**) 2.6 ML Au. The contrast only weakly decreases with Au coverage

single crystal. As LEED images show, the Au film is single crystalline and grows in the (111) orientation on Co. STM images at different coverages revealed an almost perfect layer-by-layer growth. Since the domain structure of Co is rather complex and different domains of different contrast can be observed, we carried out the measurements of the contrast as a function of Au overlayer thickness on the same magnetic domain. For this, an Au evaporator was installed such that deposition is possible under glancing incidence underneath the tip while the tip is retracted by a couple of micrometers. After deposition, the tip is approached again and the very same area can be imaged. Figure 9.15 shows the magnetic images after a series of depositions. Interestingly, the contrast is found to decay only weakly with Au thickness. Hence, imaging through a protective layer is possible, and the theoretical predictions are confirmed by our experiment. The slow decay of the spin polarization of the tunneling electrons can be explained by a spin-scattering length that usually even exceeds the mean free path of the electrons in Au on the order of several tens of nanometers [41].

9.7 Conclusions and Outlook

Magnetic imaging utilizing scanning tunneling microscopy with ferromagnetic tips has been presented. It is a technique with high lateral resolution of at least 1 nm for both topographic and magnetic information, which can be separated sufficiently well regardless of the crystallographic or electronic structure of the sample due to a modulation of the magnetization of the tip. In contrast to magnetic force microscopy, it directly measures one of the magnetization components of the sample surface, usually the perpendicular component. The technique, however, requires relatively high experimental efforts, i.e., special tips and atomically clean sample surfaces, to achieve magnetic contrast. The possibility of imaging the magnetization of a sample through a thin Au film might lead to less stringent requirements for magnetic imaging in the future. Similar to magnetic force microscopy, the use of a magnetic tip causes stray fields that can influence the sample magnetization under unfavorable conditions. However, the modulation technique allows one to check for an influence during imaging and to use this influence to locally measure the magnetic susceptibility. In the future, the development of tips of other shapes, e.g., ring-shaped tips, might allow one to purely measure in-plane magnetic components of a well-defined direction. In that case, the influence of the stray field of the tip might also be reduced.

Acknowledgement. I would like to thank Mrs. Chen Chen and especially Haifeng Ding for their patience in carrying out many of the experiments presented here, Riccardo Hertel for the micromagnetic modeling of the tip, Gerold Steierl for SEM images, Manuel Vazquez for providing the amorphous wires, and Jürgen Kirschner for his support and many hints and discussions. Furthermore, I would like to thank Olivier Fruchard, Jian Shen, Hans Peter Oepen, and Patric Bruno for helpful suggestions, and Wolfgang Kuch for his critical reading of the manuscript. Finally, I wish to thank Yoshishige Suzuki for the kind permission to show his material.

References

1. G. Binning, H. Rohrer, Ch. Gerber, and E. Weibel, Appl. Phys. Lett. **40**, 178 (1982).
2. G. Binning, H. Rohrer, Ch. Gerber, and E. Weibel, Phys. Rev. Lett. **49**, 57 (1982).
3. M. Jullière, Phys. Lett. **54A**, 225 (1975).
4. J.C. Slonczewski, Phys. Rev. B **39**, 6995 (1989).
5. T. Miyazaki and N. Tezuka, J. Magn. Magn. Mater. **139**, L231 (1995).
6. D.T. Pierce and F. Meier, Phys. Rev. B **13**, 5484 (1976).
7. M. Johnson and J. Clarke, J. Appl. Phys. **67**, 6141 (1990).
8. R. Wiesendanger, H.J. Güntherodt, G. Güntherodt, R.J. Gambino, and R. Ruf, Phys. Rev. Lett. **65**, 247 (1990).
9. S. Blügel, D. Pescia, and P.H. Dederichs, Phys. Rev. B **39**, 1392 (1989).
10. M.W.J. Prins, R. Jansen, and H. van Kempen, Phys. Rev. B **53**, 8105 (1996).
11. Y. Suzuki, W. Nabhan, and K. Tanaka, Appl. Phys. Lett. **71**, 3153 (1997).
12. H. Kodama, T. Uzumaki, M. Oshiki, K. Sueoka, and K. Mukasa, J. Appl. Phys. **83**, 6831 (1998).
13. M.W.J. Prins, R.H.M. Groeneveld, D.L. Abraham, H. van Kempen, and H.W. van Kersteren, Appl. Phys. Lett. **66**, 1141 (1995).
14. Y. Suzuki, W. Nabhan, R. Shinohara, K. Yamaguchi, and T. Katayama, J. Magn. Magn. Mater. **198–199**, 540 (1999).
15. M. Bode, M. Getzlaff, and R. Wiesendanger, Phys. Rev. Lett. **81**, 4256 (1998).
16. W. Wulfhekel and J. Kirschner, Appl. Phys. Lett. **75**, 1944 (1999).
17. J.S. Moodera, and G. Mathon, J. Magn. Magn. Mater. **200**, 248 (1999).
18. Magnetic wires were kindly provided by M.Vazquez.
19. J. Velazquez, M. Vazquez, D.-X. Chen, and A. Hernando, Phys. Rev. B **50**, 16737 (1994).
20. H. Theuss, B. Hofmann, C. Gómez-Polo, M. Vázquez, and H. Kronmüller, J. Magn. Magn. Mater. **145**, 165 (1994).
21. Y. Kuk and P.J. Silverman, J. Vac. Sci. Technol. A **8**, 289 (1990).
22. E. Kneller, *Ferromagnetismus*, Springer-Verlag, Berlin (1962).
23. J. Unguris, M.R. Scheinfein, R.C. Celotta, and D.T. Pierce, Appl. Phys. Lett. **55**, 2553 (1989).
24. A. Hubert and R. Schäfer, *Magnetic Domains*, Springer-Verlag, Berlin, p. 315 (1998).
25. A. Aharoni, *Introduction to the Theory of Ferromagnetism*, Oxford University Press, New York (1996).

26. A. Hubert and W. Rave, J. Magn. Magn. Mater. **196–197**, 325 (1999).
27. H.F. Ding, W. Wufhekel, and J. Kirschner, Europhys. Lett. **57**, 100 (2002).
28. W. Wulfhekel, H.F. Ding, and J. Kirschner, J. Appl. Phys. **87**, 6475 (2000).
29. H.F. Ding, W. Wulfhekel, C. Chen, and J. Kirschner, Material Science & Engeneering B **84**, 96 (2001).
30. A. Sakai In: T. Sakurai, Y. Watanabe (eds) *Advances in Scanning Probe Microscopy*, Springer, pp. 143 (1999).
31. M.B. Stearns, J. Magn. Magn. Mater. **5**, 167 (1977).
32. H.F. Ding, W. Wulfhekel, and J. Kirschner, J. Magn. Magn. Mater. **242–245**, 47 (2002).
33. J.S. Moodera, M.E. Taylor, and R. Meservey, Phys. Rev. B **40**, 11980 (1989).
34. J.S. Moodera, J. Nowak, L.R. Kinder, P.M. Tedrow, R.J.M. van de Veerdonk, B.A. Smits, M. van Kampen, H.J.M. Swagten, and W.J.M. de Jonge, Phys. Rev. Lett. **83**, 3029 (1999).
35. J.J. Sun and P.P. Freitas, J. Appl. Phys. **85**, 5264 (1999).
36. P. LeClair, H.J.M. Swagten, J.T. Kohlhepp, R.J.M. van de Veerdonk, W.J.M. de Jonge, Phys. Rev. Lett. **84**, 2933 (2000).
37. A. Vedyayev, N. Ryzhanova, C. Lacroix, L. Giacomoni, and B. Dieny, Europhys. Lett. **39**, 219 (1997).
38. W.S. Zhang, B.Z. Li, and Y. Li, Phys. Rev. B **58**, 14959 (1998).
39. S. Zhang, P.M. Levy, Phys. Rev. Lett. **81**, 5660 (1998).
40. J. Mathon and A. Umerski, Phys. Rev. B **60**, 1117 (1999).
41. M.K. Weilmeier, W.H. Rippard, and R.A. Buhrman, Phys. Rev. B **59**, R2521 (1999).

10
Spin-Polarized Scanning Tunneling Spectroscopy

M. Bode and R. Wiesendanger

Within recent years spin-polarized scanning tunneling spectroscopy (SP-STS) has developed into a mature tool for ultrahigh spatial resolution magnetic domain imaging. In this chapter, we will introduce the measurement principle of SP-STS and describe experimental procedures and requirements. We will discuss present data measured on different ferro- and antiferromagnetic sample systems demonstrating the particular strength of SP-STS, i.e., the ability for simultaneous observation of structural, electronic, and magnetic properties.

10.1 Introduction

As already described in the previous chapter, spin-polarized electron tunneling in planar tunnel junctions has been a well established experimental technique since the early 1970s [1,2]. The power of persuasion of these first experiments was based on an appropriate choice of electrode materials, i.e., a ferromagnetic and superconducting electrode, in combination with the use of spectroscopic techniques.

The planar tunnel junctions were fabricated by growing a thin ferromagnetic film on an oxidized, self-passivating aluminum sample. The thin layer of aluminum oxide served as a tunneling barrier. In particular, the electronic properties of the superconductor (aluminum at $T = 0.3\ K$) played an important role. Close to the Fermi level its quasiparticle density of states (DOS) is dominated by two peaks separated by a small gap with a width of a few meV. If exposed to a strong external field, the Zeeman energy leads to an additional well-defined spin splitting of both peaks leading to four peaks, two of which exhibit a polarization parallel to the magnetization of the ferromagnet and two of which exhibit an antiparallel orientation. Therefore, the spin-polarized electronic structure of the superconductor is known from elementary physical principles, and any spin-polarized contribution to the tunneling current will lead to a predictable asymmetry in the tunneling spectra.

Indeed, Tedrow et al. [1,2] observed strongly asymmetric spectra when the counter electrode was made of a ferromagnetic material, which clearly proved the spin polarization of the tunneling current. A quantitative analysis allowed the calculation of the degree of spin polarization and revealed significant discrepancies when compared with clean single-crystalline ferromagnetic surfaces. These discrepancies are, however, not surprising since the electronic structure of ferromagnetic surfaces is certainly modified by the presence of the tunneling barrier, mostly made of AlO_2, and by the roughness of the interface.

These seminal experiments performed by Tedrow et al. [1, 2] show exemplarily how to overcome the main problems of spin-polarized scanning tunneling microscopy (SP-STM), i.e., the clear identification of a spin-polarized signal and the separation of topographic, electronic, and magnetic contributions. In direct analogy to those experiments, one could use a superconducting probe tip. However, the experimental setup must be exposed to a strong external magnetic field in order to achieve a sufficiently high Zeeman splitting, which will destroy the remanent domain structure and saturate the magnetization of the sample. Therefore, the use of superconducting tips may be useful for a quantitative analysis of the spin polarization of clean surfaces, but it is certainly not practicable for imaging magnetic domains.

As originally proposed by Pierce [3], optically pumped semiconducting or ferromagnetic probe tips may be used alternatively. Actually, optically pumped GaAs is routinely used as a source of spin-polarized electrons. By making use of the spin splitting of the GaAs valence band, the illumination of a GaAs tip with circularly polarized light may lead to some spin sensitivity. This experimental setup is particularly promising since no magnetic materials are involved, which excludes any unwanted modification of the sample's magnetic domain structure. Although this approach toward spin-sensitive STM has been followed in several laboratories [4–6], magnetic imaging has not yet been demonstrated unambiguously. Possibly, the problems, at least in part, are caused by the fact that due to the geometry of the tunnel junction, which is formed by the more or less flat sample and the pyramidally shaped tip, the polarization of the incident light at the apex of the probe tip is rather unpredictable. Furthermore, it may vary depending on the actual position of the tip with respect to the sample.

Alternatively, the intrinsic spin imbalance of (ferro)magnetic materials may be used for spin-sensitive experiments. Different procedures have been proposed in the past in order to avoid the tip's magnetic stray field modifying or even destroying the domain structure of the sample to be investigated. In the previous chapter, it was shown that bulk magnetic samples can be imaged with amorphous magnetic materials that exhibit an extremely low saturation magnetization. This experimental approach is, however, probably not suitable for the investigation of thin magnetic films. Instead, we have minimized the magnetic stray field by utilizing nonmagnetic probe tips that were coated by an ultrathin layer (typically a few atomic layers) of magnetic material. In the following, we will briefly describe the experimental setup and procedures. Then we will explain the contrast mechanism of spin-polarized scanning tunneling spectroscopy by making use of the exchange-split surface state of ferromagnetic Gd(0001), which represents an analogue to the Zeeman-split DOS of a superconductor in a strong external magnetic field. Then we will demonstrate the scientific impact of SP-STS on three different test samples: The ability of SP-STS to image magnetic domains and domain walls with a spatial resolution well below

1 nm has been demonstrated on ferromagnetic Fe nanowires [7]. Furthermore, the extremely high surface sensitivity of SP-STS allows the magnetic imaging of layered antiferromagnets as, e.g., Cr(001). Finally, we will show that atomic spin resolution can be achieved on the densely packed antiferromagnetic surface of a Mn monolayer on W(110) in the constant-current mode.

10.2 Experimental Setup

All experiments described in this section have been performed under UHV conditions, i.e., neither for the preparation of the sample nor the tip has the vacuum been broken. Up until now, different kinds of scanning tunneling microscopes have been used for SP-STM, commercially available and home-built, operated at room, variable, or low temperatures. Common to all microscopes is the absence of strong permanent magnetic fields at the sample location and a reliable tip exchange mechanism that allows the use of in-situ cleaned and coated probe tips [8].

Obviously, a proper tip preparation is an essential requirement for performing SP-STM experiments. Currently, we are using tungsten tips that are coated by a thin layer of ferromagnetic material. The tungsten tips are prepared by etching a polycrystalline wire in saturated NaOH solution (8 g NaOH/100 ml H_2O). This etching procedure results in tips with a typical diameter of 20 – 50 nm. After introduction into the UHV system, the tips are heated to at least 2200 K in order to remove oxide and other contaminants. This preparation procedure does not change the general shape of the tip, but causes the apex of the tip to become blunt, as can be seen in the scanning electron microscope (SEM) images shown in Fig. 10.1.

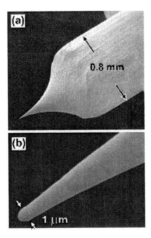

Fig. 10.1. Scanning electron micrographs of an electrochemically etched tungsten tip after flashing to $T > 2300$ K. (**a**) The overview shows the shaft of the tip (diameter: 0.8 mm) and the overall shape. (**b**) A high-resolution image reveals that the very end of the tip is blunt with a typical diameter of 1 μm

While the overall shape of the tip as displayed in the overview of Fig.10.1(a) remains almost unaffected by the high-temperature treatment, the high-resolution image of Fig.10.1(b) reveals that the tip diameter is increased to 1 μm probably due to the melting of the tip apex. In the next preparation step, the tips are coated by a thin layer of ferromagnetic material. With respect to the layer thickness several points have to be considered. On one hand, the thickness of the ferromagnetic coating should be minimized in order to reduce the tip's stray field, which might destroy the intrinsic domain structure of the sample. On the other hand, the film thickness should be well above the critical value for the onset of ferromagnetism. Additionally, in many cases, the magnetic anisotropy of a film, which determines its easy magnetization direction, varies for different film thicknesses. All these aspects have to be taken into account when choosing a suitable ferromagnetic coating. So far we have worked with probe tips that were coated by 7.5 ± 2.5 ML Fe, 8 ± 1 ML Gd, or 8 ± 1 ML GdFe. As we will show below, in general an in-plane magnetic contrast is obtained with the Fe tip, while the Gd and the GdFe tips are sensitive to the out-of-plane component of the sample's spin polarization. This fact may surprise since it might be expected that the elongated shape of the tip, as shown in Fig. 10.1(a), leads to a dominant contribution of the shape anisotropy. We have, however, shown in Fig. 10.1(b) that the tip diameter exceeds the film thickness by about three orders of magnitude. We believe that this causes material parameters like the interface and surface anisotropies to dominate the easy magnetization direction of magnetic thin film probe tips.

In our approach, tunneling spectroscopy of the local density of states is another important experimental tool since it allows the separation of structural, electronic, and magnetic sample properties. It is performed by adding a small modulation (typically 20 – 30 mV at about 2 kHz) to the applied DC bias voltage and measuring the differential conductivity dI/dU by means of lock-in technique [9].

10.3 Experiments on Gd(0001)

In the introduction of this chapter, we described the experiments performed by Tedrow et al. [1,2] on superconductor-insulator-ferromagnet planar tunnel junctions in a high external magnetic field. In the beginning, we asked ourselves whether it is possible to find a magnetic sample that exhibits an electronic structure that is equally suited to prove spin-polarized tunneling with the STM. We found that the exchange-split d-like surface state of Gd(0001) represents a good analogue to the Zeeman-split BCS-like DOS of the superconductor. While the superconductor exhibits four peaks of well-defined spin polarization, if exposed to a strong external magnetic field, the Gd(0001) surface state is exchange split into two spin parts by exchange interaction with the half-filled 4f-shell of Gd. Actually, the majority part of the surface state is occupied, and the minority part is empty, i.e., they are energetically positioned below and above the Fermi level, respectively. Since photoemission spectroscopy (PES) and inverse (I)PES are sensitive only to electronic states below or above the Fermi level, respectively, a combined PES and IPES experiment is necessary to detect both spin parts. Such an experiment has been performed by Weschke and coworkers [10]

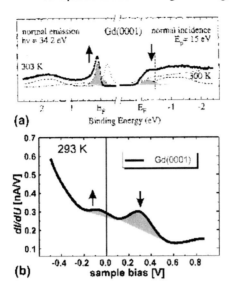

Fig. 10.2. (a) Photoemission and inverse photoemission spectrum of the exchange-split surface state of Gd(0001) showing the majority and the minority spin parts just below and above the Fermi level, respectively. (b) Tunneling spectra taken with the STM tip positioned above a Gd(0001) island. The double-peak structure represents the surface state

and is shown in Fig. 10.2(a). Since, however, the bias voltage between the sample and the tip of an STM can be tuned from positive to negative values or vice versa in a spectroscopic measurement, both empty, as well as occupied sample states, can be detected in a single experiment. Indeed, the tunneling spectrum shown in Fig. 10.2(b) exhibits two peaks at sample bias voltages of $U = -0.1$ V and $U = +0.3$ V representing the occupied and empty part of the surface state, respectively. Below the Curie temperature $T_C = 293$ K, the majority and minority characters of theses states have been confirmed by spin-resolved measurements [11, 12].

As schematically represented in Fig. 10.3(a) and (b), spin-polarized tunneling should lead to a striking asymmetry in the tunneling spectra: the dI/dU signal of the particular part of the surface state being parallel to the magnetization of the tip is expected to be enhanced, while the dI/dU signal of the peak being antiparallel to the tip is reduced. Indeed, we could observe this expected behavior experimentally with the STM. In order to minimize the field strength required for switching the magnetization direction of the sample, we have evaporated Gd on the W(110) substrate held at elevated temperature ($T = 530$ K). As can be seen in the inset of Fig. 10.3(c) this preparation procedure results in the growth of isolated Gd islands. These islands exhibit an extraordinary low coercivity of about 1.5 mT [13]. Such a small field strength can easily be reached even by coils placed outside the UHV chamber.

Since thin Gd(0001) films exhibit an in-plane anisotropy, we have used an Fe-coated probe tip for this experiment. After inserting the Fe-coated tip into the tunnel-

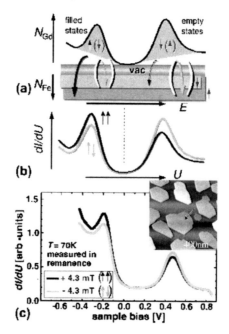

Fig. 10.3. (a) The principle of SP-STS using a sample with an exchange split surface state, e.g., Gd(0001), and a magnetic Fe tip with a constant spin polarization close to E_F. Due to the spin-valve effect (see Chap. 9), the tunneling current of the surface state spin component being parallel to the tip is enhanced compared with the opposite spin direction. (b) This should lead to a reversal in the dI/dU signal at the peak position of the surface state upon switching the sample magnetically. (c) Exactly this behavior could be observed in the tunneling spectra measured with the tip positioned above an isolated Gd island (see arrow in the inset)

ing microscope, it was magnetized by applying the maximum possible field that could be produced by the coils, i.e., $\mu_0 H \approx +10$ mT. Then the sample was inserted into the sample holder, cooled down to $T = 70$ K, and magnetized by a field of $+4.3$ mT. Subsequently, 128 tunneling spectra were measured in remanence with the tip positioned above the Gd, island marked by an arrow in the inset of Fig. 10.3(c). Then the tip was retracted from the sample surface by about 200 nm, and the direction of the magnetic field was reversed ($\mu_0 H \approx -4.3$ mT). After bringing tip and sample into tunneling distance again, another 128 spectra were measured in remanence above the identical island. This procedure was repeated several times. As can be seen in the averaged dI/dU spectra of Fig. 10.3(c), the occupied part of the exchange-split surface state is enhanced after the application of a positive field, i.e., with tip and sample magnetized parallel, while the empty part of the surface state is reduced and vice versa. This observation is in accordance with the expected behavior for spin-polarized tunneling.

Fig. 10.4(a) shows two tunneling spectra that have been measured with a Fe-coated probe tip on two adjacent Gd(0001) domains, named #1 and #2, being separated by a domain wall. Since the sample is chemically homogeneous, the differences between both spectra must be caused by spin polarized tunneling. Obviously, the

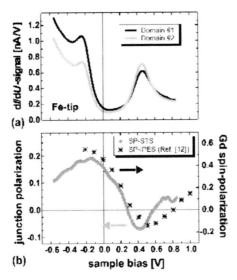

Fig. 10.4. (a) Tunneling spectra as measured with a Fe-covered probe tip above adjacent domains. An asymmetry of the dI/dU signal between the empty and filled parts of the surface state can clearly be recognized. In contrast, variations in the dI/dU signal when measured with a pure W tip are always symmetric. (b) Spin polarization of the tunneling current between an Fe-covered probe tip and the Gd(0001) surface at $T = 70$ K compared to spin-polarized inverse photoemission data of Gd(0001) measured at $T = 130$ K($*$) by Donath et al. (see [12])

largest difference between both spectra, which determines the spin polarization of the tunnel junction, does not occur at the Fermi level ($U = 0$ V), but close to the position of the Gd(0001) surface state. We can calculate the spin polarization P of the tunnel junction as formed by both magnetic electrodes, tip and sample, at any energetical position around E_F by dividing the difference of the spectroscopic signal measured on both domains through the sum:

$$P = \frac{\mathrm{d}I/\mathrm{d}U_{\#1} - \mathrm{d}I/\mathrm{d}U_{\#2}}{\mathrm{d}I/\mathrm{d}U_{\#1} + \mathrm{d}I/\mathrm{d}U_{\#2}}. \tag{10.1}$$

The result is plotted in Fig. 10.4(b). Based on the observation that the spin polarization of the tunnel junction exhibits extreme values at $U = -0.13$ V and $U = 0.42$ V, i.e., around the peak position of the minority surface state, we can conclude that – in contrast to the behavior of planar tunnel junctions with an oxide barrier for which the spin polarization was found to decrease monotonically with increasing bias voltage – surface states are of great importance for the strength of the observed magnetic signal in vacuum tunneling experiments. For comparison, Fig. 10.4(b) also shows the spin polarization of homogeneous, approximately 30 ML thick Gd(0001) films grown on W(110) as determined by means of spin-resolved inverse photoemission spectroscopy (SP-IPE) [12]. An excellent overall qualitative agreement can be recognized. Both SP-STS and SP-IPE data exhibit a positive spin polarization P on both sides of the Fermi level; P vanishes at about 300 meV and changes sign at around $U = +0.5$ V. However, the polarization of the tunnel junction is found to be about a factor of 2.5 smaller than the Gd spin polarization as determined by SP-IPE. This difference is probably caused by the finite polarization of the second electrode, i.e., the Fe-coated probe tip.

10.4 Domain and Domain-Wall Studies on Ferromagnets

In the previous section, we described our experiments on Gd(0001), a sample which is particularly suitable for the demonstration of spin-polarized scanning tunneling spectroscopy because of its extraordinary electronic properties, i.e. the existence of an exchange-split surface state. However, almost nothing was known about the magnetic domain structure of Gd(0001) thin films on W(110) before our study [14]. In order to overcome this drawback we looked for a sample with a well-defined and previously known domain structure.

We found that Fe nanowires prepared on stepped W(110) substrates fulfill this condition. Fe nanowires have been intensively studied with a large variety of experimental methods. Especially at coverages between one and two monolayers, interesting magnetic properties were reported. For example, combining longitudinal and polar Kerr-effect measurements, an onset of perpendicular magnetization was found for Fe coverages $\Theta > 1.1$ ML. Generally, the coverage range between 1.4 and 1.8 ML Fe/W(110) is characterized by magnetic saturation at relatively low external perpendicular fields combined with the absence of a hysteresis, i.e., zero remanence. These experimental results have been interpreted as the manifestation of perpendicularly magnetized Fe double-layer (DL) stripes that prefer to occupy a demagnetized ground state by antiparallel dipolar coupling, i.e., by periodically changing the magnetization direction between adjacent DL stripes [15].

Figure 10.5(a) shows an STM image of a surface that was prepared following the recipe of Elmers, Gradmann, and coworkers [15], i.e., by deposition of 1.5 ML Fe on a stepped W(110) substrate held at elevated temperature ($T \approx 520$ K). In our case, the substrate is miscut by about 1.6°, which results in an average terrace width of 9 nm. Adjacent terraces are separated by steps of monatomic height. It is well-known that at submonolayer coverage these experimental parameters lead to the Fe decoration of substrate step edges, the so-called step-flow growth mode. After the completion of the first ML, the second atomic layer grows in a similar manner. As schematically represented in Fig. 10.5(b), the substrate is finally covered by stripes of alternating Fe nanowires of ML and DL coverage. According to the model of Elmers and Gradmann, the Fe DL nanowires are alternatingly magnetized up and down, resulting in a magnetic period that amounts to twice the structural period given by the average terrace width, i.e., about 18 nm. As schematically represented in Fig. 10.5(c), this antiparallel order is a consequence of the dipolar coupling that reduces the stray magnetic field of the perpendicularly magnetized Fe double layer. At domain walls the double layer may be locally magnetized along the hard magnetic axis, i.e., in-plane. Details of the magnetic structure, however, remained unclear. Since, e.g., the typical domain wall width of 180° amounts to about 100 nm [16], it

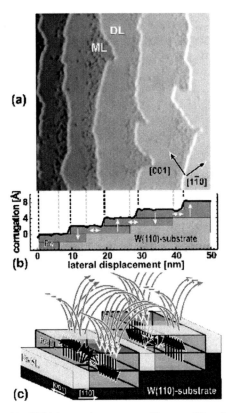

Fig. 10.5. (a) Topographic STM image (scan range: 50 nm × 50 nm) of 1.6 ML Fe/W(110) after annealing to 450 K. (b) Line section measured at the bottom edge of the STM image. The local coverage alternates between one and two atomic layers. White arrows symbolize the easy magnetization directions of the mono- and double layers, i.e., in-plane and perpendicular to the surface, respectively. (c) According to Elmers et al. [15], adjacent perpendicularly magnetized double-layer stripes exhibit an antiparallel dipolar coupling. Within domain walls, the Fe double layer on W(110) locally exhibits an in-plane magnetization

was controversially discussed whether in an ultrathin magnetic film a spin rotation can occur on a lateral scale of a few nanometers.

Before a sample surface can be studied by means of SP-STS, the electronic structure of the sample has to be investigated with non-magnetic tips. Figure 10.6(a) shows typical tunneling spectra of Fe on W(110) at mono- and double layer coverage as measured with a W tip. While the monolayer exhibits a peak at $U = +0.4$ V, an even stronger peak at $U = +0.7$ V was found to be characteristic of the double layer. Besides some variations at structural dislocation sites, we found the spectra to be identical for different mono- and double layer Fe nanowires within the accuracy of our measurement. In contrast, two different types of spectra were found for the Fe DL nanowires as soon as magnetic tips were used, as shown in Fig. 10.6(b) and

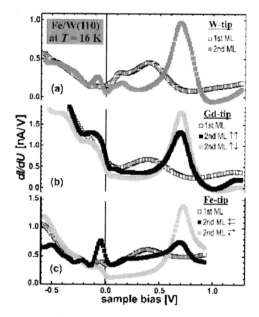

Fig. 10.6. Tunneling spectra of Fe/W(110) at mono- and double layer coverage as measured with a (**a**) nonmagnetic W tip and with a magnetic (**b**) Gd-coated and (**c**) a Fe-coated tip. With both magnetic tips we find an additional variation for the spectra of the double layer. This variation is caused by spin-polarized tunneling in magnetic domains (**b**) and domain walls (**c**)

(c) for Gd- and Fe-coated probe tips, respectively. This additional variation is caused by spin-polarized tunneling between both magnetic electrodes, i.e., tip and sample, depending on their relative magnetization directions (parallel or antiparallel). As we will show in the following, Gd tips are magnetized along the tip axis, which results in a *domain* contrast in an SP-STS experiment on the perpendicularly magnetized Fe double layer on W(110). In other words, the two different types of spectra are caused by a tip that is magnetized parallel to one domain (↑↑) and antiparallel to the other domain (↑↓). In contrast, Fe tips are magnetized perpendicular to the tip axis, which makes them sensitive to the in-plane component of the sample magnetization. Consequently, a magnetic contrast is observed when the tip is positioned above *domain walls* of the Fe double layer. Again, in the center of a domain wall the sample magnetization may be parallel or antiparallel to the tip magnetization.

Obviously, the size and the sign of the spin contrast strongly depends on the bias voltage. The Gd tip used in Fig. 10.6(b), for example, gives high contrast at the DL peak position, i.e., at $U = +680$ mV. At $U < +500$ mV and $U > +850$ mV, the contrast inverts, as indicated by the crossing of the curves of parallel and antiparallel magnetization. The data of Fig. 10.6(c), which have been measured with a Fe-coated tip, exhibit an additional contrast inversion at $U < -0.2$ V. In our experience, the voltage at which maximal spin contrast is achieved varies between different tip preparation cycles [17]. This is probably caused by the dependence of the spin-dependent electronic structure of the STM probe tips on the shape and the chemical composition of the cluster forming the apex of the tip.

It is very time-consuming to measure full tunneling spectra at every pixel of the image (typically 20 h for each image). Therefore, we have reduced the measurement time (to about 20 min) by scanning the sample at one particular bias voltage that, according to the spectra, is expected to give high magnetic contrast. Figure 10.7(a) and (b) show the simultaneously recorded topographic and spectroscopic dI/dU signal, respectively, of 1.8 ML Fe/W(110), as recorded with a Gd-coated tip at $U = 0.7$ V. Since the Fe monolayer exhibits a lower differential conductivity dI/dU at this particular bias voltage (cf. Fig. 10.5), it appears black. Due to the locally varying electronic properties, the dislocation lines in the Fe-double layer that point along the [001] direction show up as dark stripes [18]. However, there is an additional variation of the dI/dU signal along the Fe double-layer stripes. It is caused by spin-polarized tunneling between the magnetic tip and the sample. As we have verified by the application of an external magnetic field pointing along the surface normal [19], Fig. 10.7(b) shows an out-of-plane contrast, i.e., the Gd-coated probe tip is sensitive to the perpendicular component of the spin polarization. One clearly recognizes a stripe domain pattern running along the [1$\bar{1}$0]-direction. The periodicity amounts

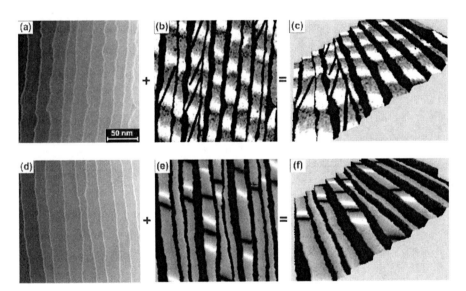

Fig. 10.7. (a) STM topograph and (b) magnetic dI/dU signal of 1.6 ML Fe/W(110), as measured with a Gd-coated tip showing the domain structure of perpendicularly magnetized double-layer Fe nanowires. The sample exhibits a stripe domain phase with domains running along the [1$\bar{1}$0]-direction. (c) Rendered perspective topographic image taken from the data in (a) combined with a gray-scale representation of the magnetic dI/dU signal taken from (b). (d)–(f) Same as (a)–(c), but measured with a Fe-coated tip being sensitive to the in-plane component of the magnetization, thereby giving domain wall contrast on this particular sample. All data were measured at a sample bias voltage $U = 0.7$ V

to approximately 50 nm. Since both the topographic, as well as the magnetic dI/dU signal have been measured simultaneously at the same position of the sample surface, we can compose a rendered three-dimensional surface contour that is superimposed by the magnetic signal in a gray-scale representation. This is shown in Fig. 10.7(c) and allows an intuitive understanding of structural and magnetic properties of the investigated surface on a single-digit nanometer scale.

So far we have shown that Fe nanowires at a total coverage just below two atomic layers exhibit a stripe domain phase, i.e., the magnetization periodically changes between up and down along a single double-layer stripe. This implies, however, that numerous domain walls must be present. Inside the domain wall the magnetization continuously rotates between either perpendicular magnetization directions. Therefore, in the center of the wall the magnetization must point along the hard axis of the Fe double layer. However, this rotation may take place in two different rotational directions, i.e., either clockwise or counter-clockwise. Is it possible to distinguish between those two cases? We found that this can indeed be accomplished by using Fe-coated tips. Figure 10.7(d) and (e) again show the simultaneously recorded topographic and spectroscopic dI/dU signals, respectively, of a similar sample as, imaged in Fig. 10.7(a)–(c). In contrast to the data measured with the Gd tip, the magnetically induced variation of the dI/dU signal is now localized to narrow lines running along the [1$\bar{1}$0]-direction. We identify these bright and dark lines as domain walls that exhibit opposite senses of rotation. Since, however, we currently cannot control the in-plane magnetization direction of our tips, we are not able to judge which part of the line, dark or bright, corresponds to (counter)clockwise rotating domain walls.

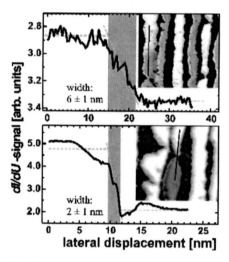

Fig. 10.8. Line sections showing the change of the dI/dU signal when crossing a domain wall located in a smooth (*upper panel*) or constricted (*lower panel*) Fe double-layer stripe. Maps of the dI/dU signal are shown as insets. The positions at which the line sections were drawn are marked by black solid lines

The development of spin-polarized STM/STS was motivated by the hope that the high spatial resolution of STM can be combined with magnetic sensitivity. Although we will show later on that even atomic resolution can be obtained by SP-STM, we would like to demonstrate here the gain in resolution made possible by SP-STS already by showing line profiles drawn across some domain walls in Fe double-layer nanowires. In Fig. 10.8, we have plotted two section lines crossing a domain wall in a smooth (top panel), as well as in a constricted Fe nanowire (bottom panel). While the width of the former domain wall amounts to $w = 6 \pm 1$ nm, the latter is much narrower ($w = 2 \pm 1$ nm). Both values of the domain wall width were well beyond the resolution limit of other magnetic imaging techniques. Although these domain walls extend over less than 20 lattice sites, we found that they can still be interpreted in the framework of micromagnetic continuum theory [20] by $w = 2(A/k)^{1/2}$, with the exchange stiffness A and the first order anisotropy constant k [21]. The mechanism that leads to the narrowing of the domain wall in the constriction has recently been described by Bruno [22].

10.5 Surface Spin-Structure Studies of Antiferromagnets

The exchange bias effect, i.e., the pinning of a ferromagnet due to exchange coupling to an antiferromagnet, has been discovered as early as 1957 [23]. As already pointed out in Chaps. 2 and 7, antiferromagnets have recently been of great economic interest due to their application in planar tunnel junctions. This has triggered numerous studies on details of the domain structure of antiferromagnetic surfaces and the coupling mechanism between the antiferromagnet and the ferromagnet [24]. However, due to the fact that the magnetic moments of antiferromagnets cancel out on lateral dimensions above the atomic scale, the measurement of their magnetic properties is particularly challenging. Therefore, a magnetic imaging technique that combines high lateral resolution with high surface sensitivity is desirable, and its availability may lead to a better understanding of antiferromagnetic surfaces and of details of the exchange-bias effect.

Antiferromagnetic chromium has been intensively investigated in the past and is still the subject of numerous experimental and theoretical studies [25, 26]. Several magnetic effects, e.g., the giant magnetoresistance effect and the interlayer exchange coupling, have been discovered in Fe/Cr multilayers. In spite of its importance, little was known about the domain structure of Cr.

Cr exhibits a transverse and a longitudinal spin-density wave below the Néel temperature $T_N = 311$ K and the spin-flip temperature $T_{sf} = 121$ K, respectively. The spin-density wave may propagate along three equivalent [001]-directions, leading to a ferromagnetic coupling within single (001)-planes, but to an antiferromagnetic coupling between adjacent (001)-planes. Although it was shown theoretically that even the (001) *surface* of Cr couples ferromagnetically and exhibits an enhanced magnetic moment [27], no net magnetic moment could be found by spin-resolved photoemission [28]. As first pointed out by Blügel et al. [29], this is caused by the fact that surfaces cannot be prepared atomically flat. Instead, a real surface always exhibits steps that separate the different atomically flat terraces from each other. Thereby, different (001)-planes with opposite magnetization directions are exposed to the surface, leading to the so-called "topological antiferromagnetism" of the Cr(001) surface. Typically, these terraces have a width of about 10 nm up to several hundreds of nanometers. Furthermore, the magnetization of the first subsurface layer points opposite to the surface. Therefore, all experimental methods with insufficient lateral resolution (> 10 nm) or surface sensitivity (> 1 ML) average over regions of opposite magnetization, which leads to their compensation.

SP-STM/STS does not suffer from these limitations. Both the lateral resolution as well as the surface sensitivity are well beyond the requirements mentioned above. Indeed, by using constant-current SP-STM Wiesendanger et al. [30] found the first experimental evidence of topological antiferromagnetism on the Cr(001) surface. In this experiment, CrO_2 tips with a high degree of spin polarization were successfully used to detect periodic alternations of the measured mono-atomic step heights in constant-current images. The deviations of the measured step height values from the topographic mono-atomic step height could be related to the effective spin polarization of the tunnel junction.

A significant drawback of this experimental approach was, however, the superposition of topographic and magnetic structure information. To solve this problem, we have performed spectroscopic measurements on Cr(001) similar to those previously described for Fe nanowires (cf. Fig. 10.6). Figure 10.9(a) shows an averaged spectrum measured with a nonmagnetic W tip. A peak which is known to be d-derived and spin polarized [31] can be recognized very close to the Fermi level ($U = 0$ V). The inset shows the topography and a map of the dI/dU signal at the peak position. The correlation of both images reveals that the differential conductivity does not change across a step edge if the measurement is performed with a nonmagnetic tip. As shown in Fig. 10.9(b), spatial variations of the dI/dU signals could only be detected after using Fe-coated tips that exhibit the required in-plane anisotropy: Due to different relative magnetization directions between the tip on one side and terraces A and B on the other side, the dI/dU signal changes at the position of the step edge.

Again, we can reduce the measurement time considerably by restricting ourselves to a single bias voltage, which gives high magnetic contrast. The data presented in Fig. 10.9(b) might suggest that a high magnetic contrast can be achieved at the Cr(001) surface state peak position, i.e., very close to the Fermi level. We have, however, to take into account that the tip-sample distance is not constant, but – as a result of the constant-current mode of operation – depends on the local differential conductivity, which is not only an intrinsic property of the sample surface but which in spin-polarized experiments is also influenced by the relative magnetization direction of tip and sample. If Cr(001), which is chemically homogeneous, is scanned in the constant-current mode with a magnetic tip at a bias voltage corresponding to the energetic position close to the surfaces state, the tip-sample distance is increased (decreased) above Cr terraces magnetized (anti)parallelly with respect to the tip. As already mentioned above, this variation of the tip-sample distance shows up as deviations of the measured step height values from the topographic step height [30, 32]. This, however, leads to a strong reduction of the variation of the differential conductivity

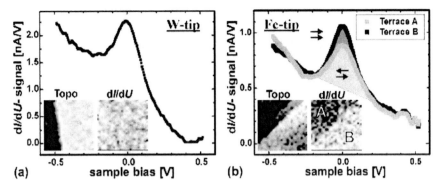

Fig. 10.9. Typical tunneling spectrum of Cr(001) as measured (**a**) with a non-coated W tip and (**b**) with a Fe-coated tip. All spectra are dominated by a strong peak at $U = -20$ mV, which represents the d-like surface state [31]. The insets show the topography (left) and maps of the dI/dU signal at the surface state peak position (right). In both cases, the topography shows two atomically flat terraces that are separated by a monoatomic step edge. While the spectra measured with the W tip are identical on both terraces, the spatially resolved dI/dU signal, as measured with the Fe-coated tip, reveals significant differences between the two terraces due to the vacuum-tunneling magnetoresistance effect

dI/dU above oppositely magnetized Cr(001) terraces. A high (magnetic) dI/dU contrast can only be achieved if the spin polarization of the (energy-integrated) tunnel current differs from the spin polarization of the electronic states at the energy that corresponds to the applied bias voltage U [35]. In our experience, on Cr(001), the highest dI/dU contrast is obtained at $U \approx \pm(250 \pm 50\,\mathrm{mV})$. Figure 10.10(a) shows the defect-free topography of a Cr(001) surface measured with a Fe-coated tip at $U \approx -290$ mV. Figure 10.10(b) reveals that the nine atomically flat terraces are separated by steps of monoatomic height. At this bias voltage, the measured step height is equal to the topographic step height for all step edges. According to Blügel's model, this topography should lead to a magnetization that alternates between adjacent terraces. This is indeed observed experimentally in the magnetic dI/dU signal that alternates from terrace to terrace between high (bright) and low (dark), thereby proving the idea of "topological antiferromagnetism" [Fig. 10.10(c)].

Figure 10.11 shows the topography of a different sample that exhibits two screw dislocations within the scan range. They are marked by arrows. In a "topological antiferromagnet", the presence of screw dislocations leads to magnetic frustrations. As can be recognized in the magnetic image of Fig. 10.11(b), in this particular case, a domain wall is created at one dislocation and annihilated at the other dislocation. Again, a three-dimensionally rendered representation of the measured data can be generated on the basis of the measured topography and dI/dU signal.

As indicated by two lines in Fig. 10.11(b), we have drawn line profiles of the dI/dU signal across the domain wall in two adjacent Cr terraces. The result is plotted in Fig. 10.12. We found that the line profiles can be fitted nicely by a tanh function, which describes a 180° wall in the framework of micromagnetic theory [20]:

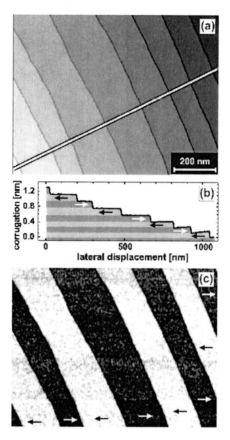

Fig. 10.10. (a) STM topograph of the Cr(001) surface (scan range: 800 nm × 600 nm). (b) The line section that has been drawn along the white line reveals that the terraces are separated by step edges of monatomic height. (b) According to Blügel's model of "topological antiferromagnetism", [29] this topography should lead to a surface magnetization that alternately switches between opposite directions from terrace to terrace. (c) In fact, using a Fe-coated probe tip alternately magnetized Cr terraces are observed by SP-STS ($U = -290$ mV)

$$y(x) = y_0 + y_{sp} \tanh\left(\frac{x - x_0}{w/2}\right), \tag{10.2}$$

where $y(x)$ is the dI/dU signal measured at position x, x_0 is the position of the domain wall, w is the wall width, and y_0 and y_{sp} are the spin-averaged and spin-polarized dI/dU signal, respectively.

As already mentioned above, the development of SP-STM was driven by the ultrahigh spatial resolution of STM, which allowed for the first time the study of structural and electronic properties of surfaces with atomic resolution in real space and over extended areas of the surface. It was an open question whether atomic resolution can also be achieved in a spin-sensitive measurement. The smallest magnetic structure

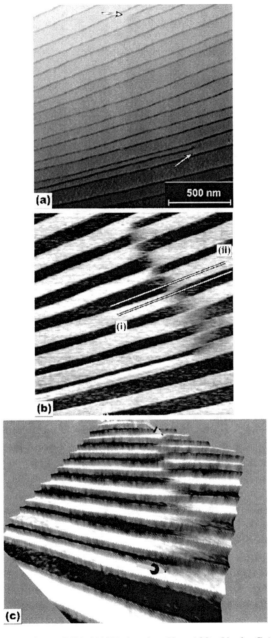

Fig. 10.11. (a) Topography and (b) dI/dU signal at $U = 190$ mV of a Cr(001) surface. The surface is magnetically dominated by a magnetization that alternates between adjacent Cr(001) terraces. Within the scan range of 2 μm × 2 μm, the sample exhibits two screw dislocations. Spin frustration leads to the formation of a magnetic domain wall between these dislocations. A line section drawn along the lines will be shown in Fig. 10.12. (c) Rendered perspective topographic image combined with a gray-scale representation of the magnetic dI/dU signal

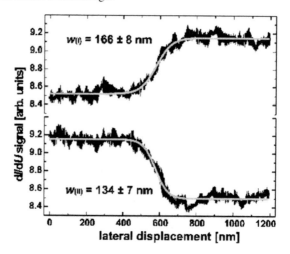

Fig. 10.12. Two domain wall profiles taken along the white lines in Fig. 10.11. The wall profiles can be fitted nicely by micromagnetic theory, resulting in domain wall widths of $w_{(i)} = 134 \pm 7$ nm and $w_{(ii)} = 66 \pm 8$ m

one can imagine is an antiferromagnetic, densely packed layer in which atoms in nearest neighbor sites exhibit opposite magnetization directions. Such a magnetic configuration was already predicted by Blügel and coworkers in 1988 [33] for several transition metal overlayers on noble-metal (111) substrates. It turned out, however, that the experimental proof of this prediction is extremely difficult. First of all, as already mentioned above, the magnetic moments of the antiferromagnetic layer cancel each other, leading to zero total magnetization on the macroscopic scale. Furthermore, the Néel temperature of that layer was unknown and probably very low. Finally, noble metals tend to intermix with transition metal adlayers, which makes their preparation almost impossible. Obviously, the first two problems can be overcome by low-temperature SP-STM. But what about the third problem? It is well-known that intermixing plays no role for refractory substrates as, e.g., tungsten (W).

In fact, in a more recent calculation, Blügel and coworkers found that a single Mn monolayer on W(110) also exhibits an antiferromagnetic ground state [34]. The typical morphology of Mn on W(110) ($\Theta = 0.68$ ML) at submonolayer coverage can be seen in the STM topograph of Fig. 10.13. It is dominated by pseudomorphically grown Mn islands of monolayer height. As long as a tungsten tip is used for the STM measurements, the experiment is not sensitive to the spin of the tunneling electron. Therefore, the opposite magnetization directions of adjacent Mn atoms cannot contribute to the tunneling current, and atomic resolution STM images should only represent the chemical unit cell. Indeed, no magnetic contribution to the experimental data can be recognized in the atomic resolution STM image of Fig. 10.14(a), which has been measured on an atomically flat Mn island on W(110) using a non-magnetic W tip. Since the electronic properties of all Mn atoms are identical, the magnetic superstructure is "ignored". The inset allows a comparison of the measured data with

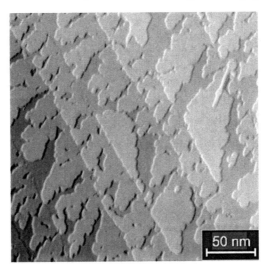

Fig. 10.13. STM topograph of 0.68 ML Mn on W(110). The morphology is dominated by pseudomorphically grown Mn islands of monolayer height

theoretically calculated atomic resolution STM images. A good qualitative agreement can be recognized.

If, however, an appropriate magnetic tip is used – due to the in-plane anisotropy of the Mn monolayer, we have used a Fe-coated tip here – the tunneling current depends on the relative magnetization direction of the tip and the sample. It has been shown theoretically for the general case [34, 35] that due to the exponential damping of large reciprocal lattice vectors, the largest magnetic superstructure will dominate the image in most cases, i.e., even if the magnetization of the tip is canted out of the plane by 80°. Figure 10.14(b) shows a high resolution STM image of 1ML Mn/W(110) obtained with a Fe tip. The scan range is the same as in Fig. 10.14(a). Instead of an atomic resolution image with a periodicity determined by the chemical unit cell, we now recognize stripes running along the [001]-direction. The periodicity perpendicular to the stripes amounts to 4.5 Å, i.e., twice the structural periodicity. The interpretation of this observation is straightforward: If the magnetic tip and the Mn rows are magnetized parallel, the tunneling current will be relatively large and, as a consequence of the constant-current mode of operation, the tip will be retracted (bright). The magnetization of all other Mn rows is antiparallel to the tip, which results in a lower tunneling current. Consequently, the tip is approached toward the surface (dark).

Nowadays, ferro- and antiferromagnetic materials are important components in high-density data storage devices. Therefore, advanced tools for microscopic characterization of the nanoscale magnetic structures are required. As pointed out previously, the investigation of antiferromagnetic surfaces is particularly difficult, as their net magnetization vanishes and many magnetically sensitive imaging techniques are not applicable. Here, spin-polarized scanning tunneling microscopy and spectroscopy

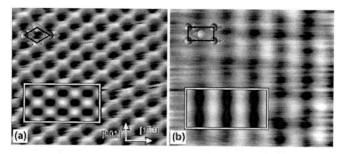

Fig. 10.14. Atomic resolution constant-current STM images measured on antiferromagnetic monolayer Mn islands on W(110) using (**a**) a nonmagnetic W tip and (**b**) a magnetic Fe tip (tunneling parameters for both images: $I = 40$ nA, $U = -3$ mV). With a W tip, the opposite magnetization direction of adjacent Mn atoms cannot be distinguished, leading to an STM image with a periodicity that is determined by the size of the chemical unit cell. In contrast, the Fe tip is sensitive to the spin of the tunneling electrons. Therefore, the periodicity of the antiferromagnetic c(2 × 2) unit cell shows up in (**b**)

close a gap and may allow significant contributions toward a better understanding of important physical phenomena, as, e.g., the exchange bias effect.

Conclusions

We have shown that nanostructured magnetic domains in ferromagnets, as well as antiferromagnets can be imaged by spin-polarized scanning tunneling spectroscopy (SP-STS) with unprecedented spatial resolution. The contrast mechanism of SP-STS is based on an additional variation of the measured differential conductivity dI/dU when a magnetic tip is used. This method allows the clear and simultaneous identification and separation of structural, electronic, and magnetic surface properties. Since STS is a near-field technique, it can be operated even in strong external fields [19]. Using the constant-current mode of operation (SP-STM), we could show that atomic spin resolution on antiferromagnetic surfaces can be obtained.

Acknowledgement. We gratefully acknowledge contributions of and discussions with S. Blügel, M. Getzlaff, S. Heinze, M. Kleiber, A. Kubetzka, X. Nie, O. Pietzsch, and R. Ravlic. Financial support has been provided by the "Bundesministerium für Bildung und Forschung" and by the "Deutsche Forschungsgemeinschaft".

References

1. P.M. Tedrow, R. Meservey, and P. Fulde, Phys. Rev. Lett. **25**, 1270 (1970).
2. P.M. Tedrow and R. Meservey, Phys. Rev. B **7**, 318 (1973).
3. D.T. Pierce, Physica Scripta **38**, 291 (1988).
4. S.F. Alvarado and P. Renaud, Phys. Rev. Lett. **68**, 1387 (1992).
5. Y. Suzuki, W. Nabhan, and K. Tanaka, Appl. Phys. Lett. **71**, 3153 (1997).

6. R. Jansen, R. Schad, and H. van Kempen, J. Magn. Magn. Mat. **198–199**, 668 (1999).
7. M. Pratzer, H.J. Elmers, M. Bode, O. Pietzsch, A. Kubetzka, and R. Wiesendanger, Phys. Rev. Lett. **87**, 127201 (2001).
8. O. Pietzsch, A. Kubetzka, D. Haude, M. Bode, and R. Wiesendanger, Rev. Sci. Instr. **71**, 424 (2000).
9. R. Wiesendanger, *Scanning Probe Microscopy and Spectroscopy*, Cambridge University Press (1994).
10. E. Weschke, C. Schuessler-Langeheine, R. Meier, A.V. Fedorov, K. Starke, F. Huebinger, and G. Kaindl, Phys. Rev. Lett. **77**, 3415 (1996).
11. D. Li, J. Pearson, S.D. Bader, D.N. McIlroy, C. Waldfried, and P.A. Dowben, Phys. Rev B **51**, 13895 (1995).
12. M. Donath, B. Gubanka, and F. Passek, Phys. Rev. Lett. **77**, 5138 (1996).
13. M. Farleand and W.A. Lewis, J. Appl. Phys. **75**, 5604 (1994).
14. M. Bode, M. Getzlaff, and R. Wiesendanger, Phys. Rev. Lett. **81**, 4256 (1998).
15. H.J. Elmers, J. Hauschild, and U. Gradmann Phys. Rev. B **59**, 3688 (1999).
16. H.P. Oepen and J. Kirschner, Phys. Rev. Lett. **62**, 819 (1989).
17. M. Bode, O. Pietzsch, A. Kubetzka, and R. Wiesendanger, J. Electr. Spectr. Relat. Phenom. **114–116**, 1055 (2001).
18. M. Bode, R. Pascal, M. Dreyer, and R. Wiesendanger, Phys. Rev. B **54**, 8385 (1996).
19. O. Pietzsch, A. Kubetzka, M. Bode, and R. Wiesendanger, Science **292**, 2053 (2001).
20. A. Hubert and R. Schäfer, *Magnetic Domains*, Springer (1998).
21. O. Pietzsch, A. Kubetzka, M. Bode, and R. Wiesendanger, Phys. Rev. Lett. **84**, 5212 (2000).
22. P. Bruno, Phys. Rev. Lett. **83**, 2425 (1999).
23. W.H. Meiklejohn and C.P. Bean, Phys. Rev. **105**, 904 (1957).
24. J. Stöhr, A. Scholl, T.J. Regan, S. Anders, J. Lüning, M.R. Scheinfein, H.A. Padmore, and R.L. White, Phys Rev. Lett. **83**, 1862 (1999).
25. E. Fawcett, Rev. Mod. Phys. **60**, 209 (1988).
26. H. Zabel, J. Phys.: Cond. Matter **11**, 9303 (1999).
27. C.L. Fu and A.J. Freeman, Phys. Rev. B **33**, 1755 (1986).
28. L.E. Klebanoff, S.W. Robey, G. Liu, and D.A. Shirley, Phys Rev B **30**, 1048 (1984).
29. S. Blügel, D. Pescia, and P.H. Dederichs, Phys. Rev. B **39**, 1392 (1989).
30. R. Wiesendanger, H.-J. Güntherodt, G. Güntherodt, R.J. Gambino, and R. Ruf, Phys. Rev. Lett. **65**, 247 (1990).
31. J.A. Stroscio, D.T. Pierce, A. Davies, R.J. Celotta, and M. Weinert, Phys. Rev. Lett. **75**, 2960 (1995).
32. M. Kleiber, M. Bode, R. Ravlic, and R. Wiesendanger, Phys. Rev. Lett. **85**, 4606 (2000).
33. S. Blügel, M. Weinert, and P.H. Dederichs, Phys. Rev. Lett. **60**, 1077 (1988).
34. S. Heinze, M. Bode, A. Kubetzka, O. Pietzsch, X. Nie, S. Blügel, and R. Wiesendanger, Science **288**, 1805 (2000).
35. D. Wortmann, S. Heinze, Ph. Kurz, G. Bihlmayer, and S. Blügel, Phys. Rev. Lett. **86**, 4132 (2001).

11

Magnetic Force Microscopy: Images of Nanostructures and Contrast Modeling

A. Thiaville, J. Miltat, and J.M. García

Magnetic force microscopy is a scanning technique, derived from atomic force microscopy, that maps the magnetic interaction between a sample and a magnetic tip. It is simple to use (most microscopes operate in ambient conditions) and provides images with a resolution down to 20 nm in best cases. The images do not, however, correspond directly to the sample magnetization, because of the long range of magnetic forces, their complex sources, as well as the potentially perturbing effect of tip-sample interaction. The emphasis of this chapter is on contrast modeling, especially in nanostructures.

11.1 Introduction: The Magnetic Force Microscope

The invention of scanning tunneling microscopy (STM, 1982) and subsequently of atomic force microscopy (AFM, 1986) caused a revolution in surface science. They enabled observation in real space of surface structures down to the atomic scale. In fact, a new generation of microscopes was born, the near-field scanning probe microscopes. Their key ingredients are:

- an actuator working at the atomic scale: this is the piezoelectric tube, where typically 1 V results in 1–10 nm displacements;
- a sharp tip whose apex can be brought in close proximity of the surface and interact with it, producing a signal that depends strongly on the tip-surface distance;
- a feedback system to control tip vertical position, owing to the detected signal and given a specific operating point (often called set point).

In STM, the signal is the tunnel current between tip and sample, both being conductive. It depends exponentially on the tip-sample distance, with a characteristic length of 0.1 nm. In AFM and a variety of derived techniques, the force between tip and sample is sensed by the deflection of a soft cantilever carrying the tip. The variety of techniques derived from the AFM expresses the variety of forces that can exist between tip and surface. Beside atomic forces (van der Waals, chemical interactions, etc.), electrostatic and magnetic forces are mainly detected. Thus, a magnetic force microscope (MFM) is, simply speaking, an AFM in wich the tip can sense magnetic forces [1, 2]. Note that other scanning probe microscopes can be used for magnetic imaging without detecting forces. These are the scanning Hall probe microscope [3,4] and the scanning SQUID microscope [5,6], where either the sample magnetic stray field or magnetic flux is measured inside a region with an area about one square micrometer. These techniques are covered in Chap. 13 of this book. The detection of a magnetic signal in an STM, using the spin-polarized tunnel effect, is covered in Chaps. 9 and 10.

Several reviews on MFM have already appeared. The first [7] is fairly representative of the status reached by the initial, homemade microscopes. The main modes of operation are well described, and many images on different samples are shown. The theory of MFM imaging in the regime of no perturbation is also discussed. The second [8] updates the apparatus description by presenting that of the current commercial machines. Imaging under an applied field is described in detail, with the demonstration of field effects on the tip, and specific examples of patterned elements imaging are shown. A recent encyclopedia article [9] is also helpful as a rapid introduction. Aside from these comprehensive texts, a number of papers with a review character appeared during the development of MFM [10–12]. Finally, some pages are devoted to MFM in the book on magnetic domains by Hubert and Schäfer [13]. Nevertheless, for the sake of legibility, the basis of MFM will be recalled briefly first. The core of the chapter will be devoted to the interpretation of MFM images and to their modeling. The emphasis will be laid on soft magnetic materials, i.e., those materials in wich the magnetic structures are dominated by magnetostatic energy (in contrast to hard materials in which the anisotropy energy dominates). Indeed, soft magnetic materials are more prone to perturbations by the MFM tip field and have been more resilient to observation by MFM. The following chapter, also devoted to MFM, proposes, on the other hand, a detailed discussion of the factors that determine the resolution of this microscope.

11.2 Principle of MFM

11.2.1 MFM Layout

The most common layout of an MFM is shown in Fig. 11.1 (atmospheric AFM operating in the vibrating mode). The piezo tube displaces the sample in the x and y directions in order to build, point by point, an image. The regulation couples deflection signals S, piezo z voltage, and tip vibration drive parameters A_d and ω_d. The signal at every point is derived from these three channels, as explained later. In fact, the various modes of AFM and MFM differ mainly in their regulation schemes.

The tip sits at the end of a flexible lever (called a cantilever), which is part of a larger chip that can be manipulated. Forces on the tip affect the position of the cantilever, whose instantaneous position is measured by reflection of a laser beam onto a position-sensitive detector. The cantilever is driven close to its resonance frequency by a piezo bimorph.

11 Magnetic Force Microscopy: Images of Nanostructures and Contrast Modeling

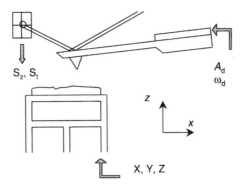

Fig. 11.1. Schematic of a MFM (vibrating mode). The various parts (piezo and sample, tip, tip chip, light detector) are not drawn on the same scale. The piezo tube moves the sample in x-, y-, z-directions under high voltages applied to separate electrodes. The small piezo bimorph, on which the tip chip is fixed, drives tip oscillations with amplitude A_d and angular frequency ω_d. The signal (cantilever deflection) is measured by the Poggendorf method: a laser beam is reflected onto a split photodiode. The difference of light intensity in the sectors gives access to tip vertical deflection (S_Z) or torsion (S_T)

Comments about the main parts of the microscope, in order to explain their specificity to MFM, appear worthwhile at this stage (see also Chap. 12 for instrumentation specific details):

1. A single piezo tube is used. This allows large sample displacements (more than 100 μm) if the tube is long enough, which is very convenient for magnetic structures as they involve different scales, from nanometers to tens of micrometers. Software aimed at correcting as much as possible the deficiencies of a piezo tube as an x-y actuator is actually required at large displacements.
2. Tips have evolved from etched magnetic wires to batch fabricated tip-cantilever chips prepared by micro-lithography [14], coated by a thin magnetic film. This ensures a great reduction of the tip stray field and tip magnetic volume, which favors a less invasive, higher resolution microscope operation. Additionally, the mechanical characteristics of the cantilevers are much better controlled. The usual cantilever materials are single crystalline silicon, or silicon nitride. Restoring force constants of the cantilevers span the $0.1-100\,\text{Nm}^{-1}$ range, and resonance frequencies vary typically from 10 kHz to 1 MHz. The magnetic coating of the tip has a typical 5–50 nm thickness. It is often magnetically hard (large coercivity) so that the tip magnetic state is not changing during imaging. A very soft coating may, however, also be used for special purposes [15].
3. Cantilever motion is commonly measured by optical beam deflection [16], which is easy to adjust, flexible but space consuming. The other main technique, optical interferometry [17], is more compact and can be fitted inside the piezo tube in an arrangement where the tube carries the tip, so that one half-space is left for the sample and its environment (e.g., an electromagnet). Both techniques enable sub-angstrom motion detection.

11.2.2 Modes of Operation

While atomic forces are short-range (the van der Waals interaction energy between induced point dipoles falls off as $1/r^6$), magnetic forces are long-range (interaction between magnetic permanent dipoles falls as $1/r^3$). If magnetic contrast is to predominate, the tip has to fly at some distance from the surface. Experimentally, this distance is on the order of 10–100 nm. One of the reasons for the widespread use of MFM is its ability to image surfaces that are not flat on the nanometer scale. In that case, separation of magnetic forces from varying atomic forces due to topography requires some care. A simple procedure consists in measuring first the topographic profile for each scan line. By retracing the measured topography with an added vertical distance, the tip flies at a constant height above the surface. This is the "lift mode" introduced by Digital Instruments [18], and therefore, the tip flying height is sometimes also called the lift height. The magnetic signal is measured on the second pass, in which the topographic signal is close to constant if the lift height is large enough. Note that larger heights decrease tip stray fields, hence also the sample perturbation, but lead to lower signals and degraded resolution.

Two cases, therefore, need to be considered. The so-called no perturbation regime corresponds to tip and sample magnetizations that do not change when imaging is performed (rigid magnetizations). On the other hand, when the tip or sample magnetization changes during imaging, one speaks of perturbations. Section 11.5 is devoted to this second regime.

Another distinction, of an instrumental nature, separates operation modes in which the tip undergoes forced oscillations from those where it does not.

11.2.2.1 Static (DC) Mode

In this simplest mode, the force on the cantilever is measured by its instantaneous deflection. Under the assumption of rigid magnetization in tip and sample, the magnetic force on the tip is the gradient of the magnetostatic interaction energy

$$E_{int} = -\mu_0 \iiint_{tip} \vec{M}_{tip} \cdot \vec{H}_{sam} = -\mu_0 \iiint_{sam} \vec{M}_{sam} \cdot \vec{H}_{tip}. \qquad (11.1)$$

Note the two equivalent formulations that exchange the roles of tip and sample (symbol: sam). The force components are thus

$$F_i = \mu_0 \iiint_{tip} \vec{M}_{tip} \cdot \frac{\partial \vec{H}_{sam}}{\partial x_i} = \mu_0 \iiint_{sam} \vec{M}_{sam} \cdot \frac{\partial \vec{H}_{tip}}{\partial x_i}. \qquad (11.2)$$

This relation indicates that tips magnetized in different directions will be sensitive to the corresponding components of the stray field. This was verified for hard magnetic coatings [19]. As, due to the tip shape, the tip moment is stronger and more stable for axial magnetization, tips are most often magnetized along their axis. The cantilever deflection senses the force mainly in one direction, which is close to the normal of the sample surface (the cantilever is mounted with an inclination of about 10° so that only the tip comes into contact). The lateral force produces a torsion of the cantilever. Torsion sensitivity is smaller than deflection sensitivity, roughly by the ratio of tip height to cantilever length (typically 10 over 200 µm). Thus, the main force is the vertical one, normal to the surface, F_Z. The minimal detectable force is limited by the thermal vibrations of the cantilever. With the usual spring constant $k = 1\ Nm^{-1}$, this force amounts to $\sim 10^{-11}$ N at room temperature [2].

11.2.2.2 Dynamic (AC) Modes

In this mode, the cantilever is driven close to resonance, and the gradient of the magnetic force comes into play. The description of the cantilever as harmonic oscillator, with quality factor Q and spring constant k, gives the following response to forced oscillations of amplitude A_d at an angular frequency ω_d [20]. The tip vertical position is $z = A\cos(\omega_d t + \varphi)$ with an amplitude

$$A = A_d \omega_r^2 \Big/ \sqrt{\left(\omega_r^2 - \omega_d^2\right)^2 + (\omega_r \omega_d / Q)^2} \tag{11.3}$$

and a phase

$$\varphi = \arctan\left(\omega_r \omega_d \big/ \left[Q\left(\omega_r^2 - \omega_d^2\right)\right]\right). \tag{11.4}$$

To first order, the resonant angular frequency ω_r is related to the free oscillation value ω_0 and the gradient of the magnetic force by

$$\omega_r = \omega_0 \sqrt{1 - (1/k)(\partial F_z/\partial z)} \sim \omega_0(1 - (1/2k)(\partial F_z/\partial z)). \tag{11.5}$$

Usual relative frequency shifts are small, on the order of 10^{-4}, justifying the expansion of the square root. In the same regime, where (11.1) is valid, the force gradient reads

$$\frac{\partial F_z}{\partial z} = \mu_0 \iiint_{tip} \vec{M}_{tip} \cdot \frac{\partial^2 \vec{H}_{sam.z}}{\partial z^2} = \mu_0 \iiint_{sam} \vec{M}_{sam} \cdot \frac{\partial^2 \vec{M}_{tip.z}}{\partial z^2}. \tag{11.6}$$

A positive force gradient produces a frequency decrease, usually coded black. As forces go to zero with increasing distance to the sample, a positive force gradient means that the tip is attracted toward the sample. For reasons of acquisition speed and insensitivity to external interferences, it has been recognized [20] that one should rather measure the cantilever oscillation (angular frequency ω_r, phase φ) than the amplitude A. As resonant frequencies are $10^4 - 10^5$ Hz and minimal detectable frequency shifts (due to thermal noise) are below 10^{-2} Hz, the force gradient limit is below $10^{-6}\ Nm^{-1}$. For the sake of comparing with the static mode, consider a power-law variation of the magnetic force. One then has $\partial F/\partial z \propto F/z$, showing that at $z = 10$ nm a $10^{-6}\ Nm^{-1}$ gradient corresponds to a 10^{-14} N force, well below the static limit. This explains the popularity of the AC technique when compared to the DC mode. The comparison of both modes is also developed in Chap. 12, in terms of spatial resolution.

The finite quality factor Q results from friction in air and the interaction with the adsorbed water layer. It reaches about 100 in air and rises to 10^4–10^5 in vacuum [20, 21], but at the expense of ease of use. It is an important parameter because the minimum detectable frequency shift is again limited by thermal noise and is inversely proportional to the square root of Q.

The AC mode, disregarding amplitude detection, is split in two versions, wherein the signal is the phase (Eq. 11.4) or the frequency (Eq. 11.5). Phase imaging is simpler, as it requires only driving the cantilever at a constant frequency. However, one sees from Eq. 11.4 that this signal is a mixture of force gradient (through ω_r) and oscillation damping (through Q). The quality factor is not at all constant. As it depends mainly on hydrodynamic damping, it varies with tip height above the sample [22]. Therefore, the preferred AC mode is that in which the resonance frequency is tracked (through an additional feedback loop set at $\varphi = \pi/2$). Indeed, one sees from Eq. 11.4 that independently of Q, the condition $\varphi = \pi/2$ is reached when $\omega_d = \omega_r$. A recently developed AC mode takes advantage of the magnetic dependence of Q [23] to produce a magnetic dissipation image [24]. In the simplest arrangement with the frequency feedback operating, the resonance amplitude is $A = A_d Q$ so that the Q factor, and the dissipated energy, can be calculated from the measured A and ω_r [24]. Contrary to previous modes, a closed form expression for the dissipated power or the Q factor cannot be given. Indeed, many dissipation processes exist such as magneto-elastic coupling to phonons and eddy currents [25], irreversible wall motion, and damping of magnetization rotation [24].

11.3 Gallery of Nanostructures MFM Images

Nanostructures made out of soft magnetic materials, which are easily modified by applied fields if shape permits it, are nowadays the subject of intense research. Images demonstrating the MFM capabilities when applied to soft magnetic nanostructures are presented below.

11.3.1 Ultrathin Films

With ultrathin films, a sensitivity (in terms of a detectable magnetic moment) down to layers less than a nanometer thick, with domains as small as 200 nm, has been demonstrated. Figure 11.2 shows domains in cobalt films sandwiched by gold. Due to interface anisotropy, the magnetization stays perpendicular up to about 1.5 nm in thickness. The different contrasts correspond to up- and down-oriented domains, with respect to the (axial) tip magnetization direction. As the stable sample state is saturation, finite domains were produced by relaxation under fields slightly below coercivity. They have a ramified structure, the larger ones extending over several micrometers (see also Fig. 11.10, discussed later).

Domains in Cu/Ni/Cu ultrathin films were also observed with a magnetic thickness as small as 2 nm [28]. As shown below, ultrathin films are also interesting samples for studying the imaging process in the MFM.

11 Magnetic Force Microscopy: Images of Nanostructures and Contrast Modeling   231

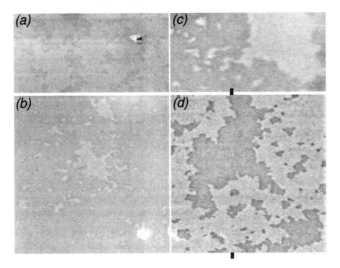

Fig. 11.2. Domains in Au/Co/Au films with ultrathin Co layers [26, 27]. Cobalt thickness is 6.2 Å **(a)**, 8.2 Å **(b)** and 13.9 Å **(c,d)**. Image widths are 30 μm **(a,b)** and 15 μm **(c,d)**. Images **(a,b,d)** were acquired with a tip covered with 200 nm CoCr, while the coverage was reduced to 50 nm in **(c)**. Lift height is 15 nm **(a,b,d)** and 20 nm **(c)**. Domains appear with a rather uniform contrast that increases, in comparison to noise, with cobalt thickness at constant tip coverage or with tip coverage at constant sample thickness

11.3.2 Nanoparticles

For samples where all dimensions are nanometric, MFM has also proved to be appropriate. Out of the many papers that appeared on this subject, Fig. 11.3 displays

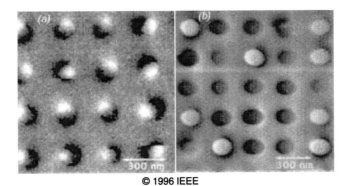

Fig. 11.3. MFM images of cobalt cylinders patterned by interference lithography [29]. Diameters are 100 nm **(a)**, 70 nm **(b)**, while heights are 40 nm **(a)**, 100 nm **(b)**. Shape anisotropy results in an in-plane **(a)** or out-of-plane magnetization **(b)**, with very different contrasts matching those expected from in-plane- and out-of-plane-oriented dipoles, respectively. Notice the switching of the bottom right particle in **(a)** during scanning

MFM images of patterned cobalt "nano-pillars" (where all dimensions are between 50 and 100 nm), with in-plane (a) and-out-of plane (b) easy-axis magnetization [29]. The contrasts (a) and (b) are very different, so that in both cases one can "see" the orientation of the magnetization in each particle.

Moreover, the individual switching of particles has been monitored, by performing imaging under an applied field [30]. With this tool, the static magnetic properties of each particle can be determined. As an example, thermally activated reversal in individual nanoparticles has been monitored with an MFM [31].

11.3.3 Nanowires

An extreme case of magnetic nanostructures is afforded by electrodeposited nanowires, where the diameter may be brought down to ~ 35 nm while the length is as large as 20 μm. Observation of domain structures along such wires has been performed nearly exclusively by MFM, on wires deposited on a substrate [32, 33]. The detailed interpretation of the contrast becomes somewhat complex, as the tip does not at all move on a planar surface due to the wire topography. Consequently, as MFM does not directly sense the sample magnetization, it cannot be told from the images alone what the magnetization orientation in each part of the wire is, except in the case of single domain nanowires. This reveals one fundamental ambiguity of stray field imaging (see also Sect. 11.4.3).

11.3.4 Patterned Elements

An enormous wealth of magnetic structures is accessed when considering samples patterned out of a soft thin film, with lateral dimensions larger than the characteristic length of micromagnetics, so that they can sustain nonuniform structures [13].

As an example, systematic observations of the magnetic configuration as a function of the element size and aspect ratio (rectangular shape) were performed by R.D. Gomez [8, 34]. This allowed the classification of the possible domain structure in such elements (comparison with observations in the transmission electron microscope – Lorentz and Foucault modes, see Chaps. 4 and 5 – established that the

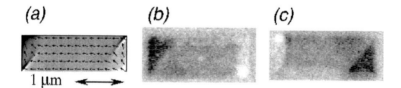

Fig. 11.4. A rectangular permalloy element (2.1 μm × 0.7 μm, thickness 16 nm) in the high remanence S state. (**a**) Calculated magnetic structure (2-D program, 5 nm mesh) with a grayscale according to the magnetic charges (11.7). (**b,c**) MFM images for two opposite tip (axial) magnetizations (tip coverage 15 nm CoCr, lift height 20 nm). Note the variation in size of the end regions between (**b**) and (**c**), due to tip-induced perturbations (see Fig. 11.16) [35]

structures were faithfully imaged). The contrasts pertaining to various types of walls (180°, 90° Néel walls, cross-tie wall, vortex) were also identified. Figure 11.4 shows two images of a rectangular NiFe element displaying a non-solenoidal configuration known as the S-state. Contrast is qualitatively similar to that of Fig. 11.3a, as the element is close to a single domain structure. However, due to its larger size, the charges at the ends are allowed to spread and form quasi-closure domain structures.

Figure 11.5 contains beautiful images of two cross-tie walls [36]. Adequate imaging conditions could only be obtained after the tip was demagnetized in order to gain a low moment, thus avoiding the perturbation of the structure. It is characterized as one 180° Néel wall, whose polarity alternates [13]. The transition regions are vortices ($S = +1$ circular Bloch lines) and cross lines ($S = -1$ Bloch lines), the latter at the origin of Néel walls of lower magnetization rotation angle extending perpendicular to the main wall.

Figure 11.6 shows the best observation, up until now, of vortex cores by MFM [37]. This observation was attempted by many people, but only recently met with success. One of the reasons is the small size of the vortex core. Indeed, micromagnetics shows that it extends over a few exchange lengths (5 nm for permalloy) [13]. Thus, a good resolution is necessary, as well as a low corrugation of the background. The absence of background signal requires circular edges of a good quality (otherwise small angle

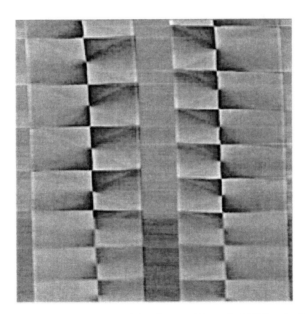

Fig. 11.5. Cross-tie walls in two 35-nm-thick, 12-μm-wide vertical NiFe strips [36], as imaged with a low moment, demagnetized tip (image size 30 μm). The secondary Néel walls perpendicular to the main wall extend to the strip edges, and contrast is distributed accordingly. With a standard Digital Instruments "MESP" tip, the central wall was displaced close to the strip edge and ripple-like structures were seen perpendicular to it, indicating a strong perturbation. Reprinted from [36] with permission from Elsevier Science

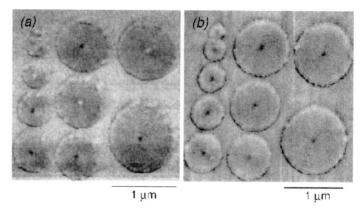

Fig. 11.6. Permalloy disks (thickness 50 nm) with diameters between 0.1 and 1 μm displaying the vortex structure [37]. After in-plane saturation at 1.5 T, different core orientations are seen (**a**). After perpendicular saturation, all cores are aligned (**b**). A "low-moment" CoCr-coated tip was used. The core contrast extends over about 50 nm. Reprinted with permission from [37]. Copyright American Association for the Advancement of Science

domain walls, or a diffuse contrast akin that in Fig. 11.4, is observed). Second, the tip should hardly displace the vortex during scanning, both in the magnetic attraction and repulsion regimes. This requires low tip fields, and small disk diameters that stabilize the vortex efficiently at the center.

11.4 MFM Contrast in Absence of Perturbations

This section examines the idealization of the MFM as a perfect measuring apparatus that does not change the state of the object while measuring it. This means that, although the tip applies a field to the sample, the reaction of the sample to this field will be neglected (see Sect. 11.5 for tip field evaluations). We also suppose that the tip is not affected by the sample field. Then, the only term of magnetic energy of the "sample + tip" system that changes with tip position is the interaction energy (11.1). Its derivatives with respect to tip height z give the force and force gradient. If one keeps in mind the idea of measuring the sample, these expressions are seen as integrals over tip volume of the sample stray field or its z-derivatives.

A first step in image interpretation is to assume that the tip is magnetically punctual. The natural model for a thin film-coated tip magnetized axially is that of a point monopole at the tip apex, with distributed monopoles of opposite sign at the bottom of the tip pyramid. Considering the large pyramid height, these remote monopoles are often disregarded. For a tip magnetized transverse to its axis, the point monopole approximation is insufficient and the point dipole model applies.

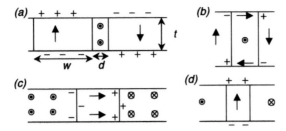

Fig. 11.7. Magnetic structures that enable 2-D stray field calculations. The film thickness has to be smaller than the exchange length in order to suppress magnetization variation across sample thickness. (**a**) A thin film with perpendicular anisotropy. The surfaces carry opposite charges. (**b**) Volume charges can appear on the walls if they transform to Néel wall, a process possible at thicknesses larger than the exchange length, or if the wall magnetization reverses at some places (presence of lines). (**c**) A thin film with easy-plane and planar magnetization. Volume charges exist around the Néel walls. (**d**) Surface charges appear when the magnetization is forced to become perpendicular, at vortex cores

11.4.1 Two-Dimensional Case

The evaluation of the stray field for an arbitrary magnetic structure has to be done numerically. However, in the case where the problem can be restricted to a two-dimensional one, analytical formulae can be obtained that are very helpful in understanding the results. The 2-D non-trivial situation is that of a (thin) film whose magnetic structure results in a 2-D charge distribution. For a magnetic structure, the magnetic poles (charge) density [13] consists of a volume term,

$$\rho = -\operatorname{div} \vec{M} = -M_s \operatorname{div} \vec{m}, \qquad (11.7)$$

and a surface term (stands for the local surface normal, oriented outwards)

$$\sigma = \vec{M} \cdot \vec{n} = M_s \vec{m} \cdot \vec{n}. \qquad (11.8)$$

The 2-D situation is realized for structures that do not vary across the film thickness, so that it is limited to thin film samples. This occurs in two main cases:

1. In films with perpendicular anisotropy, thin compared to the exchange length, walls are of pure Bloch-type, hence uncharged. One is left with only two opposite charge densities at the film surfaces (Fig. 11.7a). This case was considered in detail by H. Hug [28] in connection with experiments on ultrathin nickel films deposited on copper.
2. In films with in-plane magnetization that are thinner than the exchange length, a volume charge is present on both sides of the Néel walls (Fig. 11.7c).

In this case, the stray field is easily obtained in Fourier space. The transfer function approach that is appropriate for the description of the microscope in this regime is discussed in Chap. 12.

11.4.2 One-Dimensional Case

An even simpler situation is the 1-D case, in which fully analytical expressions of the stray field can be obtained. Such expressions are quite helpful in evaluating MFM contrast as a function of its parameters: film thickness t, tip flying height z, domain width w, and domain wall thickness d. The 1-D situation is that of a periodic domain structure, with parallel domain walls of infinite length (Fig. 11.7a).

Consider first a charge distribution at the sample top surface. For $z \neq 0$, the stray field reads

$$H_x = (1/2) M_s \sin(kx) \exp(-k|z|). \tag{11.9a}$$
$$H_z = (1/2) \operatorname{sign}(z) M_s \cos(kx) \exp(-k|z|). \tag{11.9b}$$

The stray field for a given structure is obtained by superposition of such expressions, via decomposition of the magnetization in a Fourier series $k = (2n+1)\pi/w$. As H in Eq. 11.9 is a complex exponential, the summation of the series is possible. In the case of perpendicular magnetization *with zero wall thickness*, one finds after some algebra

$$H_x = \frac{M_s}{\pi} \left[\operatorname{arctanh}\left(\frac{\sin(\pi x/w)}{\cosh(\pi z/w)}\right) - \operatorname{arctanh}\left(\frac{\sin(\pi x/w)}{\cosh(\pi(z+t)/w)}\right) \right], \tag{11.10}$$

$$H_z = \frac{M_s}{\pi} \left[\arctan\left(\frac{\cos(\pi x/w)}{\sinh(\pi z/w)}\right) - \{z \to z+t\} \right], \tag{11.11}$$

$$\frac{\partial H_x}{\partial z} = -\frac{M_s}{w} \left[\frac{\sin(\pi x/w) \sinh(\pi z/w)}{\cosh^2(\pi z/w) - \sin^2(\pi x/w)} - \{z \to z+t\} \right], \tag{11.12}$$

$$\frac{\partial H_z}{\partial z} = -\frac{M_s}{w} \left[\frac{\cos(\pi x/w) \cosh(\pi z/w)}{\sinh^2(\pi z/w) - \cos^2(\pi x/w)} - \{z \to z+t\} \right], \tag{11.13}$$

These expressions have apparently not appeared in the literature yet. They are built as the difference of two contributions, arising from the opposite charge distributions at the two surfaces. This is indicated in the equations that follow Eq. 11.10, by the substitution rule $\{z \to z+t\}$; Eq. 11.10 being fully written.

A finite domain wall thickness may be modeled in a first approximation as a linearly varying surface charge located on the sample surfaces. Then, using the same procedure as above, one gets, for example,

$$\frac{\partial H_z}{\partial z} = \frac{M_s}{\pi d} \left[\operatorname{arctanh}\left(\frac{2\cosh(\pi z/w) \sin(\pi d/2w) \cos(\pi x/w)}{\sinh^2(\pi z/w) + \sin^2(\pi d/2w) + \cos^2(\pi x/w)}\right) - \{z \to z+t\} \right]. \tag{11.14}$$

Similar expressions can be constructed for the in-plane case with volume charges only. Indeed, the z derivative of the fields in the volume charge case is proportional to the fields of the surface charge case. These expressions are however less useful, because 1-D domain structures are very rare for planar magnetization samples.

Finally, one should not forget that even in the no perturbation regime these expressions give the MFM contrast only for a point monopole or dipole tip. An integration over the tip is necessary in general.

One application of these analytical formulas concerns ultrathin films with perpendicular magnetization. In Fig. 11.8a, $\partial H_z/\partial z$ above a 1 nm film with 500-nm-wide domains is plotted at several tip heights $z = 10$, 50, and 100 nm (as explained in Sect. 11.5, this derivative is proportional to the frequency shift for a point monopole tip). At the low flying height, contrast is localized on the domain walls, and depends on domain wall thickness (the maximum signal (normalized to M_s) is 1.8×10^{-3} nm^{-1} for $d = 0$, and 1.0 at $d = 30$ nm). At large flying height, a domain-like contrast appears, which does not depend on d, as the latter is small compared to z. The intermediate case shows a contrast at domain center that is non-zero, but much smaller

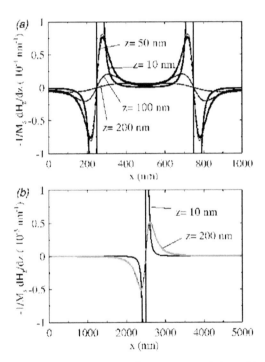

Fig. 11.8. Calculated MFM contrasts for a 1-D pattern, consisting of large domains in an ultrathin film 1 nm thick, with perpendicular magnetization. (**a**) Domain width $w = 500$ nm, tip flying height $z = 10$, 50, 100, and 200 nm, wall width $d = 0$ and 30 nm. (**b**) Domain width $w = 5,000$ nm, tip flying height $z = 10$ and 200 nm, wall width $d = 0$ and 30 nm. Domain contrast may appear at the smaller period, but completely vanishes for the large one, even if z is raised to 200 nm. The domain wall width of 30 nm has an influence only on the profiles with small flying heights (for example, in (**a**) at $z = 50$ nm, the thin curve is for $d = 0$ nm, while the thick one with symbols is for $d = 30$ nm)

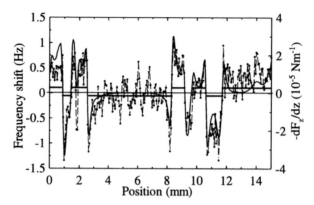

Fig. 11.9. Experimental line profile (dots) from Fig. 11.2d, and corresponding profile calculated for a point monopole tip using the 2-D approach. The data are averaged over five columns of the image, indicated by markers on Fig. 11.2d (the whole image contains 256 lines). The average contrasts measured on 20×20 pixels boxes, five for each contrast, placed in large areas of uniform contrast, are also indicated (thick broken horizontal lines; the variance on these values is 0.13 Hz). The calculation corresponds to a point monopole located at 180 nm above the sample that creates a maximum induction at sample level of 56 mT (field 45 kA m^{-1})

(1/20) than the maximum contrast at the domain walls. For $z = 200$ nm, the contrast maximum moves to the domain center. Such calculations cast some light on the puzzling observation of domain-like contrast in ultrathin films with perpendicular magnetization (Fig. 11.2). Indeed, as the films are ultrathin, the stray field is almost entirely concentrated inside the film. Outside, it should exist only above the domain walls. These exact calculations show that at large flying heights contrast can appear inside the domains if they are not too large (compare Fig. 11.8a and b). Nevertheless, if real images have to be fitted, the 2-D approach is necessary.

Figure 11.9 shows a line scan of Fig. 11.2, compared to a calculation for a point monopole tip located at 180 nm from the surface. The 2-D calculation was performed on a domain structure obtained by assigning dark and light pixels in Fig. 11.2d to up- and down-oriented domains with zero wall thickness (as first done by Hug [28]). The experimental trace is well reproduced, given the noise of the image. However, this recessed monopole position may appear too large, even for a 200 nm tip-coating thickness. Also, the average domain contrasts are still larger than those calculated (but their difference is on the order of statistical noise). This called for more precise experiments.

Figure 11.10 is a large scale image (60 μm) of a region with a square domain of 40 μm size. The domain was "written" on the saturated layer by MFM with a tip having a thick coating ($Co_{80}Cr_{20}$ alloy, 30 nm) that produces a large stray field. The sample is a cobalt ultrathin film (1.4 nm) sandwiched by platinum (see [38] for a general presentation of the sample properties). Imaging was performed with a tip covered by sputtering half the "writing" thickness, actually the same value used for imaging patterned permalloy elements (see Sect. 11.5.3). The very straight domain

11 Magnetic Force Microscopy: Images of Nanostructures and Contrast Modeling 239

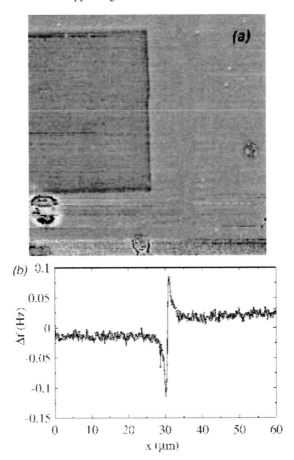

Fig. 11.10. MFM image of a 40 μm square domain in a Pt/Co(1.4 nm)/Pt ultrathin film (**a**). The domain was "written" first by imaging several times this square with a tip having a thick magnetic coverage. The image is 60 μm in size, and the tip lift height is 25 nm. A contrast profile (**b**) obtained by averaging over 150 lines (the image contains 512 lines) shows, on top of a domain wall contrast (0.2 Hz), a domain contrast tens of micrometers away from the wall (0.03 Hz)

wall (Fig. 11.10a) allows for an averaging over many line profiles. As a result, a small but clear contrast can be seen very far (tens of micrometers away) from the wall (Fig. 11.10b). If this data is compared to Eq. 11.14 at an infinite domain width, the contrast step between the two domains cannot be reproduced. This definitely calls for further investigation of MFM imaging of ultrathin films.

The formulas above are also helpful in the discussion of MFM spatial resolution. Even if the probe is as small as a point-like magnetic structure (monopole, dipole), the tip flying height profoundly affects the images. This is exemplified in Fig. 11.11, drawn for parameters appropriate to FePd ordered alloys films. These samples have

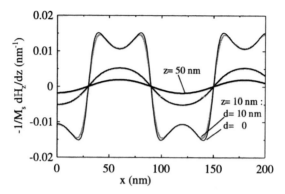

Fig. 11.11. Calculated MFM contrasts for a 1-D pattern with fine domains. Parameters are those of FePd films ($t = 30$ nm, $w = 60$ nm, $d = 10$ nm). Flying heights $z = 10, 30$, and 50 nm are compared. Only at the smallest value can one say that the structure is really resolved (the curve with zero domain wall width differs from that at $d = 10$ nm)

shown the smallest regular domains (of width 60 nm) up until now [39]. One sees that usual flying heights only give sinusoidal contrast profiles, because they are too large compared to the domain width.

11.4.3 MFM as a Charge Microscopy

It was recognized early [19] that the MFM images of soft elements with in-plane magnetization were similar to those of the magnetic charge distribution. This is vividly illustrated here by Figs. 11.3a and 11.4, where the charges considered are volume charges (11.7). For samples with perpendicular magnetization, the similarity is also apparent if one considers the surface charges (11.8) on the upper sample surface (Figs. 11.2, 11.3b). As the magnetic charges are the sources of the stray field (there is no stray field if they are zero everywhere), it is also quite natural to see the MFM images as charge pictures. This conception was formalized by A. Hubert and coworkers [40]. The interaction energy (11.1), considering its second form in which integration is performed over the sample, can be rewritten with the help of the magnetostatic potential ϕ_{tip} from which the tip field derives ($\vec{H}_{tip} = -\vec{\nabla}\phi_{tip}$) as an integral over the volume and surface charge densities

$$E_{int} = \mu_0 \iiint_{sam} \rho \phi_{tip} + \mu_0 \iint_{sam} \sigma \phi_{tip}. \tag{11.15}$$

We have thus alternative expressions of the force, in terms of magnetic charges

$$F_z = \mu_0 \iiint_{sam} \rho_{sam} H_{tip,z} + \mu_0 \iint_{sam} \sigma_{sam} H_{tip,z}$$

$$= \mu_0 \iiint_{tip} \rho_{tip} H_{sam,z} + \mu_0 \iint_{tip} \sigma_{tip} H_{sam,z}. \tag{11.16}$$

An interpretation of these expressions is that the suitable derivative of the tip potential is the response function of the charge microscope: Its magnitude determines the signal, its localization sets the resolution. Profiles of the potential and its derivatives were computed in [40] for various tip shapes: They roughly extend over the thickness of the tip coating, when the flying height is below this thickness. These profiles were obtained under a specific assumption about the tip magnetic structure, but it can be accepted that for any tip such a more or less localized potential will exist. Potential profiles were also determined experimentally from the analysis of images on point-source like magnetic objects [41].

This conception of MFM is rather efficient for a quick survey of the images. For a more thorough understanding, however, a number of difficulties arise. Consider for example the ultrathin film case discussed before (Figs. 11.2, 11.9). In the domains, one has two equal and opposite charge densities at the surfaces, spaced 1 nm apart. Thus, seen from a distance of tens to hundreds of nanometers, it is meaningless to assume that the top surface charge determines the contrast. These two opposite charge densities just cancel, and contrast should only remain in the vicinity of domain walls. Figure 11.8a was drawn for periodic stripe domains of width $w = 500$ nm. For a width 10 times larger, the contrast becomes clearly localized at the walls (Fig. 11.8b). The charge picture is here inadequate: It leads to the expectation of a domain contrast that converges to a constant non-zero value at infinite distance from domain walls, which is not found by exact calculations. And if one just reduces the film to a purely 2-D structure, all charges disappear (walls are Bloch walls, i.e., the divergence of magnetization is zero) and no force signal is expected.

A second example is afforded by the vortex structure. As schematically drawn in Fig. 11.7d, it features a perpendicularly magnetized core at the center of a circularly rotating in-plane magnetization pattern. A strictly in-plane vortex would bear no charge (it is a flux closure structure), and the center magnetization would be zero. Due to the exchange interaction, a real vortex has a finite width core with perpendicular magnetization. The surface charges of the core are conducive to a deviation of the magnetization from the circular rotation, so that both volume and surface charges occur. They are drawn in Fig. 11.12, together with the magnetization. The charge pattern is complex, as volume charges opposite to surface charges extend below the latter. The resulting contrast is not easy to figure out. The partial compensation of surface by volume charges may explain why vortex core observations by MFM are so rare. One can anticipate that this will be possible only in a certain range of thickness. In that case, as the structure is not as simple as the idealization, the charge pattern is complex and in some respect unknown. The rather wide contrasts in Figs. 11.4 and 11.5 testify to the ability of charges to spread out. Thus, for a given sample the precise charge distribution should be computed. Then, the effort is not much larger if one computes the stray field simultaneously.

A final remark should be made about the inversion problem (i.e., given an MFM image, find the magnetic structure). It has been proved that the information supplied by the MFM is not sufficient for this purpose [43]. A simpler inverse problem, namely, the recovery of the magnetic charges distribution, was also proved to be hopeless [44]. In fact, surface and volume charges play equivalent roles. In the case of films, MFM

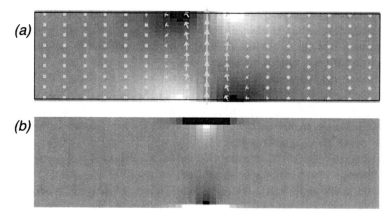

Fig. 11.12. Numerically computed 3-D structure of a vortex in a 200 nm diameter, 50-nm-thick permalloy disk, with a mesh size of $4 \times 4 \times 2$ nm. The structures are displayed in a vertical cross section that contains the disk axis. The magnetization pattern (**a**) was coded in gray scale according to the radial component, whose presence means that the divergence-free situation is not realized. The magnetic charges (sum of volume and surface contributions for the surface cells) are drawn in (**b**) with a gray scale representation. Volume charges are seen to be opposite to the nearest surface charges linked to the core magnetization of the vortex. Computations were performed with the OOMMF software [42]

images can always be related to equivalent surface charges (effective charges [44]). Thus, an external source of knowledge is always necessary in order to interpret MFM images.

11.5 MFM Contrast in the Presence of Perturbations

It was recognized early that, for soft samples, the MFM image of the magnetic structure may appear strongly distorted [45]. The replacement of etched wires by thin-film coated tips led to a great reduction of tip stray fields and allowed for much more faithful images. Nevertheless, complete avoidance of tip stray field is not the solution, as from Eq. 11.1 no contrast would then exist. Therefore, one has to accept the presence of this field. Ideally, it would be tailored so as to give, for the sample considered, the lowest detectable contrast. More sensitive microscopes are thus still needed.

11.5.1 Tip Stray Field Values

Several experiments have been devised to measure the tip stray field magnitude and spatial distribution. The most direct determination was obtained by measuring the deflection of high energy electrons passing close to the tip in a STEM [46], see Chap. 4. A typical value for Digital Instruments standard "MESP" probes is a maximum axial induction of 40 mT (field of 32 kA m^{-1}) at about 50 nm from the tip apex, the lateral FWHM being 0.25 μm. A more indirect technique involved deconvoluting maps of Hall voltage versus tip position, using micron-size Hall sensors [47]. Maximum values were found to be 10–20 mT (8–16 kA m^{-1}) at 100 nm from the tip apex.

Additionally, the tip stray field can be fitted to experimental data, provided the sample structure is well known. Such is the case of ultrathin Ni layers (10 nm) on Cu, with perpendicular magnetization, narrow Bloch walls and large domains [48]. Fields determined for tips covered with about 10 nm of iron by evaporation had maximal values at apex of 10 mT (8 kA m^{-1}) and a FWHM less than 0.1 μm.

Whatever the exact result of these measurements, the values found are not small for soft samples when compared to coercive fields on the order of 100 A m^{-1} (in-plane) and anisotropy fields on the order of 1 kA m^{-1} (in-plane). However, tip fields are sharply localized, and their effect cannot be guessed from that of an equivalent uniform field. In recent years, micromagnetic calculations have been applied to this problem. In some cases, results could be compared in detail to experiments, as shown below.

11.5.2 Forces in the Case of Perturbation

This discussion is restricted first to the case where the tip magnetic structure remains unaffected. This approximation is reasonable for the experimental configuration of a magnetically hard tip and a soft sample. The total magnetic energy, dropping the constant tip internal energy, is now

$$E_{mag} = E_{int} + E_{sam}, \tag{11.17}$$

where the interaction term is Eq. 11.1. In the frame of micromagnetics [13], the local sample magnetization has a constant magnitude M_s and a direction vector (unit vector). The two contributions, therefore, read

$$E_{int} = -\mu_0 M_s \iiint_{sam} \vec{m} \cdot \vec{H}_{tip}, \tag{11.18}$$

$$E_{sam} = \iiint_{sam} A(\nabla \vec{m})^2 + KG(\vec{m}) - (\mu_0 M_s/2)\vec{m} \cdot \vec{H}_{sam}. \tag{11.19}$$

The energy contributions to the sample energy are called exchange, anisotropy, and magnetostatic energies, respectively.

Here arises an essential question: Which energy should one consider in order to evaluate the magnetic force on the tip? It has sometimes been assumed ([49], for example) that the sole interaction energy (Eq. 11.1) is appropriate, and, in fact, a direct extension of the no perturbation case. Let us consider the question at a basic level. As there are no forces other than magnetic and elastic, the "sample + tip-cantilever" system is isolated and its energy is constant. Therefore, the tip moves under the gradient of the total magnetic energy. In the case of no perturbation (no change of the sample and tip magnetization configurations), the sample and tip energies do not change, so the force results from the sole interaction term. This is, however, no longer true if the sample magnetization is allowed to change.

Typical MFM tips, because of their mechanical softness, vibrate at frequencies below 100 kHz. These are low compared to natural frequencies of magnetic structures, which are on the order of GHz [50]. Therefore, a first approximation is to consider the quasi-static limit of the problem: At each tip position (in x, y, and z) the sample magnetization is allowed to reach equilibrium by minimizing E_{mag} with respect to $\vec{m}$. This calculation is just what micromagnetic codes can perform.

The force on the tip is the sum of an external contribution (the tip field at sample level varies with tip height) and of an induced contribution, as the sample magnetization reacts to tip field. Therefore, we write the total derivative (symbol d) with respect to the tip height as the sum of the partial derivative (derivative over the terms that explicitly contain z_{tip}, symbol ∂) and the "motion" term (through the variation of the equilibrium magnetization with tip height)

$$F_z = -\frac{dE_{mag}}{dz_{tip}} = -\frac{\partial E_{mag}}{\partial z_{tip}} - \frac{\delta E_{mag}}{\delta \vec{m}} \cdot \frac{D\vec{m}}{Dz_{tip}}. \qquad (11.20)$$

For the variation of the energy with the magnetization, we use the functional derivative symbol (δ) customary to functionals that involve the gradient of the variable [13]. The evolution of the sample magnetization with tip height is denoted by yet another symbol, D. As we assumed that the sample magnetization is in equilibrium at every position of the tip, the energy is stationary with respect to variations of $\vec{m}$ (i.e., $\iiint (\delta E_{mag}/\delta \vec{m}) \cdot \Delta \vec{m} = 0$ for any variation $\Delta \vec{m}$ of the magnetization), so that one is left with the explicit term, which reads

$$F_z = \mu_0 M_s \iiint_{sam} \vec{m} \cdot \frac{\partial \vec{H}_{tip}}{\partial z_{tip}}. \qquad (11.21)$$

In this expression, the magnetization distribution $\vec{m}$ is that which is at equilibrium under the tip field associated with the tip position considered; it is generally different from the magnetization in the absence of the tip field, denoted by $\vec{m}$. Note that if we had evaluated the gradient of the interaction energy (Eq. 11.1), since $\iiint (\delta E_{int}/\delta \vec{m}) \cdot \Delta \vec{m} \neq 0$, an additional term would be present in the integral (namely, $\vec{H}_{tip} \cdot D\vec{m}/Dz_{tip}$). It is now important to understand that Eq. 11.21 remains the same when the tip magnetization distribution is also changing due to perturbation, again because of stationary conditions. One *only* has to use the modified tip field, not that of the tip without the sample.

Turning now to the dynamic mode, we calculate the force gradient from the force (11.21). Here, nothing cancels out, and one obtains

$$\frac{dF_z}{dz_{tip}} = \mu_0 M_s \iiint_{sam} \frac{D\vec{m}}{Dz_{tip}} \cdot \frac{\partial \vec{H}_{tip}}{\partial z_{tip}} + \vec{m} \cdot \frac{\partial^2 \vec{H}_{tip}}{\partial z_{tip}^2}. \qquad (11.22)$$

A third term should be added in the general case to express the modification of the tip field due to tip position-dependent perturbation.

11 Magnetic Force Microscopy: Images of Nanostructures and Contrast Modeling   245

The literature contains many experimental demonstrations of the existence of perturbations, even for the standard tips used nowadays. Trivial cases are those where the structure changes during scanning (see, e.g., Fig. 11.3a bottom right corner dot), or grossly differs from what is known, either by another technique or by our knowledge of magnetic structures. These cases can sometimes be avoided by using low moment tips and large flying heights. More subtle perturbations, which only change the details of the image, also exist. A check consists of reversing tip magnetization (see Figs. 11.4 and 11.13). From Eq. 11.21, one can express the sum of two images with opposite tip magnetizations as

$$F_z(+) + F_z(-) = \mu_0 M_s \iiint_{sam} \frac{\partial \vec{H}_{tip}(+)}{\partial z_{tip}} \cdot \left[\vec{m}(+) - \vec{m}(-)\right]. \quad (11.23)$$

If the magnetization is unperturbed, $\vec{m}(+) = \vec{m}(-) = \vec{m}_0$ and the result is zero. Otherwise, there is some perturbation. Obviously, the tip has to be reversed many times so as to check for reproducibility of its structure. This approach was first used in [51, 52].

11.5.3 Perturbations in Patterned Permalloy Elements

Figure 11.13 shows the structure of a 2 μm permalloy square exhibiting the flux-closure structure with four domains. The MFM images correspond well to this schematic pattern, however, with a small deformation. Comparing the 2 images taken with opposite tip magnetizations, a curvature of the domain walls is apparent.

Another feature of the images is the global attractive force (negative frequency shift) when the tip is above the element, with respect to the tip above the nonmagnetic substrate. This attraction is present irrespective of tip magnetization. A mechanism for tip attraction was put forward by J.J. Saenz [53] in the case of perpendicularly magnetized samples. It may, however, be generalized. A magnetic material (except when saturated in the direction of the field) has a non-zero susceptibility χ. Idealizing

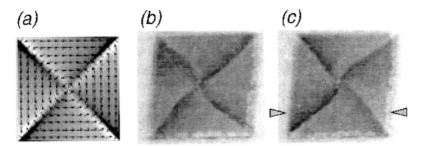

Fig. 11.13. Images of a 16-nm-thick permalloy square with 2 μm edge length: (**a**) calculated structure in zero field, with gray scale shading according to the magnetic charge (mesh 5 × 5 × 16 nm); (**b,c**) MFM images with opposite tip polarities (tip coating 20 nm CoCr, lift height 20 nm)

it by a paramagnet, one gets $E_{\text{mag}} = -\mu_0 \chi (\vec{H}_{\text{tip}})^2 / 2$, which gives rise to tip attraction. How this mechanism applies locally for a nonuniform magnetization distribution and in an inhomogeneous tip field has to be studied by micromagnetic calculations.

The occurrence of apparent wall curving is also best studied by such calculations. A 2-D micromagnetic program developed earlier (see [54,55] and references therein) was modified to include a tip field in wich the tip can be located at any position [56]. As a first approximation, the field of an effective monopole was used. In order to find the parameters (pole strength and vertical position) of the effective monopole, a finite elements micromagnetic calculation of a 3-D typical tip was performed [56, 57]. For each distance of a sample plane to the tip apex, the computed field profile was found to be close to that of a monopole. The monopole is only an approximation, however, as the fitted pole strength and position were found to vary systematically with distance to the tip apex [57], such that the effective monopole has a higher strength and is located farther inside the tip as the sample plane recesses from the tip apex. The calculated effect of the tip field on the sample micromagnetic structure is shown in Fig. 11.14 for several positions of the tip and both tip polarities. When the tip is above one domain (Fig. 11.14a,d), one sees clearly the appearance of an induced charge in the sample. This charge is opposite to that of the tip monopole

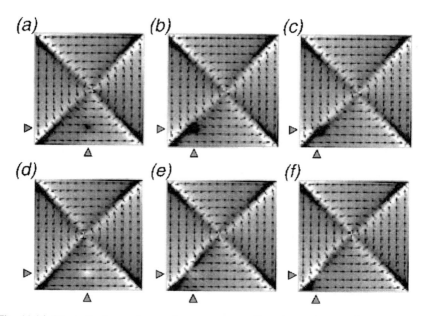

Fig. 11.14. Magnetization patterns of a square permalloy element exposed to the field of a negative (**a–c**) and positive (**d–f**) monopole tip. Gray-scale coding relates to the volume charge (this is a 2-D calculation). The monopole position is indicated by arrows. The sample is 1000 nm wide, 16 nm thick, and the mesh size is $4 \times 4 \times 16$ nm. The effective monopole of the tip is located at 60 nm from the top surface and creates a maximum axial field of 32 kA m^{-1} (induction 40 mT) in the film center. The induced charge is clearly seen, as well as its interaction with the wall charges

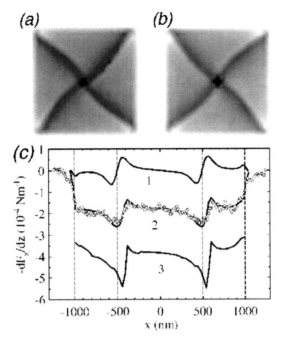

Fig. 11.15. Contrast calculations corresponding to the experimental results of Fig. 11.13. Calculated images **(a,b)** for a 1000-nm-wide, 10-nm-thick sample (home-developed 2-D program, mesh size 32 × 32 × 10 nm), monopole tip at 60 nm from sample mid-plane, creating a maximum field of 32 kA m^{-1} (induction 40 mT). The gray scale encoding is such that black corresponds to a positive force gradient, as in the experiments. Note that the mesh size does not allow one to compute the vortex core, and that it is a 2-D calculation, so a contrast due to vortex core cannot appear. For the true sample size, **(c)** shows a line profile (dots) from Fig. 11.13c (markers) superposed with calculation results for this line. The calculation uses a sample 2000 nm wide, 16 nm thick (mesh size 16 × 16 × 16 nm) and a monopole tip at 53 nm from sample mid-plane, creating at sample level a maximum field of 24 kA m^{-1}. The three lines are (1) no perturbation model; (2) calculations according to Eq. 11.22; (3) calculation of the force as the gradient of the interaction energy. The outer dashed lines indicate the sample edges, the inner dotted lines point to the wall positions in zero field. Frequency shifts were converted into force gradients using a cantilever stiffness $k = 5.2$ Nm^{-1}

and can be viewed as the image charge of the tip monopole in an imperfect shield (the permalloy sample). When the tip moves toward the walls, the induced charge follows and interacts with the wall structure. As the walls are of Néel type, they bear two opposite charge sheets on their sides. Thus, the tip in the top figures always pushes down the left wall to the left, while the opposite tip (bottom figures) pushes it to the right. This explains the characteristic wall deformation in the MFM images. Figure 11.15a,b display computed images (force gradient) for a 1 μm square; a very good qualitative agreement is found with the experiment. Figure 11.15c provides a quantitative comparison for the 2 μm square, for one scan line. The monopole

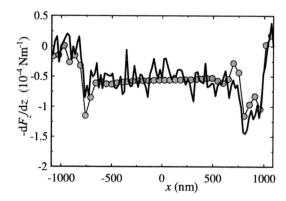

Fig. 11.16. Profile along the center line of Fig. 11.4 superposed to numerical simulation results. Calculation parameters are: mesh size $28 \times 28 \times 16$ nm, monopole tip at 53 nm from sample mid-plane, creating a maximum field of 11.1 kA m^{-1}, cantilever stiffness $k = 6.5$ Nm^{-1}

strength was slightly adjusted for a best fit. The global attractive force is clearly seen both in experiments and calculations. Wall position in the images is also correctly reproduced, as well as wall contrast. Note that the calculation, taking into account only the interaction energy, would predict twice the overall contrast and sharper walls (curve labeled 3).

Another calculation was performed for the rectangular element shown in Fig. 11.4. The phenomena are similar, and perturbation is visible as unequal sizes of the two quasi-closure domains at the ends. Figure 11.16 shows contrast profiles (experimental and calculated) along the central scan line, again with a nice agreement.

One should not deduce from the previous examples that all perturbation processes are best discussed in terms of induced charges. In other cases, thinking simply of the effect of the tip field on the sample magnetization is more direct. This is adequate, for example, in a rectangle with four domains and one vortex on the central wall when focusing on vortex motion [58].

Summarizing, it appears that micromagnetic calculations can, in some cases, reproduce MFM images in the presence of perturbations with a high accuracy. They also give insight into what happens in the sample during observation. For soft samples, the calculations show that the charge density in the sample can rearrange under tip fields, which calls for caution when considering the MFM as a microscope that "just maps" the sample charges.

11.6 Conclusion and Perspectives

Magnetic force microscopy has now become a standard magnetic microscope, operating in conditions similar to optical microscopes, but with a more than ten times better resolution. However, the signal is not related in a simple way to the magnetization direction in the sample just below the tip apex. Because magnetic forces are long range, the signal contains a contribution from regions far from the tip. Moreover, such a signal is related to magnetic charges, rather than to the magnetization itself.

We have seen in this chapter that, in most cases, a qualitative image interpretation is readily obtained. But quantitative values, fine details, or even magnetization orientation in complex cases can remain obscure. Micromagnetic calculations of the sample have been shown to help in this respect, although one is still far from a complete modeling of the microscope.

Advanced magnetic force microscopes are still possible. Tips should be improved so that their magnetic structure is better controlled and their stray field strength tailored to each sample. More sensitive force detection schemes will allow for lower tip fields, hence lower sample perturbation and more faithful images. Finally, new contrast modes are continuously appearing, enriching the capabilities of this versatile microscope (for example see [59, 60]).

Acknowledgement. This chapter could not have been written without the work of Laurent Belliard [26] and Dalibor Tomáš [54] for their Ph.D., and of Denis Bourgeois for his masters theses. The work of JMG at Orsay was supported by an individual Marie Curie fellowship from the E.C.

References

1. Y. Martin and H.K. Wickramasinghe, Appl. Phys. Lett. **50**, 1455 (1987).
2. J.J. Saenz, N. Garcia, P. Gruetter, E. Meyer, H. Heinzelmann, R. Wisendanger, L. Rosenthaler, H.R. Hidber, and H.J. Guentherodt, J. Appl. Phys. **62**, 4293 (1987).
3. A.M. Chang, H.D. Hallen, L. Harriott, H.F. Hess, H.L. Kao, J. Kwo, R.E. Miller, R. Wolfe, J. van der Ziel, and T.Y. Chang, Appl. Phys. Lett. **61**, 1974 (1992).
4. A. Oral, S.J. Bending, and M. Henini, J. Vac. Sci. Technol. **B14**, 1202 (1996).
5. J.R. Kirtley, M.B. Ketchen, K.G. Stawiasz, J.Z. Sun, W.J. Gallagher, S.H. Blanton, and S.J. Wind, Appl. Phys. Lett. **66**, 1138 (1995).
6. J.R. Kirtley, M.B. Ketchen, C.C. Tsuei, J.Z. Sun, W.J. Gallagher, L.S. Yu-Jahnes, A. Gupta, K.G. Stawiasz, and S.J. Wind, IBM J. Res. Develop. **39**, 655 (1995).
7. P. Gruetter, H.J. Mamin, and D. Rugar, in *Scanning Tunneling Microscopy II*, edited by R. Wiesendanger and H.J. Guentherodt (Springer Verlag, Berlin, 1992).
8. R.D. Gomez, in *Magnetic Imaging and Its Applications to Materials*, edited by M. de Graef and Y. Zhu (Academic Press, San Diego, 2001).
9. J. Miltat, and A. Thiaville, in *Encyclopedia of Materials: Science and Technology*, edited by K.H.J. Buschow, R.W. Cahn, M.C. Flemings, B. Ilschner, E.J. Kramer, and S. Mahajan (Elsevier Science, Amsterdam, 2001).
10. D. Rugar, H.J. Mamin, P. Guethner, S.E. Lambert, J.E. Stern, I. McFadyen, and T. Yogi, J. Appl. Phys. **68**, 1169 (1990).
11. U. Hartmann, T. Goeddenhenrich, and C. J. Magn. Magn. Mater. **101**, 263 (1991).
12. S. Porthun, L. Abelmann, and J.C. Lodder, J. Magn. Magn. Mater. **182**, 238 (1998).
13. A. Hubert and R. Schaefer, *Magnetic Domains* (Springer, Berlin, 1998).
14. P. Gruetter, D. Rugar, H.J. Mamin, G. Castillo, S.E. Lambert, C.J. Lin, and R.M. Valletta, Appl. Phys. Lett. **57**, 1820 (1990).
15. P.F. Hopkins, J. Moreland, S.S. Malhotra, and S.H. Liou, J. Appl. Phys. **79**, 6448 (1996).

16. G. Meyer and N.M. Amer, Appl. Phys. Lett. **53**, 1045 (1988).
17. D. Rugar, H.J. Mamin, and P. Guethner, Appl. Phys. Lett. **55**, 2588 (1989).
18. K. Babcock, M. Dugas, S. Manalis, and V. Elings, Mat. Res. Soc. Symp. **355**, 311 (1995).
19. W. Rave, L. Belliard, M. Labrune, A. Thiaville, and J. Miltat, IEEE Trans. Magn. **30**, 4473 (1994).
20. T.R. Albrecht, P. Gruetter, D. Horne, and D. Rugar, J. Appl. Phys. **69**, 668 (1991).
21. M. Dreyer, R.D. Gomez, and I.D. Mayergoyz, IEEE Trans. Magn. **36**, 2975 (2000).
22. G. Lévêque, P. Girard, S. Belaidi, and G. Cohen Solal, Rev. Sci. Instrum **68**, 4137 (1997).
23. P. Gruetter, Y. Liu, and P. LeBlanc, Appl. Phys. Lett. **71**, 279 (1997).
24. R. Proksch, K. Babcock, and J. Cleveland, Appl. Phys. Lett. **74**, 419 (1999).
25. Y. Liu and P. Gruetter, J. Appl. Phys. **83**, 7333 (1998).
26. L. Belliard, Ph.D. thesis, Orsay University, 1997.
27. L. Belliard, A. Thiaville, S. Lemerle, A. Lagrange, J. Ferré, and J. Miltat, J. Appl. Phys. **81**, 3849 (1997).
28. H.J. Hug, B. Stiefel, A. Moser, I. Parashikov, A. Kliczmik, D. Lipp, H.J. Guentherodt, G. Bochi, D.I. Paul, and R. C. O'Handley, J. Appl. Phys. **79**, 5609 (1996).
29. A. Fernandez, P.J. Bedrossian, S.L. Baker, S.P. Vernon, and D.R. Kania, IEEE Trans. Magn. **32**, 4472 (1996).
30. S. Gider, J. Shi, D.D. Awschalom, P.F. Hopkins, K.L. Campman, A.C. Gossard, A.D. Kent, and S. von Molnar, Appl. Phys. Lett. **69**, 3269 (1996).
31. M. Lederman, S. Schultz, and M. Ozaki, Phys. Rev. Lett. **73**, 1986 (1994).
32. L. Belliard, J. Miltat, A. Thiaville, S. Dubois, J.L. Duvail, and L. Piraux, J. Magn. Magn. Mater. **190**, 1 (1998).
33. Y. Henry, K. Ounadjela, L. Piraux, S. Dubois, J.M. George, and J.L. Duvail, Eur. Phys. J. **B20**, 35 (2001).
34. R.D. Gomez, T.V. Luu, A.O. Pak, K.J. Kirk, and J.N. Chapman, J. Appl. Phys. **85**, 6163 (1999).
35. J.M. Garcia, A. Thiaville, and J. Miltat, J. Magn. Magn. Mater. **249**, 163 (2002).
36. H. Joisten, S. Lagnier, M.H. Vaudaine, L. Vieux-Rochaz, and J.L. Porteseil, J. Magn. Magn. Mater. **233**, 230 (2001).
37. T. Shinjo, T. Okuno, R. Hassdorf, K. Shigeto, and T. Ono, Science **289**, 930 (2000).
38. J. Ferré, in *Spin Dynamics in Confined Magnetic Structures I*, edited by B. Hillebrands and K. Ounadjela (Springer, Berlin, 2002).
39. V. Gehanno, Y. Samson, A. Marty, B. Gilles, and A. Chamberod, J. Magn. Magn. Mater. **172**, 26 (1997).
40. A. Hubert, W. Rave, and S.L. Tomlinson, phys. stat. sol. (b) **204**, 817 (1997).
41. T. Chang, M. Lagerquist, J.G. Zhu, J.H. Judy, P.B. Fischer, S.Y. Chou, IEEE Trans. Magn. **28**, 3138 (1992).
42. OOMMF is a public micromagnetic program developed at the NIST, USA, by M.J. Donahue and coworkers. It is available at http://www.math.nist.gov. We used version 1.2α2.
43. I.A. Beardsley, IEEE Trans. Magn. **25**, 671 (1989).
44. B. Vellekoop, L. Abelmann, S. Porthun, and C. Lodder, J. Magn. Magn. Mater. **190**, 148 (1998).
45. H.J. Mamin, D. Rugar, J.E. Stern, R.E. Fontana, and P. Kasiraj, Appl. Phys. Lett. **55**, 318 (1989).
46. S. McVitie, R.P. Ferrier, J. Scott, G.S. White, and A. Gallagher, J. Appl. Phys. **89**, 3656 (2001).
47. A. Thiaville, L. Belliard, D. Majer, E. Zeldov, and J. Miltat, J. Appl. Phys. **82**, 3182 (1997).

48. P.J.A. van Schendel, H.J. Hug, B. Stiefel, S. Martin, and H.J. Guentherodt, J. Appl. Phys. **88**, 435 (2000).
49. S.L. Tomlinson and E.W. Hill, J. Magn. Magn. Mater. **161**, 385 (1996).
50. J. Miltat, G. Albuquerque, and A. Thiaville, in *Magnetism and Synchrotron Radiation*, edited by E. Beaurepaire, F. Scheurer, G. Krill, and J.P. Kappler (Springer, Berlin, 2001).
51. S. Foss, R. Proksch, E.D. Dahlberg, B. Moskowitz, and B. Walsh, Appl. Phys. Lett. **69**, 3426 (1996).
52. W. Rave, E. Zueco, R. Schaefer, and A. Hubert, J. Magn. Magn. Mater. **177–181**, 1474 (1998).
53. J.J. Saenz, N. Garcia, and J.C. Slonczewski, Appl. Phys. Lett. **53**, 1449 (1988).
54. D. Tomas, Ph.D. thesis, Charles University, 1999.
55. G. Albuquerque, J. Miltat, and A. Thiaville, J. Appl. Phys. **89**, 6719 (2001).
56. J. Miltat, L. Belliard, A. Thiaville, D. Tomas, and F. Alouges, contribution HA01 at the 7th Joint MMM-Intermag Conference, 1998 (unpublished).
57. J.M. Garcia, A. Thiaville, J. Miltat, K.J. Kirk, J.N. Chapman, and F. Alouges, Appl. Phys. Lett. **79**, 656 (2001).
58. J.M. Garcia, A. Thiaville, J. Miltat, K.J. Kirk, and J.N. Chapman, J. Magn. Magn. Mater. **242–245**, 1267 (2002).
59. R. Proksch, P. Neilson, S. Austvold, and J.J. Schmidt, Appl. Phys. Lett. **74**, 1308 (1999).
60. M.M. Midzor, P.E. Wigen, D. Pelekhov, W. Chen, P.C. Hammel, and M.L. Roukes, J. Appl. Phys. **87**, 6493 (2000).

12

Magnetic Force Microscopy – Towards Higher Resolution

L. Abelmann, A. van den Bos, C. Lodder

In this chapter, magnetic force microscopy is treated in detail with an emphasis on high resolution and hard magnetic materials such as recording media. The chapter starts with basic MFM operation, instrumentation, and a frequency domain theory of image formation using transfer functions. Subsequently, the limits of resolution in MFM are discussed, and the concept of critical wavelength as a measure for resolution is introduced. To achieve high resolution, the tip-sample distance has to be small, and methods of tip-sample distance control are discussed. Finally, generations of MFM tips are treated, including a full silicon micromachined design, which eventually might take the resolution of MFM below 10 nm.[1]

12.1 Principle of MFM

The technique of magnetic force microscopy has been discussed extensively in literature [8, 12, 18, 20], so we will restrict ourselves to a short description. The principle of magnetic force microscopy is very much like that of atomic force microscopy – some even dare to assert that *MFM is just an AFM with a magnetic tip*, much to the dislike of MFM developers, because in an MFM much smaller forces are measured. In essence, it is true, however, and every MFM is capable of AFM as well (the other way around is not true in general).

In an MFM, the magnetic stray field above a very flat specimen or sample is detected by mounting a small magnetic element, the tip, on a cantilever spring very close to the surface of the sample (Fig. 12.1). Typical dimensions are a cantilever length of 200 μm, a tip length of 4 μm, a diameter of 50 nm, and a distance from the surface of 30 nm. The force on the magnetic tip is detected by measuring the displacement of the end of the cantilever, usually by optical means. The forces measured in typical MFM applications are on the order of 30 pN, with typical cantilever deflections on the order of nanometers.

---

[1] This chapter is accompanied by a Web page http://www.el.utwente.nl/smi/mfmchapter/, where you can find errata and links for further reading.

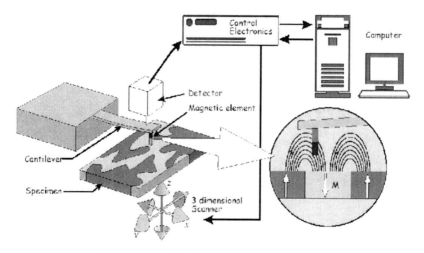

Fig. 12.1. Principle of magnetic force microscopy

An image of the magnetic stray field is obtained by slowly scanning the cantilever over the sample surface, in a raster-like fashion. Typical scan areas are from 1 up to 200 μm, with imaging times on the order of 5–30 minutes.

12.1.1 Mode of Operation

The force F exerted on the tip by the stray field of the sample has two effects on the cantilever deflection. In the first place, the cantilever end is deflected toward or away from the sample surface by a distance Δz:

$$\Delta z = F_z/c \quad [\text{m}], \tag{12.1}$$

with c the cantilever spring constant in z-direction [N/m]. This deflection can be measured using soft, usually SiN, cantilevers with spring constants on the order of 0.01–0.1 N/m. When measuring the deflection, we speak about *static mode* MFM.

For small deflections, the cantilever can be considered as a damped harmonic oscillator, which can be modeled by an ideal spring c [N/m], mass m [kg], and damper D [Ns/m] [3]. When we apply an external oscillating force $F_z = F_0 \cos(\omega t)$ to the cantilever, the resulting displacement is harmonic as well, but has a phase shift for $\omega > 0$, $z = z_0 \cos(\omega t + \theta(\omega))$. This force can be applied directly to the end of the cantilever, for instance, by electrostatic means. The most commonly used method is, however, to apply a force to the cantilever holder by means of a piezo mounted underneath the holder.

It is convenient to describe the relation between force and displacement in the Laplace domain[2]

---

[2] We use the definition commonly used in mechanical engineering textbooks $\hat{F}(s) = \int_{-\infty}^{+\infty} f(t) e^{-st} dt$

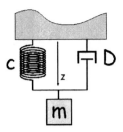

Fig. 12.2. The cantilever can be modeled as a damped harmonic oscillator

$$\frac{\hat{Z}}{\hat{F}} = \frac{1}{c + sD + ms^2} \quad [m/N] . \tag{12.2}$$

Using the natural resonance frequency ω_n and damping factor δ

$$\omega_n = \sqrt{c/m} \quad [2\pi/s] \tag{12.3}$$

$$\delta = \frac{D}{2\sqrt{mc}} = \frac{D\omega_n}{2c} \tag{12.4}$$

we can rewrite Eq. 12.2 as

$$\frac{\hat{Z}}{\hat{F}} = \frac{1}{m\omega_n^2 + 2\delta\omega_n ms + ms^2} \quad [m/N] . \tag{12.5}$$

For underdamped systems ($\delta < 1$), this system has poles at

$$s_{1,2} = -\delta\omega_n \pm i\omega_n\sqrt{1-\delta^2} = -\delta\omega_n \pm i\omega_d . \tag{12.6}$$

The term $\omega_d = \omega_n\sqrt{1-\delta^2}$ is referred to as the *damped natural frequency*. For MFM in dynamic mode, cantilevers with very small damping ($\delta \ll 0.01$) are used and $\omega_d \approx \omega_n$. In MFM it is customary to talk about the *quality factor* of resonance Q, instead of the damping factor. Q is proportional to the ratio between the energy stored in the cantilever and the energy lost per cycle:

$$Q = 2\pi \frac{\text{Energy stored in cantilever}}{\text{Energy lost per cycle}} = 2\pi \frac{\frac{1}{2}cz_0^2}{\pi D z_0^2 \omega_n} = \frac{c}{D\omega_n} = \frac{1}{2\delta} . \tag{12.7}$$

Using Q, Eq. 12.5 becomes:

$$\frac{\hat{Z}}{\hat{F}} = \frac{1}{m\omega_n^2 + \frac{\omega_n m}{Q}s + ms^2} \quad [m/N] . \tag{12.8}$$

From Eq. 12.8 we can calculate the amplitude z_0 of the cantilever vibration when driven at a frequency ω [2]

$$z_0 = \frac{F_0/m}{\sqrt{(\omega_n^2 - \omega^2) + (\omega\omega_n/Q)^2}} \tag{12.9}$$

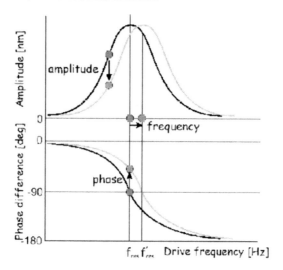

Fig. 12.3. A change in the magnetic force on the tip results in a change in resonance frequency of the cantilever, which can be detected in different ways

and the phase shift θ between the force and the deflection (see also Fig. 12.3)

$$\theta = \tan^{-1}\left(\frac{\omega \omega_n}{Q\left(\omega_n^2 - \omega^2\right)}\right). \tag{12.10}$$

In MFM, the force on the magnetic tip increases when it approaches the sample, so it is as if there is a second spring with a spring constant of $\partial F/\partial z$ attached to the cantilever. In the case that the cantilever deflection is so small that $\partial F/\partial z$ can be considered a constant, this results in a change in natural resonance frequency $f_n = \omega_n/2\pi$ of the cantilever:

$$f'_n = f_n\sqrt{1 - \frac{\partial F_z/\partial z}{c}} \quad [\text{Hz}] \tag{12.11}$$

$$\Delta f = f'_n - f_n \approx -\frac{f_n}{2c}\frac{\partial F_z}{\partial z} \quad [\text{Hz}]. \tag{12.12}$$

The approximation is accurate for $\Delta f \ll f_n$, which is always the case in MFM. Note the sign of Δf. In the above equations it is assumed that the positive z-direction is pointing away from the surface. When the tip is attracted toward the sample, the force, therefore, is *negative* and the force derivative is positive. So for *attracting* forces, the resonance frequency of the cantilever *decreases*. Please note that Eq. 12.11 is only valid for small vibration amplitudes. When $\partial F/\partial z$ cannot be considered constant, the vibration contains higher harmonics and more elaborate, and even numerical methods, are needed to calculate the resonance frequency shifts.

When we measure the resonance frequency of the cantilever, we speak about *dynamic* mode MFM (indicated by *frequency* in Fig. 12.3). In this mode, the cantilever is usually forced to resonate at an amplitude of 10–30 nm, so that an accurate detection of the very small frequency shifts is possible (typically 3 Hz on 80 kHz). In this mode, a control circuit is needed that matches the beat frequency of the actuator that drives the cantilever (e.g., a piezo), with the actual resonance frequency. Very often a phase locked loop (PLL) circuit is used, which keeps the phase difference between the driving signal and the measured deflection of the cantilever at approximately 90°. This control circuit adds additional noise to the measurement signal. For small signals it is, therefore, sometimes preferable to fix the frequency of the driving signal to f_n and measure the phase difference between the driving signal and the measured cantilever deflection (indicated by *phase* in Fig. 12.3). The phase shift, which is on the order of a few degrees, strongly depends on the damping of the cantilever, which on its turn is a function of many parameters. The phase signal is, therefore, not really suitable for quantitative analysis.

An even simpler dynamic detection mode is to drive the cantilever off-resonance. A change in resonance frequency will result in change of the vibration amplitude (*amplitude* in Fig. 12.3). Even though this amplitude mode works fine for AFM, it gives very poor results for MFM because the amplitude variations are small compared to the noise. Moreover, the response of the cantilever to a change in force is slow when the quality factor of resonance is high, which is, for instance, the case in vacuum [2]. Therefore, this mode is not used very often.

Fundamentally, there is no difference in sensitivity between the static mode and dynamic mode, because both modes use the same measurement geometry. For a number of reasons, to which we will come back later, the dynamic mode often gives better results however.

12.1.1.1 Image Formation

To calculate the force on the magnetic tip, we have to start with the calculation of the energy U of the tip-sample system. The gradient of this energy then gives us the force vector. For MFM, we are particularly interested in $\partial U/\partial z$.

We have two ways to calculate U. One can either calculate the energy of the magnetic tip in the presence of the sample stray field or the energy of the magnetic sample in the presence of the tip stray field [12]. In both cases, we have to integrate the inner product of magnetic field and magnetization over the area where the magnetization is not zero:

$$U = -\mu_0 \int_{\text{tip}} \vec{M}_{\text{tip}} \vec{H}_{\text{sample}} \, dV = -\mu_0 \int_{\text{sample}} \vec{M}_{\text{sample}} \vec{H}_{\text{tip}} \, dV \ . \tag{12.13}$$

Which method is more convenient depends on the problem that is to be analyzed. One usually takes the form where the stray field calculation is easier to perform.

When discussing image formation and resolution, it is convenient to do so in the spatial frequency domain – a method commonly used in magnetic recording theory.

We, therefore, decompose the sample magnetization $\vec{M}$ in the sample plane (x, y) into Fourier components, leaving the z component untransformed:

$$\hat{M}(k_x, k_y, z) = \int_{-\infty}^{\infty} \int_{-\infty}^{\infty} \vec{M}(x, y, z) e^{-i(xk_x + yk_y)} \, dx \, dy \, . \qquad (12.14)$$

The relation between the wavelength of a certain component and the Fourier components is

$$\vec{k} = (k_x, k_y) \, , \qquad (12.15)$$

$$k_{x(y)} = \frac{2\pi}{\lambda_{x(y)}} \, . \qquad (12.16)$$

The stray field of the sample generated by this magnetization distribution can be calculated by means of a Laplace transform. For a thin film with thickness t, one can obtain with some patience [33]

$$\begin{pmatrix} H_x(k_x, k_y, z) \\ H_y(k_x, k_y, z) \\ H_z(k_x, k_y, z) \end{pmatrix} = \begin{pmatrix} -ikx/|\vec{k}| \\ -iky/|\vec{k}| \\ 1 \end{pmatrix} \frac{1}{2} \left(1 - e^{-|\vec{k}|t} \right) e^{-|\vec{k}|z} - \sigma_{\text{eff}}(\vec{k}) \, , \qquad (12.17)$$

where $\sigma_{\text{eff}}(\vec{k})$ is an *effective* surface charge distribution. It expresses the property of the Laplace transformation that the stray field at height z above the sample surface is fully determined by the stray field at height $z = 0$. The effective surface charge distribution can be seen as a sheet of charges at the sample surfaces, which causes the same stray field as the more complex charge distribution within the sample itself.[3]

For a sample with perpendicular magnetization ($M_x = 0$, $M_y = 0$), $\sigma_{\text{eff}}(x, y)$ simply equals the surface charge density σ.

$$\sigma_{\text{eff}}(\vec{k}) = \hat{M}_z(\vec{k}) = \hat{\sigma}(\vec{k}) \qquad (12.18)$$

For a sample with an in-plane magnetization (so $M_z = 0$), we only have volume charges $\varrho(x, y)$. If the magnetization is constant over the film thickness ($\partial M_x/\partial z = 0$, $\partial M_y/\partial z = 0$), the effective surface charge distribution becomes

$$\hat{\sigma}_{\text{eff}}(\vec{k}) = -\frac{i\vec{k}}{|\vec{k}|} \hat{M}(\vec{k}) = \frac{\hat{\varrho}(\vec{k})}{|\vec{k}|} \, . \qquad (12.19)$$

For more complex situations, every magnetic charge in the sample has to be transformed, which results in rather lengthy expressions, and this method loses its power. In that case, it might be easier to calculate the stray field from the tip.

Assuming a known effective surface charge distribution, we can now calculate the energy of the tip-sample system by combining Eqs. 12.13 and 12.17. The only thing

---

[3] The σ_{eff} introduced here differs by a factor of 2 from the effective surface charge distribution defined in [21, 33]

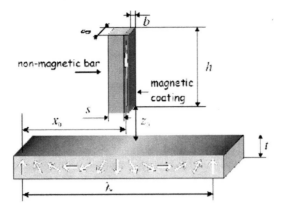

Fig. 12.4. The ideal MFM tip has a bar shape and a magnetization fixed along its axis

left unknown is the magnetization distribution in the MFM tip, which can be very complex. We will restrict ourselves, however, to the bar-type tip with a magnetization fixed along the z-axis (Fig. 12.4), in the first place because this is the ideal MFM tip shape [21] and in the second place because it results in very illustrative closed-form equations. The procedure to obtain the force F_z involves a simple integral of the stray field over the rectangular tip volume and taking $\partial U/\partial z$ [21]:

$$\hat{F}_z(\vec{k}, z) = -\mu_0 M_t \cdot b \operatorname{sinc}\left(\frac{k_x b}{2}\right) \cdot S \operatorname{sinc}\left(\frac{k_y S}{2}\right) \quad (12.20)$$
$$\times \left(1 - e^{-|\vec{k}|h}\right)\left(1 - e^{-|\vec{k}|t}\right) e^{-|\vec{k}|z} \hat{\sigma}_{\text{eff}}(\vec{k}),$$

where M_t is the tip magnetization [A/m], $b \times S$ the tip cross section, h the tip height, t the film thickness, and z the tip-sample distance (all in [m]).[4] This relation between the force and the effective surface magnetization is often called the *tip transfer function* (TTF).

Although Eq. 12.20 is complex, it is not difficult to understand. We see that the force is proportional to the tip magnetization and the magnetic charge density (σ) in the sample, combined with a number of geometrical loss factors. For most situations in magnetic data storage research, the films under investigation are thin and the film thickness loss term $(1 - e^{-|\vec{k}|t})$ will dominate over the tip height term $(1 - e^{-|\vec{k}|h})$. If we further assume that the tip cross section ($b \times S$) is much smaller than the smallest features of the charge distribution in the film, we get a very simple expression for the TTF:

$$\hat{F}_z(\vec{k}, z) = -\mu_0 M_t b S \left(1 - e^{-|\vec{k}|t}\right) e^{-|\vec{k}|z} \hat{\sigma}_{\text{eff}}(\vec{k}). \quad (12.21)$$

This approximation can be called the *monopole* approximation, because we assume that all magnetic charges ($M_t b S$) are located at one point at the end of the tip,

---

[4] $\operatorname{sinc}(x) = \sin(x)/x$

and that the other charges are very far away from the sample surface. The only loss terms that remain are the film thickness loss and the tip-sample distance loss, which usually is dominant. This immediately shows that for a good signal-to-noise ratio (SNR) in the image, the tip-sample distance has to be as small as possible.

When the details in the image start to approach the tip dimensions, the sinc functions have to be considered as well. For this bar-type tip, the force becomes zero when the wavelength of the surface charge distribution equals the tip size. This is analogue to the situation in magnetic recording, where at the *gap zero* the bit size is half the gap size of the recording head.

The force we calculate in Eq. 12.20 is measured by means of a cantilever deflection or change in resonance frequency. For the latter case, $\partial F_z/\partial z$ has to be calculated, which in the Fourier domain is simply:

$$\frac{\partial \hat{F}_z(\vec{k}, z)}{\partial z} = -|\vec{k}|\hat{F}_z(\vec{k}, z) \ . \tag{12.22}$$

A typical example of TTFs for the static and dynamic modes and the resulting deflection and resonance shifts, calculated from Eqs. 12.1 and 12.12, is given in Fig. 12.5. In this case, we consider a sample with perpendicular magnetization and only magnetic surface charges, using the parameters from Table 12.1.

Table 12.1. Parameters used in the calculation of the tip transfer functions and noise levels of Fig. 12.5. Values in parentheses are for the dynamic mode curves

M_t	Tip magnetization	1422 kA/m
b	Tip thickness (coating thickness)	20 nm
s	Tip width	100 nm
h	Tip height	1 μm
M_s	Sample saturation magnetization	295 kA/m
t	Sample thickness	70 nm
z_0	Tip sample distance	20 nm
c	Cantilever spring constant	0.01(3) N/m
f_n	Cantilever resonance frequency	7(75) kHz

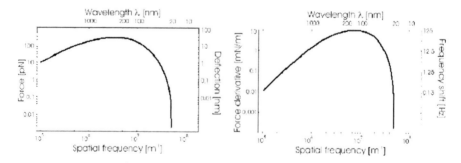

Fig. 12.5. Tip transfer functions for a typical situation (Table 12.1)

12.1.2 Instrumentation

The scope of this book does not allow for an exhaustive discussion of scanning probe microscope instrumentation. When using MFM or making decisions on the type of instrument to be used, some background information can be useful. For an excellent example of modern microscope design, see *Hug* et al. [13]. In scanning probe instrumentation, we can recognize the following subsystems:

Probe

The force on the magnetic tip is detected by means of a cantilever and a displacement sensor. The tip and cantilever will be discussed in detail in Sect. 5, because the tip size obviously determines the ultimate resolution of the MFM. But this resolution can only be obtained if the detection system is capable of measuring the very small deflections of the cantilever. Nowadays, mostly optical displacement sensors are used in high resolution MFM such as beam deflection systems or various types of interferometers. The beam deflection system is very robust and easy to align, but calibration of the sensor is only possible by indirect methods such as calibration grids. The interferometer, on the other hand, is more difficult to align, but can be easily calibrated with respect to the laser wavelength. By using a fiber interferometer [25], the active size of the interferometer can be made extremely small, which is beneficial to the instrument stability. Whether the beam deflection or interferometer detection system is more sensitive depends very much on the implementation, but in principle the interferometer is a factor of 4 more sensitive [7]. The detection limits are estimated to be as low as 10^{-15} m/$\sqrt{\text{Hz}}$ [30].

Positioning

The positioning system brings the sample close to the tip (coarse positioning) and scans the tip over the sample area to obtain an image (fine positioning). The coarse positioning system has the difficult task to achieve millimeter displacements with stability better than 1 nm (when switched off). Ordinary manual or motorized screw positioners are used very often, sometimes in combination with levers. These systems suffer from drift: It can sometimes take hours before the displacement relaxes. Hysteresis is also a big problem. For vacuum systems (see next section), these screw positioners are not really useful, since they require greasing, although some vacuum-compatible systems do exist. Therefore, recent microscope designs resort to the use of piezo-actuated stepper actuators. Principles such as slip-stick, inertia, and walking beetle motions are used. The big advantage of these systems is that they are extremely stable once switched off. The disadvantage is that they are much more difficult to operate and sometimes do not move at all, which is a nightmare for instrument designers: You have a perfect MFM, and you cannot get the tip close to the sample.

The fine positioning system has the difficult task to achieve an as big as possible scan range with sub-nanometer stability, and, on top of that, scan as fast as possible. For speed and stability, it is advantageous to use a small scanner, but, of course, this is contradictory to the scan-range requirement. In most microscopes, so-called tube scanners are used. The major disadvantage of the tube scanner is the coupling between the xy- and z-motions, because the sample is tilted as well as translated. A very nice tutorial on piezo-translators can be found on the Web site of Physik Instrumente.[5]

Housing

Although seemingly trivial, the environment in which the MFM instrumentation is mounted is of crucial importance to the image quality. The MFM works at extremely small tip-sample distances of 30 nm and less, but the tip never touches the sample surface. The mechanical path from tip to sample is, therefore, several centimeters, six to seven orders of magnitude larger than the tip-sample distance. Therefore, the MFM is extremely sensitive to temperature variations, mechanical and acoustic vibrations, and airflow. For high resolution, the MFM is mounted on a vibration isolation table (Fig. 12.6) in an environment that shields acoustic noise and airflow. Air-operated instruments are put in noise isolation cupboards, which enclose both the MFM and the vibration isolation table. To eliminate electromagnetic noise, the cupboard can be equipped with wire mesh to provide a Faraday cage. Sometimes ionizing units are mounted inside the cupboard to prevent static charge on the sample.

Fig. 12.6. Noise isolation cupboard in the authors' laboratory featuring nonparallel side-walls, two-layer sound isolation, a Faraday cage, and ionizing unit (not visible)

The ultimate acoustic noise isolation is obtained when the MFM is mounted in vacuum. A vacuum environment has many other important advantages. Impingement of air molecules, which counteracts cantilever movement, and squeezing of air between the cantilever and the sample dominate damping of the very small cantilever. The quality factor of resonance Q goes up by several orders of magnitude when

---

[5] http://www.physikinstrumente.com/tutorial/

the air is removed. This increase in quality factor strongly increases the instruments sensitivity and, therefore, the resolution (see Sect. 12.3.2). Next to this, Brownian motion of the air molecules excites the cantilever as well and causes additional noise. Finally, the vacuum removes a large part of the gases, which stick to the sample surface, such as water. Especially the highly mobile molecules on the surface cause a problem, since they move toward the tip and form a meniscus that pulls the tip into the sample. A vacuum of about 0.1 Pa (10^{-3} mbar) is sufficient to benefit fully from the effect of vacuum. Care must be taken, however, that the pressure is not too high; the conditions for electric breakdown between the piezo-electrodes are optimal for pressures between 1 Pa and 40 kPa (10^{-2} mbar and 0.4 bar) [19]. Since 1 Pa is difficult to reach with a rotary pump only, two-stage pumping systems are, therefore, almost a necessity.

Control

The actual acquisition of an image requires a complicated control system, which takes care of tip-sample positioning, both for scanning and for controlling the tip-sample distance, data acquisition and visualization, and possibly other control loops such as phase locking in dynamic mode measurements. With the gradual progress in digital signal processing (DSP), more and more tasks are being taken over by digital control systems. (This is how one of the biggest scanning probe microscope manufacturers got its name). Still high quality analog electronics, such as the pre-amplifiers and high voltage amplifiers for the piezos, are crucial and should not be neglected.

In the design of control electronics, two issues play an important role. First of all, the noise caused by the control systems should be minimal. This means that high quality components such as powerful DSPs and low-noise, high-bandwidth DA converters have to be used. As a result, for microscopes operated in air, the price of the control system usually equals that of scanning probe microscope hardware. In the second place, the control system should provide a user-friendly interface to the microscope system with a lot of flexibility in scanning modes and parameter tuning. Unfortunately, every single scanning probe microscope manufacturer has his own dedicated front-end, using proprietary code. Until now, there have been no attempts to standardize, not even in the format of the output data files. This seriously hampers the transfer of data between different research laboratories, which often use different brands of microscopes. We believe it is the task of the academic community to improve this situation and have proposed such a standard (www.spml.org).

12.2 MFM in Magnetic Data Storage Research

The MFM has found a widespread application in magnetic recording research, mainly because it is a relatively cheap high resolution imaging technique that does not require sample preparation. Moreover, in magnetic recording one designs media and heads in such a way that the surfaces are very smooth and the external stray fields are very high, both of which are beneficial to MFM. And, finally, recording media are very resistant to the influence of external fields, so tip-sample interaction is not really a problem. It is, therefore, not coincidental that reference samples for MFM originate from the magnetic recording community (such as the NIST hard disk [23] or the magneto-optic disk used for the CAMST reference sample [1]).

The magnetic recording industry is, however, on an incredibly steep roadmap. Each year, the data density more than doubles. Progress in MFM technology has been made mainly on the ease and reliability of use of the instrumentation. In the early 1990s, an MFM measurement could take a week. The authors painfully remember that it could take a day to prepare an MFM tip, but seconds to break it. Much has been done by the microscope industry to improve this situation. The use of batch fabricated tips, auto-approach, and digital signal processing has brought the total time from sample mount to the first image down to 10 min. The resolution has, however, not improved dramatically. Figure 12.7 shows a more or less representative selection of MFM resolution since its invention. Already in 1991, a resolution as low as 40 nm was obtained [8], and it took 10 years to bring this down to 25 nm. In the meantime, the bit length in experimental hard disks has decreased from 700 nm in 1990 to 33 nm in 2001. With the instrumentation and tips we have today, it is very difficult to measure these small bits, and very soon the use of MFM in magnetic recording research will become limited.

Unfortunately (or fortunately, depending on which side one is on), there is no reasonable alternative to MFM in magnetic recording research. Electron microscope techniques (such as Lorentz microscopy) require elaborate sample preparation and are not really suitable for perpendicular recording media. Spin-polarized STM measurements require extremely clean surfaces, which can up to now only be achieved by in-situ preparation. A very interesting question, therefore, is the ultimate limit of MFM resolution. Can we do much better than the 20–30 nm of today, and what will be needed to achieve that? This is discussed in the remainder of this chapter.

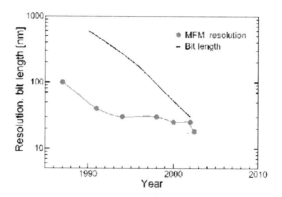

Fig. 12.7. The progress of magnetic recording has been much faster than the progress in MFM resolution, which might cause a problem in the near future. (The bit length is taken from hard disk laboratory demonstrations, averaged for a 30% annual growth.) MFM resolution is taken from literature [1,6,10,14,17,26], the last point is an unpublished result by H.J. Hug at Intermag 2002

12.3 Limits of Resolution in MFM

The resolution that can be achieved by MFM is determined by the combination of many factors, of which the tip geometry, tip-sample distance, and instrument sensitivity are the most important. In this paragraph, we set out on a dangerous but important task: To determine the ultimate resolution that can be achieved by MFM.

12.3.1 Critical Wavelength

Although everyone knows what a high-resolution image is, to quantify resolution is not trivial. When an imaging system has high resolution, we usually mean that it is able to separate two closely spaced objects. We do not mean that it can detect one single small object: This has to do with sensitivity. We can define resolution as the minimum spacing between two objects that can still be observed. Instead of two, we can, of course, take a number of equally closely spaced objects. If we take an array, we can even define resolution in a certain direction. By using the array, the definition of resolution can quite naturally be transferred to the spatial frequency domain as the *minimum spatial wavelength that can still be observed*. Using the theory of Sect. 1.2, we see that the signal strongly decreases for high frequency components or small wavelengths. Analogous to magnetic recording theory, we can define a certain upper limit on the spatial frequency, above which we call the signal "non-detectable." This is usually done with respect to the background noise level: The signal-to-noise ratio (SNR) should exceed a certain value. For magnetic recording, the SNR should be about a factor of 10 (20 dB), for imaging we consider an image still acceptable at an SNR as low as 2 (3 dB) (Fig. 12.8).

In principle, the SNR can be made arbitrarily high by increasing the measurement time, because that decreases the measurement bandwidth. This, however, assumes that the background noise is white, i.e., it has a flat frequency spectrum. In practice, the noise strongly increases when the frequency becomes very low. This is usually referred to as $1/f$ noise and is caused by drift, for instance, by temperature variations or piezo creep. As an illustration, let us take the schematic noise spectrum of Fig. 12.9. The measurement bandwidth ΔB is related to the measurement time between the

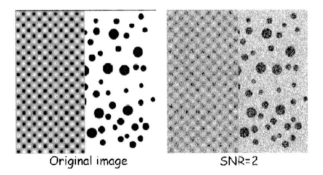

Fig. 12.8. Image quality can be quite acceptable at low signal-to-noise ratios

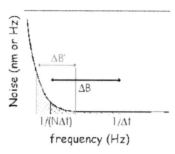

Fig. 12.9. Increasing the measurement time can increase low frequency noise

pixels Δt and the total number of pixels N, the combination of which determines the total time to measure an image,

$$\Delta B = \frac{1}{\Delta t} - \frac{1}{N\Delta t} \approx \frac{1}{\Delta t}. \tag{12.23}$$

We see that the bandwidth is inversely proportional to the measurement time. But at long measurement times, the bottom of the measurement band will enter the $1/f$ noise region, which is indicated by $\Delta B'$ in Fig. 12.9. So if we exceed a certain measurement time, the total noise (the area under the curve) will start to increase and cancel out the effect of a smaller bandwidth. In MFM, this situation results in a "stripey" image, where the individual scan lines have noticeable offsets with respect to each other. Image processing can partly eliminate this effect (remove line average), but only if the low frequency noise periods are much longer than the time between scan lines. (One should be careful, however, because with imaging processing of this kind one enters the twilight zone of image manipulation). So for reasons of image quality and, of course, for practical reasons, we always have to take a limited measurement time into account. A minimum bandwidth of 200 Hz is reasonable. This allows us to link noise in the frequency domain to noise in the spatial domain through the scan speed. Now we can find the resolution in figures like Fig. 12.10 by

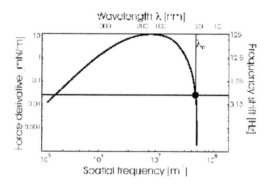

Fig. 12.10. Using the background noise level, we can define the limit of resolution by means of a critical wavelength (λ_c)

calculating the wavelength where the signal drops below the noise level multiplied with the desired SNR. Because the minimum SNR is quite low and usually the tip transfer function is very steep around this point, we simply set the SNR to unity. We call this point the *critical wavelength* or λ_c.

12.3.2 Thermal Noise Limited Resolution

Using our simple bar-type tip model and the critical wavelength, it must be possible to predict the resolution of MFM if we know the background noise level. Noise is, of course, originating from many sources: thermal agitation of the cantilever, electronic crosstalk, mechanical vibration, etc. Many noise sources can be eliminated by proper design of the detection system and operation in vacuum. In the ideal situation, we are only left with the thermal noise in the cantilever. This provides us with a kind of fundamental limit to MFM resolution. Whether or not MFM will ultimately reach this limit depends on many factors, but it can serve as a benchmark against which to compare progress.

Thermal Noise

The atoms within the cantilever vibrate around an equilibrium position and exert small forces on the cantilever. On average, these forces average out, but at the small scales where we measure in MFM, we can actually observe small fluctuations in the cantilever position with time. When we assume that the cantilever has only one free degree of motion, the total thermal energy in the system is $\frac{1}{2}kT$. This energy is equivalent to the average mechanical energy. So we can express the average thermal energy $\langle U_{th} \rangle$ in the average amplitude of vibration of the cantilever end $\langle \Delta z_{th} \rangle$, through the spring constant c:

$$\langle U_{th} \rangle = \frac{1}{2}kT = \frac{1}{2}c \langle \Delta z^2 \rangle \tag{12.24}$$

with

$$\langle \Delta z^2 \rangle = \frac{1}{T} \int_0^T (\Delta z(t))^2 \, dt = (\Delta z_{rms})^2 \quad [\text{m}^2] \, . \tag{12.25}$$

The time average deflection of the cantilever is, of course, an interesting value, but it does not give us the required noise frequency spectrum. For this, we move into the frequency domain using Fourier analysis.

$$\hat{z}(\omega) = \frac{1}{2\pi} \int_{-\infty}^{\infty} z(t) e^{-i\omega t} \, dt \tag{12.26}$$

$$z(t) = \int_{-\infty}^{\infty} \hat{z}(\omega) e^{-i\omega t} \, d\omega \, . \tag{12.27}$$

The mean square of the displacement translated to the Fourier description becomes

$$\langle z^2 \rangle = \int_{-\infty}^{\infty} \hat{z}(\omega)^2 \, d\omega \quad [\text{m}^2] \, . \tag{12.28}$$

Like in Eq. 12.9, we can now calculate the cantilever displacement from the force using the cantilever transfer function:

$$\hat{z}(\omega) = G(\omega)\hat{F}(\omega) \quad [\text{m/Hz}] \tag{12.29}$$

$$G(\omega) = \frac{1/m}{\sqrt{(\omega_n^2 - \omega^2)^2 + (\omega\omega_n/Q)^2}} \quad [\text{m/N}] \, . \tag{12.30}$$

We assume that the force over the range of frequencies we are interested in ($f < 10\,\text{MHz}$) the "thermal force" of the combined action of all atoms is equally distributed over all frequency components (so $\hat{F}(\omega) = \hat{F}_{\text{th}}$). This "thermal force" is usually called the thermal white noise drive. The cantilever vibration spectrum, therefore, simply equals $G\hat{F}_{\text{th}}$. In a cantilever with a high quality factor, the noise will predominantly focus around the resonance frequency and the noise at DC will be low.

In the Fourier domain, we can relate the mechanical energy to the thermal white noise drive:

$$\hat{U}_{\text{th}}(\omega) = \frac{1}{2}c\hat{z}_{\text{th}}^2 = \frac{1}{2}cG^2\hat{F}_{\text{th}}^2 \, . \tag{12.31}$$

Integrating over all spectral components yields the thermal energy, which was equal to $\frac{1}{2}kT$, so we can calculate the frequency-independent thermal white noise drive:

$$\langle U_{\text{th}} \rangle = \int_{-\infty}^{\infty} \hat{U}_{\text{th}}(\omega) \, d\omega = \frac{(\Delta F_{\text{th}})^2 Q \pi \omega_n}{2c} = \frac{1}{2}kT \quad [\text{J}] \tag{12.32}$$

$$\rightarrow \Delta F_{\text{th}} = \sqrt{\frac{kTc}{\pi \omega_n Q}} = \sqrt{\frac{kTm\omega_n}{\pi Q}} = \sqrt{\frac{kTD}{\pi}} \quad \left[\text{N}\sqrt{\text{rad/s}}\right] \, .$$

The thermal white noise drive results in a frequency-dependent cantilever movement, which can now simply be calculated from the transfer function $\Delta z_{\text{th}}(\omega) = G(\omega)\Delta F_{\text{th}}$. For static mode MFM, the bandwidth is usually far below the cantilever resonance frequency and we can approximate the thermal noise to $\Delta z_{\text{th}}(\omega) = \Delta F_{\text{th}}/c$.

The average noise in a measurement bandwidth from 0 to $\Delta \omega$ equals

$$\langle \Delta F_{\text{th}}^2 \rangle = \int_{-\Delta\omega}^{\Delta\omega} \Delta F_{\text{th}}(\omega) \, d\omega = \frac{2kTc\Delta\omega}{\pi\omega_n Q} \quad [\text{N}^2] \tag{12.33}$$

$$(\Delta F_{\text{th}})_{rms} = \sqrt{\frac{4kTc\Delta B}{\omega_n Q}} = \sqrt{kTD\Delta B} \quad [\text{N}] \, . \tag{12.34}$$

with $\Delta B = 2\pi \Delta \omega$ [Hz]. In the last expression, we can appreciate the analogy with the thermal white voltage noise (Johnson noise) in an electric resistance $\sqrt{4kTR\Delta b}$ [V].

For dynamic mode MFM, we have to analyze the noise in a self-oscillating system with positive feedback. This has been done by Albrecht [2], who obtained

$$\left(\Delta \left(\frac{\partial F}{\partial z}\right)_{th}\right)_{rms} = \sqrt{\frac{4kTc\Delta B}{\omega_n Q \langle z_{osc}^2 \rangle}} \tag{12.35}$$

where $\langle z_{osc}^2 \rangle$ is the mean square amplitude of the self-oscillating cantilever, to which the minimum detectable force derivative is inversely proportional.

Critical Wavelength

Using Eqs. 12.34 and 12.35 for the noise levels, we can now calculate the critical wavelength for the bar tip model (Eq. 12.20). To obtain simple expressions, we set the bar tip cross section square, so that we have the same resolution in x- and y-directions and we restrict ourselves to the x-direction. Furthermore, we assume that the tip height h is much larger than the film thickness t, and that the film thickness loss factor can be ignored for wavelengths close to the critical wavelength. These assumptions are not so strict, but to reduce the number of parameters even further, we also assume that the tip-sample distance loss ($e^{-k_x z}$) can be ignored. This is approximately true for $z < 0.1/k_x$ or $z < 0.02\lambda_c$. To achieve such small tip-sample distances is not trivial, and we come back to that later. Doing so, however, we obtain a fairly simple expression for the force:

$$\hat{F}_z(\vec{k}, z) = -\mu_0 M_t b^2 \operatorname{sinc}\left(\frac{k_x b}{2}\right) \operatorname{sinc}\left(\frac{k_y b}{2}\right) \hat{\sigma}(\vec{k}) \quad [\text{Nm}^2]. \tag{12.36}$$

To calculate the critical wavelength, we assume a perpendicular medium with a harmonic magnetization distribution with identical wavelength in x- and y-directions: $M_z \sin(2\pi x/\lambda) \sin(2\pi y/\lambda)$ (see Fig. 12.8, for instance). We then obtain for the force amplitude simply:

$$\Delta F_z(\lambda) = \mu_0 M_t M_z b^2 \operatorname{sinc}^2\left(\frac{\pi b}{\lambda}\right) \quad [\text{N}]. \tag{12.37}$$

For a fixed wavelength λ, the force has a maximum amplitude at $b = \lambda/2$, so when the tip diameter is exactly equal to half the wavelength of the stray field. This is logical, because for smaller wavelengths, charges with opposite polarity will be beneath the tip and reduce the force. The tip dimension b can now be eliminated, leaving

$$(\Delta F_z(\lambda))_{\max} = \mu_0 M_t M_z \left(\frac{\lambda}{\pi}\right)^2 \quad [\text{N}]. \tag{12.38}$$

This force should be larger than the base noise level we obtain for the critical wavelength

$$\lambda_c = \pi = \sqrt{\frac{\sqrt{4kTD\Delta B}}{\mu_0 M_t M_s}} . \qquad (12.39)$$

With Eq. 12.39, we have a lower fundamental limit to the resolution of MFM, in other words, *this is the resolution that is theoretically possible*. Let us take static mode MFM in air at room temperature. With $c = 0.03$ N/m, $\omega_n = 2\pi 10$ kHz, $Q = 30$, and a bandwidth ΔB of 200 Hz, we get a thermal noise level of about 1.5 pN. Taking both M_s and M_t equal to 1 MAm^{-1}, we get a critical wavelength of 3.4 nm, which means feature sizes of 1.7 nm. This extremely low value is achieved with a tip of 1.7 nm in diameter and at a tip-sample distance of less than 0.07 nm. Of course, this tip-sample is impossible to achieve in static mode MFM. Our first problem will be that we have contamination layers on the tip and sample. Suppose that we solve this by moving to UHV and in-situ preparation of tip and sample, then at this distance we will have strong Van der Waals forces, so that the tip will be pulled into the sample if the spring constant is too low. Furthermore, we assumed in our calculations that the magnetic charges are at the tip and sample surface. Of course, they are not, we expect the charges to be distributed over the exchange length, which is a few nanometers in general. What we can conclude, however, is that even for the simple case of static mode MFM in air we can do much better than the 30 nm resolution (60 nm critical wavelength) that is observed today.

So let us be realistic and assume that we can achieve a tip-sample distance z_{min} of at least 10 nm. The signal is now reduced by a factor $\exp(-2\pi z_{min}/\lambda)$ and

$$(\Delta F_z(\lambda))_{max} = \mu_0 M_t M_z e^{-2\pi z_{min}/\lambda} \left(\frac{\lambda}{\pi}\right) \quad [\text{N}], \qquad (12.40)$$

which is kind of awkward to solve analytically for λ. Numerically, we obtain Fig. 12.11, which shows that we can achieve resolution better than 20 nm in air (10 nm feature size). If we move to vacuum and increase the quality factor to 10^4, for instance, the noise level decreases to 0.08 pN and the critical wavelength is smaller than 12 nm, with minimum of 0.8 nm.

We can perform the same exercise for dynamic mode MFM, using the thermal noise background of Eq. 12.35. The oscillation amplitude is another parameter, which is set equal to its maximum value of the tip-sample distance z_{min} (so it almost hitting the sample). In principle, we cannot use the theory of Sect. 1.1 for dynamic mode MFM in this case, because the force derivative will not be constant over the cantilever oscillation. But let us regard the calculated (wrong) values as an upper limit, then if we take, for example, a typical MFM cantilever with $\omega = 2\pi 75$ kHz and a spring constant $c = 3$ N/m we obtain the curves of Fig. 12.12 in air ($Q = 300$) and vacuum ($Q = 40\,000$). The deterioration of resolution for very small tip-sample distances is caused by the fact that the noise levels increase faster than the signal. In fact, for very small distances, static mode MFM has a lower theoretical resolution than dynamic mode MFM. For larger tip-sample distances, the dynamic mode is better, although

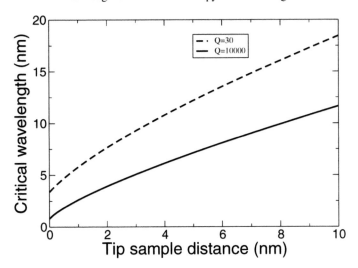

Fig. 12.11. Thermal noise limited critical wavelength for static mode MFM in air ($Q = 30$) and vacuum ($Q = 10\,000$), using $c = 0.03$ N/m, $\omega_n = 2\pi\,10$ kHz, and a bandwidth ΔB of 200 Hz

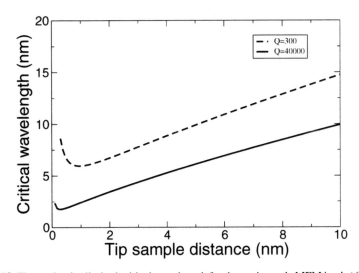

Fig. 12.12. Thermal noise limited critical wavelength for dynamic mode MFM in air ($Q = 300$) and vacuum ($Q = 40\,000$), using $c = 3$ N/m, $\omega_n = 2\pi\,75$ kHz, and a bandwidth ΔB of 200 Hz

the differences are not dramatic. In general, we can conclude that when we move to vacuum, the theoretically achievable resolution of MFM is improved by a factor of 2.

Beware!

Although the above analysis of λ_c is useful in a sense, its meaning for practical MFM is limited. We emphasize again that the analysis above gives a theoretical upper limit to the resolution, under the condition that the only noise source is thermal noise in the cantilever and that the imaging process can be modeled by the bar tip model. Of course, there are other noise sources, such as laser power fluctuation in the case of beam deflection or interferometry, $1/f$ noise in the system caused by temperature drift and crosstalk from high-voltage positioning signals for piezos and external sources. Even in the quality factor, which determines the thermal noise level, there is an uncertainty. Damping caused by air squeezing between the cantilever and the sample, meniscus formation, and eddy currents in the sample caused by the time varying magnetic stray field of the tip will reduce the quality factor if the tip is approached to the sample. It is, therefore, wise to measure the quality factor of resonance at the same tip-sample distance used during the measurement.

In our calculations, we used, on one hand, the thermal noise, which gives a lower limit on the noise level. On the other hand, we used the idealized bar tip model, which gives an upper limit to the signal level. Usually we find, however, much smaller signals. There are many possible reasons for this. One reason might be that the magnetic charge density will not be on the uttermost front surface of the tip or at the top surface of the medium. It is likely that the charges will be distributed over a volume with a thickness of at least the exchange length, which can be several nanometers. Fortunately, the area over which the charges are distributed decreases with the tip dimension. Figure 12.13 shows the result of two micromagnetic calculations on tips of different sizes (using MagFEM3D, a large-scale micromagnetic calculation package [22]). The magnetization in the large tip (20×100 nm) is able to rotate away from the front surface, with the result that only 50% of the charges are on the tip front surface. In the much smaller tip (7×20 nm) 98% of the charges are on the front surface, simply because there is no more space for the magnetization to rotate away.

There are also other reasons besides the charge distribution that create a difference between the magnetic separation and physical separation of the tip and the sample surface. Unless carefully prepared and operated in UHV, the sample surface will have a surface surfactant layer of absorbed molecules. Also, the magnetic tip will have such a layer, and the magnetic tip might not extend completely to the end of the tip, for instance, because of an oxide layer. All these effects result in an effective tip-sample distance, which is larger than the physical tip-sample distance. Therefore, we usually find a considerably lower signal than predicted by the bar tip model. For the CAMST reference sample, for instance [1], we typically find signals on the order of 1–3 Hz, whereas the bar tip model predicts 10–100 Hz.

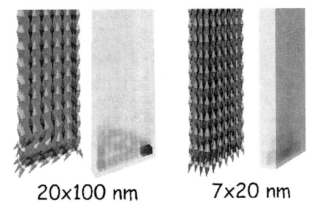

Fig. 12.13. Micromagnetic simulation of equilibrium magnetization in a bar-type tip with a dimension of 20 × 100 nm (left) and 7 × 20 nm (right). The images show the magnetization vectors and the volume charge distribution

In conclusion, we see that in reality the noise level will be higher and the signal level lower than the values assumed in the theory outlined above. The limit on resolution presented here can, however, be used to benchmark progress against. The theory also shows three important areas in which this progress can be made:

- decrease noise,
- reduce tip-sample distance,
- improve tip shape and size.

The latter two will be discussed in the following.

12.4 Tip-Sample Distance Control

There are at least two reasons why we want to measure and control the tip-sample distance. In the first place, we want to be as close as possible to the sample surface to obtain high resolution. Sample tilt, drift in the microscope positioning, and surface roughness will cause a variation in tip-sample distance over the image, and we can never get closer than the highest point on the surface. In the second place, we will get a stronger contribution from nonmagnetic forces as we approach closer and closer to the sample, and topographic information will be superimposed on the image. If we keep tip-sample distance constant, the contribution of those forces will be more or less constant as well, and we obtain a topography-free image.

There are various ways to achieve constant tip-sample distance, all differing in complexity, precision, and stability. In the following, we will try to position these approaches in a structure as in Fig. 12.14. Probably the ideal situation would be to use a *dedicated detector* optimized for tip-sample distance measurement. STM seems to be a good candidate, because it will enable us to control the tip-sample distance

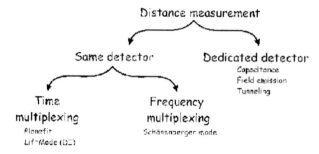

Fig. 12.14. Different ways to control tip-sample distance

with angstrom precision. The disadvantage might be that we only get reasonable tunneling currents at very low tip-sample distances, too low to allow for sufficiently soft cantilevers in static mode MFM or for sufficiently large vibration amplitudes in dynamic mode MFM. In the future, STM control might become an option however. At higher tip-sample distances, we can still use the STM technique, but then in the field emission mode. This requires higher tip-sample voltages, and the resolution will not be as good as in STM. Another option might be to measure the capacitance between the cantilever or tip and the sample.

All these options require complex integrated probes with bond pads and wires running down the cantilever. That is why currently practically all methods to measure the tip-sample distance use the *same detector* as the one used for measuring the magnetic force (beam deflection, interferometer, piezo-resistive, etc.). Since we use the same measurement channel, we have to multiplex the magnetic and topographic signals. Multiplexing can be done in the time domain. This is the most commonly used method because it allows one to use AFM to measure the tip-sample distance. Time multiplexing can be done on various scales:

Image by image:

An example of this method is the real-time plane fit (such as, for instance, used in control systems manufactured by RHK), which is often used in static mode MFM. First, we measure an AFM image of the surface and determine the slope of the image. Then we add a correction signal to the z-piezo and make sure that the AFM image appears flat: the average tip-sample distance is constant. More sophisticated systems use a polynomial fit to the surface in which bowing is taken into account. This is especially useful for large-range scans using tube scanners. In principle, it is not necessary to measure the complete AFM image; depending on the order of the polynomial, just a few strategically placed scan lines are sufficient. A big advantage of this mode is that there is no contact between the tip and the sample during image formation. Especially for soft magnetic samples, in which the domain image is easily disturbed by the MFM tip stray field, this mode has an advantage over the other two time-multiplexing modes discussed hereafter. In principle, the plane-fit method can be extended to measuring a complete AFM image and do the distance control on a pixel-by-pixel basis, but instrument drift during the time it takes to scan an image has to be less than the distance between the pixels. This type of accuracy is only found in low temperature microscopes.

Line by line:

This method is called LiftMode by Digital Instruments and is used with dynamic mode MFM. Every line in the image is scanned twice. The first scan is in tapping mode AFM with amplitude detection (see Sect. 12.1.1). The traced topography is stored in memory and used to control the tip-sample distance in the subsequent MFM trace, using phase or frequency detection. Since the time per line trace is relatively short, this method works quite well. Moreover, the topographic and magnetic information are obtained simultaneously, which makes mapping of topographic features onto the MFM image very easy. The method is less suitable for static mode AFM/MFM, because then very soft cantilevers are needed that snap into the sample. To "unsnap" the cantilever, the z-piezo has to be withdrawn over quite a distance (sometimes more than a micrometer). Piezo hysteresis and nonlinearity then make it difficult to approach the sample accurately again.

Pixel by pixel:

This mode can be compared with the spectroscopy mode used in STM. For every pixel we make a force distance measurement (F_z as function of z). From the snap-in or contact point we can calculate the tip-sample distance, and we can make a map of the magnetic force for different tip heights. Of course, this mode is very slow and suffers heavily from $1/f$ noise. The extra information obtained is useless; one can easily calculate the MFM signal at a height z from images taken at lower height. One might think that by scanning at different tip-sample distances one can determine the volume distribution of the magnetic charges in the sample, but the nature of the magnetic stray field simply prohibits that [33]. This mode, however, does allow for the smallest tip-sample spacing, especially in static mode MFM. (Some skeptics, however, just see this "spectroscopy" mode as a way to circumvent the DI patent on the LiftMode.)

The last branch in Fig. 12.14 is the reciprocal to time multiplexing, which is frequency multiplexing. This means that we measure topographic and magnetic information in different frequency bands. To achieve this, we need to modulate the magnetic force or the force responsible for the topographic signal (or both). Contact mode AFM is not possible in this frequency-multiplexing mode, because we cannot measure the magnetic forces simultaneously. So we need a non-contact measurement of the topography. Non-contact mode AFM might be a possibility, but has, to our knowledge, never been used before. It will be very hard, if not impossible, to modulate the forces measured during AFM (such as Van der Waals forces), so we have to modulate the magnetic force. This can be achieved by applying an external magnetic field, which modulates the magnetization in the tip. The same field would also act on the sample, however, and we will probably only be able to measure hard magnetic samples.

Another way to measure the topography in non-contact mode is to use an electric field. This has been demonstrated by Schönenberger et al. [28], who applied a modulated voltage between the cantilever and the sample. The frequency of modulation is well below the resonance frequency of the cantilever, so that the cantilever will start vibrating. The amplitude of vibration is a measure for the tip-sample distance and is kept constant. This method can be applied to static mode as well as dynamic mode MFM. In principle, the method is similar to the measurement of the tip-sample capacity, the electric force and the change in capacity being related by:

$$F_{el} = \frac{\partial C}{\partial z} V^2 .\qquad(12.41)$$

12.5 Tips

12.5.1 Ideal Tip Shape

Most scanning probe microscope tips are pointed and very sharp: The STM tip and AFM tips are atomically sharp; most SNOM tips are tapered optic fibers. One might, therefore, assume that an MFM tip should also have the shape of a sharp needle. Techniques such as STM, AFM, and SNOM, however, measure very short-range effects, whereas MFM measures long-range magnetostatic forces. (The same holds for electric force microscopy (EFM) and electrostatic forces.) In STM and AFM only the last atom determines the signal, but in MFM a much larger part of the tip takes part in the image formation. Therefore, the optimum tip shape is not a sharp needle but a bar or cylinder with a flat front end [21], analogue to a hard disk head for perpendicular magnetic recording.

The fact that the bar shape is the optimum shape for MFM can perhaps be most easily understood by considering the magnetic charge distribution in a tip with an ideal uniform magnetization (see Fig. 12.15). In a bar-type tip, all magnetic charges are located at the front surface of the bar and as close as possible to the sample surface. Therefore, all magnetic charges contribute equally well to the signal. In a pointed

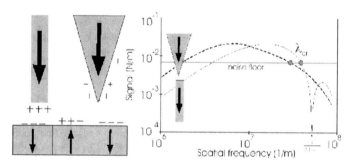

Fig. 12.15. Point-sharp versus bar-type MFM tips, graph from [21]. The width of the base of the triangular tip is adjusted in such a way that the maximum signals of both tips are equal

MFM tip, the charge is distributed over the complete tip. The charges located further away from the sample still contribute to the signal, but their effect is stronger at longer wavelengths. If we set the maximum signal of a pointed tip and a bar-type tip equal, the bar-type tip will perform better at high spatial frequencies and, therefore, show a better resolution. Of course, when the wavelengths approach the gap zero of the bar tip, the pointed tip performs better. But we can always tailor the bar type tip in such a way that this happens at wavelengths above the critical wavelength, as indicated in Fig. 12.15 on the right. One can argue that we may allow the pointed tip to have higher signal at low spatial frequencies by increasing the tip length. This does, however, not significantly increase the signal at high spatial frequencies. In all cases, the signal close to the critical wavelength will be highest for the bar-type tip.

The optimum tip shape is one thing, making such a tip another. When realizing MFM tips, we need such small feature sizes that one cannot rely on conventional lithography (as used in the semiconductor industry). Instead, we have to use tricks to obtain such small dimensions, and we enter the area of nanotechnology. The first scanning probe microscope tips were handmade in a one-by-one fashion. Still, the best tips available today are made by hand. The batch-fabricated Si and SiN tips, however, strongly improved the ease of use the MFM. In the following we briefly mention a number of MFM tips, starting with handmade tips and continuing with batch-fabricated tips.

12.5.2 Handmade Tips

The first MFM cantilevers were very thin Co or Ni wires that were etched down to a sharp point. The wires were bent around a razor blade edge, and the laser spot was deflected from the resulting "knee" – far from stable operation. The tips obtained this way were, however, not very sharp, and contained a lot of magnetic material [14]. Much better resolution could be obtained with tungsten tips coated with a magnetic layer from the side [24]. The resulting magnetic thin film was, however, not always smooth, due to the surface roughness on the tungsten point caused by the etching process.

A much smoother tip could be obtained by using contamination needles, which were at that time already used for AFM. These contamination needles were grown in an SEM, which was accidentally or deliberately contaminated with organic gases. On the side of these carbon tips, a thin magnetic layer is deposited. Some call these tips electron beam deposited (EBD) tips [5], but to avoid confusion with electron beam evaporation, which is a layer deposition technique, others call these tips electron beam induced deposited (EBID) tips [26, 27]. Especially the tips made by Skidmore et al. are impressive, with a tip radius of less than 7 nm. Strangely enough, the resolution of the images is not better than that obtained with conventional tips [29]. Instead of depositing a layer on the side of the EB(I)D tip, one can also use the contamination as an etch protection mask [16]. In this way, one can leave a tiny disk of magnetic material at the top of an AFM tip. This MFM tip does not have the ideal shape for

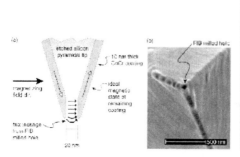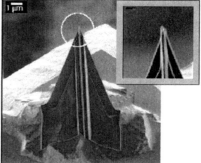

Fig. 12.16. MFM tips prepared by FIB. Left: perforated tip [6], right: Co needle [17]

high resolution, but can, for instance, be used in those situations where a low tip switching field is necessary.

The arrival of the focused ion beam (FIB) instrument opened new possibilities for MFM tip preparation. The two techniques known to the authors start with a magnetic coating on a commercial AFM cantilever. Folks et al. [6] start with a commercial MFM tip, then FIB etch a small hole at the apex of the tip. The tip is then magnetized in plane (parallel to the cantilever), so that magnetic charges occur at the edges of the tip. In this way, one obtains an in-plane tip, which is only sensitive to stray fields parallel to the sample in the direction of the tip magnetization (x- or y-directions). In general, one would like to measure the z-component, from which the x- and y-components of the fields can easily be derived. Such a tip is prepared by Philips [17], who starts with a thin Co film on a commercial AFM cantilever and then etches away the unnecessary material. In this way, an 8-μm-long 50-nm-wide needle could be prepared.

12.5.3 Coating of AFM Tips

The workhorse of the MFM cantilevers is a Si batch-fabricated cantilever with a tip optimized for AFM, which is coated with a magnetic material. There is a large variety in magnetic coatings, and if we tried to give an exhaustive list, we would certainly fail. We can, however, indicate different categories. The most common magnetic coating is an alloy based on Co with additions such as Cr and Pt, based on the pioneering work of Grütter et al. [9, 10]. These alloys are often derived from hard disk recording layer materials and have a high coercivity. The high coercivity makes the tip resistant to reversal by the medium's stray field. To achieve a high coercivity, the material is prepared in such a way that one gets magnetically separated grains. The exact location of these grains on the tip apex varies from tip to tip, which explains why some tips have higher resolution than others. Companies sell tips with different coercivities and film thicknesses (and give them meaningful names such as "high coercivity tip" or "low moment tip").

For some applications, the coercivity of the layers on the tips is not high enough, for instance, when measuring stray fields from recording heads or permanent magnets. In that case, an alternative can be found in very soft magnetic coatings, which switch in very small external fields. As a result, however, the tip is always attracted, and one looses information on the polarity of the field. Materials used are, for instance, Co [10], NiFe [10], Fe [31], and granular $Fe(SiO_2)$ films [11].

The perfectly soft MFM is a paramagnetic such as, for instance, a Pt-coated tip. Of course, the magnetic moment is much lower than in ferro- or superparamagnetic tips, and the signal is very low [32].

Although magnetically coated AFM tips are widely used in MFM, they are a bad approximation to the ideal tip shape. In the following, we discuss a new method to realize batch-fabricated tips, in a method optimized for MFM.

12.5.4 Tip Planes: The CantiClever Concept

A magnetic tip that is suitable for high resolution MFM should have lateral dimensions in the nanometer regime. One would like these dimensions to be controllable and variable to be able to customize the MFM tip for different types of measurements or samples. The CantiClever design [4] discussed in this section accomplishes this by defining both lateral dimensions by thin film deposition techniques. The magnetic tip is made by deposition of magnetic material on the side of a free-hanging, very thin layer called the tip plane. The width and thickness of the magnetic tip are defined by the thickness of the tip plane and the magnetic layer, respectively. The length of the tip is defined using photolithography. A schematic drawing of the structure is shown in Fig. 12.17.

Such a structure is very difficult to make, however, when the approach used to make conventional cantilevers is used. During fabrication, the conventional cantilevers are situated such that the oscillation direction is perpendicular to the surface of the substrate. Using this approach would need the very thin tip plane to be fabricated as a freestanding layer perpendicular to the substrate surface. Instead, the free-hanging tip plane is made in a completely new approach. During fabrication, the cantilevers are tilted 90° compared to the conventional cantilevers, creating a lateral oscillating cantilever with its oscillation direction parallel to the substrate surface, as shown in Fig. 12.18.

This approach makes the fabrication of the cantilever more difficult compared to conventional cantilevers, but also enables precise control over the cantilever resonance

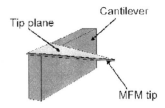

Fig. 12.17. The cantilever with the tip plane

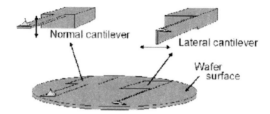

Fig. 12.18. Tip plane on the cantilever for both cantilever orientations. Left: conventional approach with horizontally vibrating cantilever. Right: new approach with laterally vibrating cantilever

frequency. The resonance frequency is given by [15]:

$$f_0 = \frac{1}{2\pi} C_0^2 \frac{t}{l^2} \sqrt{\frac{E_{Si}}{12\varrho_{Si}}}, \qquad (12.42)$$

where C_0 is the eigenvalue of the system corresponding to its fundamental frequency, having a value of 1.875, t the thickness of the cantilever, l its length, E_{Si} the Young modules of the silicon (150 GPa) and ϱ_{Si} its density (2330 kgm^{-3}).

In contrast to the fabrication method used for conventional cantilevers, this approach allows both dimensions of the cantilever, which determine the resonance frequency, the cantilever thickness t, and the length l, to be controlled and varied, as both are defined parallel to the substrate surface. A single substrate can carry a large number of cantilevers with different resonance frequencies suitable for different applications. Furthermore, standard deposition and etching techniques can be used to define the tip plane, as it is also oriented parallel to the substrate surface. The result is a reproducible manufacturing process that incorporates both the cantilever and the magnetic tip and allows for batch fabrication of the probes.

Two factors play an important role in making a high resolution MFM tip that resembles the ideal tip shape, as described above, as close as possible. First, the freestanding tip plane should be very thin, as this defines the width of the tip, but strong enough for contact imaging and at the same time should have low stress to prevent bending. Second, the tip plane should have a well-defined and very sharp cut-off corner, as illustrated in Fig. 12.19. This ensures that the tip end is flat. How such a sharp cut-off corner is realized is described in the next section.

The cantilevers are fabricated from (110) silicon wafers using KOH wet anisotropic etching. The (110)-oriented silicon wafers have a $(-1, 1, 1)$ and a $(1, -1, 1)$ plane perpendicular to the wafer surface. The etch rate of the family of (111) planes in KOH solution is so low that these planes can be considered etch stop planes. Precise alignment of the cantilevers to these planes will, therefore, enable lateral cantilevers to be made with a thickness that only depends on the dimensions of the mask because almost no underetching will occur. The crystal plane etch stop also ensures very smooth surfaces, suitable for interferometric deflection detection. The mask material used during KOH etching are thin silicon nitride layers, deposited in a low stress LPCVD process. The SiN layers are not damaged by the KOH etching process. This

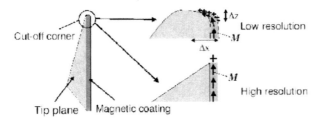

Fig. 12.19. Influence on resolution of the cut-off corner of the tip plane

property and the low stress present in the layers make them also perfectly suitable as a material for fabricating the tip planes. The SiN masking layer used during KOH etching is patterned to form the tip planes. The sharp cut-off corner of the tip plane that is required is obtained by etching away the blunt tip end at the end of the fabrication process. As a final step, a thin magnetic coating is evaporated on the front side of the tip plane. Precise alignment is important to reduce the amount of magnetic material deposited on the sides of the tip plane.

An SEM photograph of a CantiClever is shown in Fig. 12.20. The free-hanging tip plane and the smooth sides of the cantilever itself can clearly be distinguished.

The probe depicted in Fig. 12.20 features a 50-nm-thick tip plane. The inset shows a high magnification image of the end of the tip plane. The cut-off corner is indeed very sharp, with a radius of around 50 nm. An MFM scan made using a CantiClever probe on a CAMST reference sample is depicted in Fig. 12.21.

For this measurement a CantiClever with a 60 kHz resonance frequency, a 50-nm-thick tip plane coated with 50 nm of cobalt was mounted in a commercial MFM from Digital Instruments. On this sample, magnetic features as small as 25 nm could

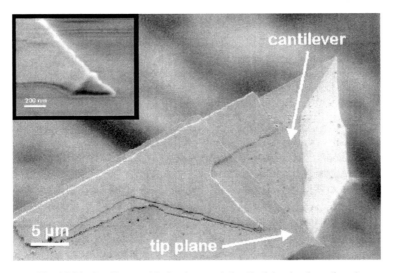

Fig. 12.20. Cantilever with tip plane and detail of the tip plane (inset)

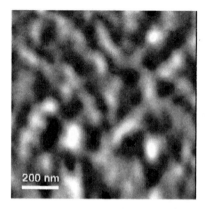

Fig. 12.21. MFM measurement on a CAMST reference sample

be distinguished. The resolution of the tips made using the CantiClever concept can be improved by decreasing the thickness of the supporting tip plane and the magnetic coating. It is expected that the lateral dimensions of the magnetic tip can be reduced below 10 nm. Finally, although it is beyond the scope of this text, the CantiClever concept described in this section has another advantage. Because of the planar nature of the fabrication process, a number of different sensors, besides an MFM tip, can be integrated fairly easily on the tip plane such as a (giant-) magnetoresistive element, making high-resolution magnetoresistance microscopy possible.

Acknowledgement. The authors wish to thank R. Ramaneti for his literature study on MFM tip coatings and C.S. Doppen for converting the original LaTeX source into Microsoft Word format.

References

1. L. Abelmann, S. Porthun, M. Haast, C. Lodder, A. Moser, M.E. Best, P.J.A. Vanschendel, B. Stiefel, H.J. Hug, G.P. Heydon, A. Farley, S.R. Hoon, T. Pfaffelhuber, R. Proksch, and K. Babcock, J. Magn. Magn. Mater. **190**, 135 (1998)
2. T.R. Albrecht, P. Grütter, D. Horne, and D. Rugar, J. Appl. Phys. **69**, 668 (1991)
3. D.K. Anand, "Introduction to Control Systems", Pergamon Press, Oxford, 2nd edition, 1984.
4. A.G. van den Bos, I.R. Heskamp, M.H. Siekman, L. Abelmann, and J.C. Lodder, IEEE Trans. Magn. **38**, 2441 (2002)
5. P.B. Fischer, M.S. Wei, and S.Y. Chou, J. Vac. Sci. & Technol. B **11**, 2570 (1993)
6. L. Folks, M.E. Best, P.M. Rice, B.D. Terris, D. Weller, and J.N. Chapman, Appl. Phys. Lett. **76**, 909 (2000)
7. A. Garcia-Valenzuela and M. Tabib-Azar, in Proceedings SPIE, Integrated Optics and Microstructures II, Vol. 2291, eds. M. Tabib-Azar, D.L. Polla, and K.-K. Wong, SPIE, 1994.
8. P. Grütter, H.J. Mamin, and D. Rugar, in "Scanning Tunneling Microscopy", Vol. II, eds. H.J. Günterodt and R. Wiesendanger, Springer, Berlin, Heidelberg, New York, 1992.

9. P. Grütter, D. Rugar, and H.J. Mamin, Appl. Phys. Lett. **57**, 1820 (1990)
10. P. Grütter, D. Rugar, H.J. Mamin, G. Castello, C.-J. Lin, I.R. McFadyen, R.M. Valletta, O. Wolter, T. Bauer, and J. Greschner, J. Appl. Phys. **69**, 5883 (1991)
11. P.F. Hopkins, J. Moreland, S.S. Malhotra, and S.H. Liou, J. Appl. Phys. **79**, 6448 (1996)
12. A. Hubert and R. Schäfer: "Magnetic Domains: The Analysis of Magnetic Microstructures", Springer-Verlag, Berlin, Heidelberg, New-York, 1998.
13. H.J. Hug, B. Stiefel, P.J.A. Vanschendel, A. Moser, S. Martin, and H.J. Güntherodt, Rev. Sci. Instrum. **70**, 3625 (1999)
14. Y. Martin and H.K. Wickramasinghe, Appl. Phys. Lett. **50**, 1455 (1987)
15. L. Meirovitch, "Fundamentals of Vibrations", Mc Graw-Hill, Boston, 2001
16. U. Memmert, A.N. Muller, and U. Hartmann, Measurement Science & Technology **11**, 1342 (2000)
17. G.N. Phillips, M.H. Siekman, L. Abelmann, J.C. Lodder, Appl. Phys. Lett. **81**, 865 (2002)
18. S. Porthun, M. Rührig, J.C. Lodder, in "Forces in Scanning Probe Microscopy", NATO ASI Series: Applied Sciences, 1995
19. S. Porthun, "High Resolution Magnetic Force Microscopy: Instrumentation and Application for Recording Media", Ph.D. thesis, University of Twente, Enschede, The Netherlands, 1996
20. S. Porthun, L. Abelmann, and C. Lodder, J. Magn. Magn. Mater. **182**, 238 (1998)
21. S. Porthun, L. Abelmann, S.J.L. Vellekoop, J.C. Lodder, and H.J. Hug, Appl. Phys. A **66**, 1185 (1998)
22. K.R. Ramstöck, Magfem3d, http://www.ramstock.de, 1999
23. P. Rice, S.E. Russek, J. Hoinville, and M.H. Kelley, IEEE Trans. Magn. **33**, 4065 (1997)
24. D. Rugar, H.J. Mamin, P. Guethner, S.E. Lambert, J.E. Stern, I. McFadyen, and T. Yogi, J. Appl. Phys. **68**, 1169 (1990)
25. D. Rugar, H.J. Mamin, R. Erlandsson, J.E. Stern, and B.D. Terris, Rev. Sci. Instrum. **59**, 2337 (1988)
26. M. Rührig, S. Porthun, and J.C. Lodder, Rev. Sci. Instrum. **65**, 3224 (1994)
27. M. Rührig, S. Porthun, J.C. Lodder, S. McVitie, L.J. Heyderman, A.B. Johnston, and J.N. Chapman, J. Appl. Phys. **79**, 2913 (1996)
28. C. Schönenberger, S.F. Alvarado, S.E. Lambert, and I.L. Saunders, J. Appl. Phys. **67**, 7278 (1990)
29. G.D. Skidmore and E.D. Dahlberg, Appl. Phys. Lett. **71**, 3293 (1997)
30. A.J. Stevenson, M.B. Gray, H.-A. Bachor, and D.E. McClelland, Appl. Opt. **32**, 3481 (1993)
31. K. Sueoka, K. Okuda, N. Matsubara, and J. Sai, J. Vac. Sci. & Technol. B **9**, 1313 (1991)
32. O. Teschke, Appl. Phys. Lett. **79**, 2773 (2001)
33. B. Vellekoop, E. Abelmann, S. Porthun, and C. Lodder, J. Magn. Magn. Mater. **190**, 148 (1998)

13

Scanning Probe Methods for Magnetic Imaging

U. Hartmann

Previous chapters give an introduction to novel magnetic imaging methods based on the scanning tunneling microscope (STM) or on the scanning force microscope (SFM). While the STM is sensitive to the surface density of electronic states and to its spin dependence, the magnetic force microscope (MFM), as a special variant of the SFM, detects the magnetic stray field produced by a sample, or the response of the sample to the local stray field produced by the probe. The basic setup in tunneling or force microscopy establishes a unified experimental approach as the basis of all scanning probe methods (SPM). The present chapter summarizes the basic aspects underlying this approach, analyzes achievements as well as limitations, and introduces three additional SPM.

Microscopic imaging requires suitable interactions between a probe and a sample that allow one to map a physical quantity – in the present context, a magnetic property – at a certain lateral resolution. In the case of tomographic methods, a certain depth resolution is also required. Most frequently employed probe-sample interactions in magnetic imaging involve electron exchange and the analysis of the spin polarization, the reflection and transmission of light in terms of the Kerr and Faraday effects, and magneto-static field effects, as, e.g., being used in the Bitter colloid method. All these interactions, used as the basis of classical methods of magnetic imaging, can also be implemented in scanning probe strategies. The electron spin is detected by the spin-polarized STM (SPSTM). Kerr and Faraday effects are utilized in the magneto-optic scanning near-field optical microscope (MOSNOM). Near-surface stray fields produced by ferromagnetic or superconducting samples can be analyzed using scanning SQUID (superconducting quantum interference device) microscopes (SSM) or microscopes based on other field sensors like Hall probes or magneto-resistive probes. Together with a discussion of some general aspects concerning probe-sample magnetic interactions, the MOSNOM and the SSM will be introduced in the following.

The high-resolution detection of spin resonances for imaging purposes has been the subject of considerable effort for quite some time. Also in this area classical approaches can be adapted to scanning probe strategies in order to analyze nuclear magnetic resonance (NMR), electron spin resonance (ESR), or ferromagnetic resonance (FMR) at a sub-micron scale. Some general information on the respective approaches of magnetic resonance force microscopy (MRFM) is provided as well by the following discussion.

13.1 General Strategies in Scanning Probe Microscopy

It is certainly not necessary to emphasize the enormous importance of microscopy in the natural sciences, in medicine, and in various engineering disciplines. In the past decades, this importance has been recognized repeatedly by the awarding of Nobel prizes to the inventors of a number of new and improved microscopy techniques. Today, a strong driving force for further developments results from the increasing demands related to new technologies. One of the key technologies is certainly magneto-electronics [1], where, as a consequence of the decreasing scale of many devices, high-resolution magnetic characterization methods have become fundamentally important for further developments in this area. A discipline in wich progress is directly related to the availability of powerful microscopy methods is the development of new and functional materials in general. The latter strongly relies on the characterization of materials at differing levels of resolution. Surface finish or texture, microstructure and defect geometry, as well as chemical composition and spatial distribution are important parameters determining the behavior of materials and practical applications in general, and in the case of magnetic materials in particular.

In order to qualify a certain approach as *microscopy*, the method should give spatially localized information on the microstructure and should have the potential to provide a magnified real-space image of the sample [2]. Today's materials scientists have a large number of such methods at their disposal. This is necessary because complete characterization of new materials requires the application of many different and complementary characterization methods used in combination. That again holds in particular for the characterization of magnetic materials and properties [3].

In 1981, G. Binnig, H. Rohrer, and coworkers invented the STM [4]. This instrument, which proved capable of imaging solid surfaces with atomic resolution, has revolutionized microscopy and surface analysis in an unprecedented way over the past 20 years [1, 5–9]. When looking back, it is evident that the outstanding success of the STM is not only due to the ultra-high resolution that can be achieved with this instrument. Perhaps more significantly, the experimental approach that the STM is based on has stimulated the development of a whole family of SPM. Every member of this family is experimentally set up very similarly to the STM. From this point of view, the most popular STM derivatives are the SFM [10] and the SNOM [11]. STM, SFM, and SNOM today represent a set of instruments that can be applied in many different and highly dedicated modes of operation, so that a variety of physical and chemical properties of a material can be analyzed. Apart from the inherent high resolution of the SPM, their versatility is a major strength.

Two important aspects are essential to all the SPM: scanning the probe and operating it in close proximity to the surface of the sample. While scanning is well established in microscopy techniques such as electron microscopy, operating the

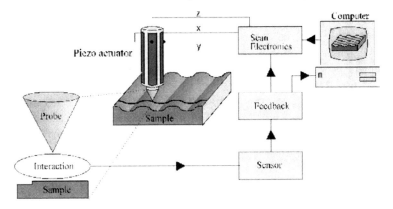

Fig. 13.1. Setup of a scanning probe microscope

probe in such a close proximity to the sample surface that it is within the *near-field regime* is certainly highly specific to the SPM discussed in the present context.

Furthermore, it is important to consider the SPM not simply as methods whereby a local probe precisely maps the topography or morphology of a surface with a high resolution. SPM should rather be considered as methods by wich the probe is used to carry out local experiments at the sample surface while raster-scanning it. The experimental data is collected, and an image of the surface reconstructed as a function of the probe's particular interaction with it. Thus, the image of the scanned surface is created using the experimental parameters chosen for the sequentially performed experiments. Consequently, different operational parameters should result in completely different images of a given sample surface.

If a sharp tip that terminates the end of an extended solid probe is in close proximity to a sample surface, as shown in Fig. 13.1, a variety of interactions can result. The particular interaction chosen for imaging is detected by an adequate probe-sensor combination. The additional components shown in the schematic diagram of Fig. 13.1 then allow one to operate the SPM in different modes such as the constant-interaction mode, the constant-height mode, or various spectroscopic modes. Which of those interactions being a priori suitable for analyzing physical properties of a sample could ultimately really be used in an SPM approach depends on the availability of an appropriate probe-sensor combination to be implemented according to Fig. 13.1. In this context, it is thus most instructive to first review relevant magnetically sensitive probe-sample interactions.

13.2 Probe-Sample Interactions Suitable for Magnetic Imaging

When discussing the applicability of SPM to high-resolution magnetic imaging, it is most important to look, on the one hand, which of the probe-sample interactions could potentially provide information on magnetic properties of a sample and which can practically be realized in any SPM approach, on the other hand. Furthermore, in the present context, the discussion should mainly be focused on ferromagnetic samples. However, as a matter of fact, some of the SPM discussed in the following turn out to be particularly useful also in the investigation of superconducting samples or spin resonance phenomena.

In magnetic imaging, the most important physical properties are the locally varying magnetization vector field, the surface density of occupied and unoccupied electronic states near the Fermi level – including its spin dependence and certain particular surface and interface states – electronic core levels, and the locally varying near-surface magnetic field produced by inner and surface magnetic charges. All magnetic imaging methods treated in this book and those not explicitly discussed here are based on the interaction of particles, waves, or a suitable external probe with these quantities. The latter are a fingerprint of the characteristic magnetic properties of the particular sample: Photoemission with the analysis of the electron spin polarization provides us with direct information on the binding energy, wave vector, and spin polarization of the electronic states in a ferromagnetic system [12]. Moreover, introducing the control of the light polarization, one observes in photoemission from the valence band and core levels magnetic dichroism effects that reflect the orientation of the magnetic moments and the symmetry of the electronic states [13]. Linear and nonlinear magneto-optics allow direct insight into the spatiotemporal evolution of dynamic processes in terms of the domain topology and its changes under the influence of an external magnetic field [1,3]. In electron microscopies, the interaction of the electrons with the near-surface stray field or the magnetization of the sample is utilized and the electrons leaving the sample are analyzed with respect to their direction, number, and spin polarization [14–18]. Neutrons can also serve as particle probes [19]. Because of their spin, neutrons do not only interact with atomic nuclei but also exhibit a magnetic scattering due to an interaction with the magnetic moments of the electron shell. The latter contribution can be used to determine magnetic configurations and fluctuations. It is, however, obvious that the practicability of neutron scattering as a tool for spatially resolved magnetic imaging is limited. The SPSTM [20] and, in particular, spin-polarized scanning tunneling spectroscopy (SPSTS) provide direct access to the spin-dependent local density of occupied and unoccupied surface electronic states. Generally, the relationship between a certain domain and micromagnetic configuration in a sample, on the one hand, and the outer stray field produced by this configuration, on the other hand, is complicated. It is a priori not possible to determine the three-dimensional magnetization vector field from the three-dimensional stray-field variation. However, in many special cases the local stray field is sufficiently closely related to its source, e.g., a domain wall or a structural defect, so that the overall magnetic configuration can be deduced from the stray-field pattern. The relationship between stray field and magnetization is the basis of the well-known Bitter colloid technique, which for decades was the most widely used method of determining domain configurations [3]. Today, the MFM, as a special variant of the SFM [21], has largely substituted the Bitter method because it produces fairly high resolution, is largely nondestructive, can be applied under various environmental conditions, and allows, to a certain degree, the obtaining of quantitative data. Even prior to the invention of the MFM, people aimed to map the external stray field produced by ferromagnetic samples by raster-scanning field-sensitive probes, like Hall sensors or inductive loops, across the sample surface [22].

The SPM treated so far in this book, namely, the SPSTM and the MFM, have turned out to be powerful tools in the investigation of magnetic materials. The SPSTM provides information about magnetic properties at the atomic scale, and the MFM is a robust working horse providing fairly high resolution in magnetic imaging at minimum experimental effort. Due to the employed probe/detector configurations and appropriate modes of operation, the instruments give us access to the spin-dependent electronic density of a sample and to its external stray field configuration by using appropriate probes, as shown in Fig. 13.2. It is also possible to measure magneto-optical properties using a dedicated SPM. The utilization of the interaction between light and a ferromagnetic sample in an SNOM is one further example of the possibility of adapting classical magnetic imaging methods, in the present case Kerr and Faraday microscopy, to the SPM. Such a strategy has also been followed in the invention of new approaches to detecting the stray field of samples by means of SPM at high resolution, but more quantitatively than possible with the MFM. The result is recent success in scanning Hall probes or SQUID at fairly high resolution across sample surfaces. Since the SQUID is the most sensitive magnetic field sensor that we have at our disposal, the SSM in which high field sensitivity is combined with high lateral resolution potentially has a broad area of application.

If NMR, ESR, or FMR could be performed at sub-micron resolution, a manifold of applications in spatially resolved chemical analysis, as well as particular applications in the investigation of ferromagnetic materials would result. The magnetic resonance force microscope (MRFM) is based on the SFM and utilizes a cantilever probe, as

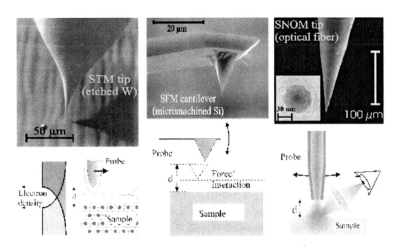

Fig. 13.2. Probes used in scanning probe methods. The left probe is used for tunneling microscopy. The one shown in the middle is for force microscopy. The probe in the right part is suitable for near-field optical imaging

shown in Fig. 13.2, to mechanically detect precession signals of spins excited by an external field.

All the SPM have in common that they are based on the same unified experimental setup. The utilization of suitable probe-sample interactions detected by appropriate probe-sensor combinations allows us to perform classical magnetic analysis with the unprecedented spatial resolution inherent to SPM. An important aspect with respect to the latter is that magnetic imaging is performed in the *near-field regime*. This is the essential advantage that makes the SPM-based approach often superior to classical ones. The near-field regime allows the STM to achieve atomic resolution independent of the probe dimensions. It allows the SNOM to break the optical resolution limit, and it allows stray-field-sensitive methods to detect individual magnetic charges instead of sole dipole contributions. In the following, the MOSNOM, SSM, and MRFM are introduced as instruments that appear somewhat more exotic than the microscopes treated in Chaps. 9–12, but that considerably extend our possibilities for high-resolution magnetic analysis.

13.3 Scanning Near-Field Magneto-optic Microscopy

Over the past decades, magneto-optic microscopy has contributed much to our understanding of magnetic structures in numerous materials and devices [3]. The inherent advantage of Kerr and Faraday microscopies is that we can obtain quantitative information on the three-dimensional magnetization vector field at the surface of a sample. Furthermore, magneto-optics is capable of providing direct access to the dynamics of magnetization reversal processes. The classical way of applying magneto-optics for microscopy purposes is to use the linear effects. However, in recent years it has been shown that nonlinear magneto-optics is extremely interesting for some special applications due to its ultrahigh sensitivity to surface and interface properties [1].

Conventional magneto-optical microscopy is obviously diffraction-limited and thus hardly produces a lateral resolution below a few hundreds of nanometers. With SNOM, it is possible to break the optical diffraction limit using a tapered optical fiber tip coated by aluminum, as shown in Fig. 13.2. If the aperture hole at the very apex of the tip is small compared with the optical wavelength, and if the probe is operated in sufficiently close proximity to the sample surface, according to Fig. 13.1, it becomes possible to image the sample surface utilizing the near field of the optical fiber probe. In this way, sub-wavelength resolution can be obtained in a straightforward way. Already quite some time ago, it was shown that the SNOM is suitable to produce magneto-optic contrast [23]. The crucial aspect is the degree of linear or circular polarization being achieved with a fiber probe like that shown in Fig. 13.2. Numerous setups have been presented in the literature [7]. A concrete experimental solution showing the major components of an MOSNOM setup is shown in Fig. 13.3. A laser serves as light source. The linear or circular polarization is adjusted using a $\lambda/2$ plate, together with a compensator. The light reflected from or transmitted by the sample is collected by suitable objectives, guided through an analyzer, and detected by a photomultiplier. The probe-sample distance is controlled by a shear-

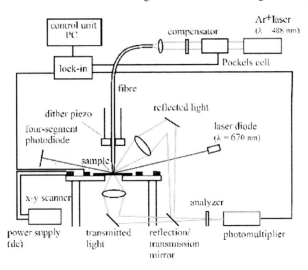

Fig. 13.3. Setup of a scanning near-field optical microscope for magnetic imaging

force sensor. The additional components required for scanning and data acquisition again correspond to those unified SPM components already discussed in the context of Fig. 13.1. Depending on the geometrical properties, i.e., on the sharpness of the fiber tip, and on the properties of the aluminum coating, the optical resolution can be better than 100 nm. Faraday as well as Kerr microscopy are possible. However, for observing the Kerr effects, it is not straightforward to adjust the fiber tip in a way that the incident and reflected light fulfill the geometrical requirements determined by

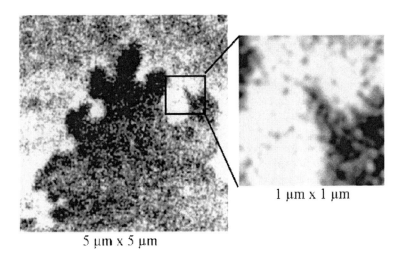

Fig. 13.4. Faraday near-field image of a Co/Pt multiplayer. The scanned areas are 5 μm × 5 μm for the left and 1 μm × 1 μm for the right image, respectively

fundamental magneto-optic aspects. Figure 13.4 shows a Faraday MOSNOM image of a Co/Pt multilayer. The total thickness of the exchange-coupled system amounts to 11 nm. The Faraday image shows the characteristic magnetic domains resulting from its perpendicular anisotropy. A detailed analysis of the magnified area shows that a resolution of about 100 nm has been obtained in this image.

With respect to its applicability, the MOSNOM has to be considered to be somehow situated in an intermediate position between the SPSTM and the MFM. On the one hand, the magnetization as a well-defined characteristic quantity is directly detected, at least under clean conditions. This makes the MOSNOM sensitive to environmental conditions, and suitable sample preparation is required in general, as with the STM. On the other hand, sufficiently inert samples can be imaged in air, and the probe-sample distance is large enough to obtain sufficiently large scanning areas, as with the MFM. The main drawback with respect to a broad application of the MOSNOM is that it is so difficult, if not impossible, to obtain adequate imaging conditions for utilizing the Kerr effects.

13.4 Scanning SQUID Microscopy

13.4.1 SQUID Basics

As already mentioned, in many applications of magnetic imaging it is of major interest to map the near-surface external magnetic field produced by a sample as sensitively and precisely as possible [3]. The phase space for such an imaging is multidimensional and includes sensitivity, spatial resolution, frequency response, linearity, and stability. Additional factors include the required proximity of the detector to the source, the detection of fields versus gradients, the need to operate in an externally applied field, the ability to reject external noise or the applied field, the ability to make measurements without perturbing the sample, and the required operating temperature of both the sample and the sensor. Among those sensors which offer the potential to quantitatively detect near-surface magnetic fields at fairly high resolutions, Hall bars, micron-sized magneto-resistors, and SQUIDs are most important. Micron-sized Hall bars have a flux sensitivity comparable to that of SQUID [24] and do not have to be operated under cryogenic conditions. However, the field sensitivity of SQUID increases linearly with the pick-up area, whereas the field sensitivity of Hall bars is nearly independent of area. Therefore, SQUID rapidly become more sensitive than Hall bars with increasing pick-up area. Furthermore, Hall bars are much more sensitive to pressure and charging effects than are SQUID. Micron-sized magneto-resistors also have flux sensitivity comparable to SQUID. However, they are typically operated with a dc current of several mA, which means that the magnetic field and the heat they generate can perturb the sample, especially for superconducting applications. There are thus good reasons to consider the SSM, in spite of its limited practicability, as a powerful tool that considerably extends the various possibilities of magnetic imaging.

When two superconductors are separated by a sufficiently thin insulating barrier, it is possible for a current to flow from one to the other, even with no voltage applied between them. In 1962, Josephson [25] predicted the existence of this supercurrent and calculated that it should be proportional to the sign of the quantum-mechanical phase difference between the two superconductors. This dependence of the supercurrent on the relative phase difference, when combined with an intimate relationship between this phase and the magnetic field, is the key to a class of sensitive magnetic field detectors called SQUID [26]. The type of SQUID used most commonly for microscopy, the dc SQUID, is composed of a superconducting ring interrupted by two Josephson junctions. The maximum zero-voltage current, i.e., the critical current, of the dc SQUID varies sinusoidally with the integral of the magnetic field through the area of the SQUID loop due to interference of the quantum phases of the two junctions. The period of modulation is the superconducting flux quantum $\Phi_0 = h/2e$. With flux modulation, phase-sensitive detection, and flux feedback, this phase can be measured to a few ppm at a one-second time integration. This makes the SQUID the most sensitive sensor of magnetic fields known [26]. The sensitivity of the SQUID magnetometer can be specified in terms of the flux noise divided by the effective pick-up area. For a typical flux-noise value of $10^{-6}\Phi_0/\sqrt{Hz}$, a pick-up area of $(10\,\mu m)^2$ corresponds to an effective field noise of $\sim 20\,pT/\sqrt{Hz}$. The practical limit of sensitivity obtained for larger pick-up areas is on the order of a few $fT/\sqrt{Hz}$ [27].

Since the SQUID detects a magnetic flux, the sensitive detection of magnetic fields requires a maximum active area penetrated by that flux. An increasing pick-up area, however, also increases the inductivity of the SQUID, which, in turn, increases the flux noise. If the SQUID is coupled to a pick-up loop, as shown in Fig. 13.5, it becomes possible to keep the SQUID loop itself small while benefiting from a large active area. An external magnetic field induces a shielding current in the pick-up loop. The corresponding flux is to a certain amount inductively coupled to the SQUID. Apart from an increase in sensitivity, the introduction of a pick-up loop allows one to remove the SQUID from the source of the magnetic field to be probed. This is of special importance for SSM setups.

Apart from dc SQUID, rf SQUID could and have been used for microscopy applications. However, due to the more complicated setup of this type of SQUID, dc setups have largely been preferred. The employment of high-temperature superconductors (HTSC) considerably reduces the cryogenic inconvenience. The first HTSC dc SQUID were already presented in 1987 [28], i.e., shortly after the discovery of

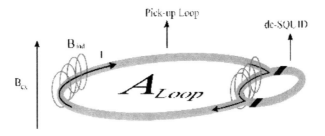

Fig. 13.5. Pick-up loop as coupled to the SQUID

the oxidic HTSC materials [29]. For low temperature, as well as for HTSC SQUID the availability of multilayer technology is of considerable benefit and has much contributed to the increase of the sensitivity specified above.

13.4.2 Conventional Approaches

In general, an SSM is any scanning SQUID setup with an active area of several millimeters or less that is used to acquire images of a magnetic field. Much of the experience with SSM gained at the millimeter scale, e.g., sample scanning, optimization of spatial resolution, and image deconvolution, is directly applicable to SSM with micron-sized critical dimensions. There are numerous SSM schemes [30] that can be conceptually divided into two classes. In the first, both sample and SQUID, are held at low temperatures, either in a common vacuum space or in a cryogenic fluid. This allows for the minimum spacing between SQUID and sample and, therefore, the best spatial resolution. If the SQUID and sample are in the same vacuum, it is possible to operate them at different temperatures, e.g., the SQUID can be at 4.2 K, while the sample is maintained at 100 K. In the second class, the SQUID is cold, but the sample is warm and can be exposed to atmosphere. In this case, the SQUID and sample are separated by some form of thermal insulation. This has the advantage of not requiring sample cooling, but entails some sacrifice in spatial resolution. In either case, a bare SQUID, a SQUID inductively coupled with a superconducting loop, or a superconducting pick-up loop integrated into the SQUID design, can be used to detect the sample magnetic field.

SSM have been used to image the magnetic fields from a wide variety of sources, ranging from the development of currents generated by the embryo in a chicken egg to the persistent currents associated with quantized flux in superconductors [30]. The majority of the effort appears to have been directed toward the development of the instruments and the demonstration of the technique, particularly for biomagnetism, corrosion science, and nondestructive evaluation.

Quantitative analyses of magnetic imaging have shown that a SQUID will have the best combination of spatial resolution and field sensitivity if the pick-up-loop diameter is approximately the same as the spacing between the loop and the sample. Similar arguments will apply to other point-detection means of measuring the magnetic field. Thus, a high-resolution instrument requires a small pick-up area to be placed close to the sample. Because SQUID microscopes usually have only a single SQUID, the separation between the scan lines must equal the desired spatial resolution, which is comparable to the loop diameter. It is important to recognize that the separation of features in a magnetic image can be enhanced with image processing that utilizes a-priori knowledge of the geometry and thus the transfer function between the source of the magnetic field and the SQUID. In this context, imaging algorithms can be viewed as inverse spatial filters and hence subject to noise-induced instability due to the amplification of high-spatial frequencies that had been preferentially attenuated with distance from the source [31].

Many SSM have been built with active pick-up areas from 4 μm to 2–3 mm. In 1983, Rogers [32] presented a setup consisting of an rf SQUID inductively coupled to a pick-up loop of 230 μm diameter that was raster-scanned across a sample surface using two stepper motor stages. The distance between the pick-up loop and the probe amounted to 50 μm, the operating temperature to 4.2 K. With this setup, a spatial resolution of about 500 μm was obtained. Also on the basis of an rf SQUID, Minama et al. [33] achieved a lateral resolution of 100 μm at a flux sensitivity of $0.1\Phi_0$. A spatial resolution of 220 μm at a field sensitivity of $18\,\text{pT}/\sqrt{\text{Hz}}$ and a bandwidth of 4 kHz were demonstrated in 1992 by Mathai et al. [34] on the basis of a low-temperature dc washer SQUID. The "washer" serves for flux focusing and consists of a superconducting plate with a tiny hole in the center. This hole is situated above the SQUID, so that the collected flux is focused directly into the active area. On the basis of this setup, a lateral resolution below 70 μm and a field sensitivity of $5\,\text{pT}/\sqrt{\text{Hz}}$ at a bandwidth of 6 kHz were obtained in 1993 [35]. In order to adjust and keep the probe-sample spacing in a convenient way, some recent approaches have been based on pick-up loops deposited on the back side of cantilevers, such as shown in Fig. 13.2 [36, 37]. The tip situated at the front of the cantilever is raster-scanned in mechanical contact across the sample surface, keeping the probe-sample distance constant. The apex of the pick-up loop can be separated from the SQUID by more than 1 mm. A lateral resolution of 10 μm has been obtained at a field noise of $40\,\text{pT}/\sqrt{\text{Hz}}$ [37]. The scheme of the setup is shown in Fig. 13.6. The SSM can also be employed on samples that do not produce remanent magnetic fields. For this purpose, eddy currents are excited in a conducting sample at an adequate frequency. A washer SQUID is then used to detect the locally varying resulting ac fields, where a resolution of 60 μm has been obtained at a sensitivity of $7\,\text{pT}/\sqrt{\text{Hz}}$ [38]. Using a 3-μm-thin thermally insulating Si_3N_4 window, separating a HTSC SQUID kept at 77 K in vacuum from ambient conditions, a resolution of about 25 μm has been obtained on biological samples [39]. A completely new approach resulting in an SSM design that is in complete accordance with the aforementioned unified SPM design criteria [40] is introduced in the following.

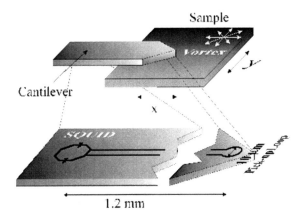

Fig. 13.6. Pick-up loop with SQUID on the back side of a cantilever

13.4.3 Flux-Guided Scanning SQUID Microscopy

The basic setup of the new type of SSM combining aspects of SPM with those of conventional SQUID microscopy is shown in Fig. 13.7. The SQUID is equipped with a 200 μm pick-up loop. Employing a pulsed excimer laser, a tiny bore is introduced right at the center of the pick-up loop. For a 1-mm-thick $SrTiO_3$ substrate, its diameter can be as small as 5 μm. A typical hole, together with the schematics of the SQUID design, are shown in Fig. 13.8. Into the bore an electrochemically prepared $(FeMoCo)_{73}(BSi)_{27}$ tip is inserted. The apex radius of curvature of such a tip can be as small as 10 nm. The soft magnetic tip serves as a flux guide. Local stray-field variations at the sample surface probed by the tip apex are guided into the pick-up loop. The most important parameters determining field sensitivity and spatial resolution of the microscope are thus the effective longitudinal susceptibility and the near-apex sharpness of the probe.

Upon scanning the probe in close proximity across the sample surface, the SQUID is operated in a flux-locked loop. The integrated modulation and compensation coil always keeps the pre-chosen point of operation in the probe's magnetization curve fixed. Thus, hysteretic and nonlinear magnetization reversal of the flux guide do not perturb SQUID operation or magnetic imaging. To produce a stray-field image, variations of the feedback signal are displayed as a function of lateral probe position. Coarse approach between probe and sample, as well as scanning are performed using suitable piezoelectric actuators.

The whole setup shown in Fig. 13.7 is based on multilayer technology. After covering the $SrTiO_3$ substrate with a 25-nm-thick Au layer, a photoresist layer is added for step-edge lithography. The Au thin film avoids diffuse scattering of light that could

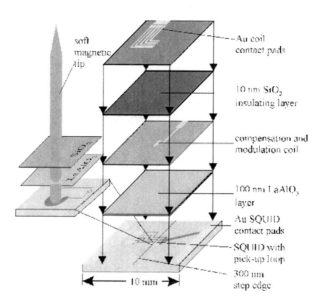

Fig. 13.7. Multilayer structure for the flux-guided detection of magnetic fields by a SQUID

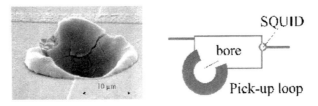

Fig. 13.8. The left image shows a typical hole into which the flux guide is inserted. The right image shows the actual arrangement of pick-up loop and SQUID

expose the resist underneath the mask. A 300-nm-deep step edge is etched by Ar^+ bombardment. For ion beams incident at 45°, rather deep step edges can be expected. A 200-nm-thick YBCO layer is deposited by laser ablation, photolithographically structured, and then Ar^+ etched in the usual way. In order to protect the junctions from oxygen diffusion, the whole arrangement is covered by a 100-nm-thick $LaAlO_3$ protection layer. The SQUID involves an outer diameter of 30 μm and an inner one of 20 μm. The junctions have a width of 4 μm with critical currents in the range of 40 μA. Outer and inner diameters of the pick-up loop are 250 μm and 200 μm, respectively (see Fig. 13.8). Such SQUID typically show a transfer amplitude of up to 36 μV peak-to-peak. Typical bias currents are 25 μA at a SQUID resistance of 2.5 Ω in the normal conducting state.

The amorphous soft magnetic probe has a base diameter of 150 μm. Its saturation magnetization is 0.5 T. Remanence and coercive field values are 2.5 mT and 3×10^{-4} kA/m. The maximum susceptibility amounts to 10^5. The probes are electrochemically etched in 0.25 molar HCl. The step-edge dc SQUID setup and the adjustment of the ferromagnetic flux guide are shown in Fig. 13.9.

The complete setup of the microscope is schematically shown in Fig. 13.10. Magnetic shielding is provided by a mu-metal cylinder. Coarse approach between probe and sample is performed by a piezoelectric motor based on shear-piezo actuators. The motor can move the sample holder over some millimeters with a step precision of about 100 nm. Fine approach as well as lateral scanning are performed by a standard tube scanner. At 77 K the scan range is 15 μm × 15 μm. The SQUID with the integrated probe is fixed at the bottom of a ceramic carrier and electrically contacted by CuBe springs. In order to avoid mechanical vibration the whole carrier is mounted via a soft damping stage to a transfer rod. Standard operation is performed in liquid nitrogen, liquid helium, or cold helium exchange gas. Furthermore, an ultrahigh vacuum variant allowing sample temperatures between 1.5 K and 300 K has been designed.

Peripheral electronics involve a standard dc SQUID setup and for piezo operation and image acquisition standard components, known from SPM. As an option for certain measurements the probe-sample distance can be kept continuously constant by measuring the tunneling current, the stray capacity, or shear forces between probing tip and sample as quantities being independent of the local magnetic stray field. In order to estimate the obtainable spatial resolution, a concentrated ferrofluid was deposited on top of a substrate. Upon drying, the colloid particles of about 10 nm

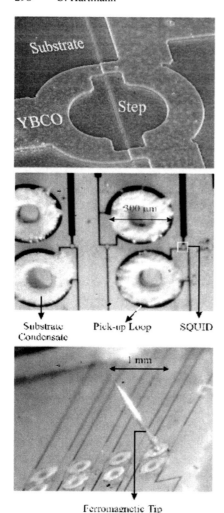

Fig. 13.9. Step-edge dc SQUID, details of the arrangement of various pick-up loop/SQUID configurations on a chip, and ferromagnetic flux guide inserted into the pick-up loop

diameter form aggregates with a range of diameters between 100 nm and several microns. In Fig. 13.11, the SQUID image clearly reveals the aggregates of various sizes. The magnified image in the upper part displays features of less than 100 nm extent. The obtained resolution is thus comparable to that achieved with the MFM [41]. However, as already emphasized, the field sensitivity of the SQUID microscope is orders of magnitude higher than what has ever been obtained by force microscopy. Yet, the actual limit remains to be determined.

The conventional SSM typically permits a spatial resolution on the order of 10 μm. The expected limit in conventional approaches is 1 μm or slightly below [37]. The spatial resolution and the field sensitivity are greatly improved with the flux-guided SSM. The presence of the ferromagnetic material in close proximity to the SQUID does not cause any notable perturbance in SQUID operation. The concept can be

13 Scanning Probe Methods for Magnetic Imaging   299

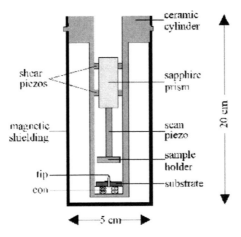

Fig. 13.10. Overall setup of the complete scanning SQUID microscope

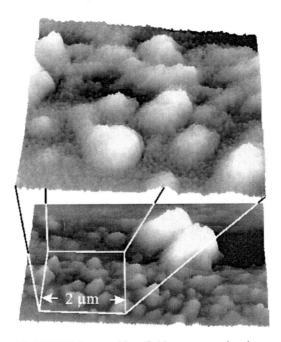

Fig. 13.11. Flux-guided SQUID image of ferrofluid aggregates showing a resolution superior to 100 nm

implemented under liquid helium, helium exchange gas, liquid nitrogen, and under ultrahigh vacuum conditions. In the last case, the sample can be at any arbitrary temperature. A considerable strength of the flux-guide concept is that it involves the possibility of a maximum spatial separation of SQUID – or superconducting pick-up loop – and sample. This indeed makes ambient-air operation straightforward. The flux guide could, e.g., be split into two parts that are magneto-statically coupled across a sufficiently thin window of the cryostat.

A maximum separation of the superconducting parts of the sensor and the sample also offers the advantage that the sample can be subjected to relatively high magnetic fields. The flux-guided SSM should thus be particularly suitable for those applications where tiny stray-field variations are measured at large bias fields. Examples of such an application are the imaging of vortices in superconductors or of one-dimensional current filaments in semiconductors, where the samples are subject to externally applied fields.

Simultaneous detection of a tunneling current or a force between probe and sample is particularly interesting in order to relate the locally detected magnetic field to topographic, morphologic, or electronic properties of a sample. The enormous application potential resulting from such simultaneously performed magnetic and nonmagnetic analysis methods is well-known from the conventional SPM. Due to its capability of achieving high field sensitivity at sub-100 nm spatial resolution, it is obvious that the new SSM approach will be of some value for various fields in solid-state and soft-matter research.

13.5 Magnetic Resonance Force Microscopy

13.5.1 Basic Concepts

Magnetic resonance in terms of NMR and ESR is a widely used and powerful method for molecular-structure analysis. FMR, a less widely used phenomenon, is at least of importance in certain niche applications that are, of course, most relevant in the present context. NMR, ESR, and FMR are thus a priori interesting candidates to be implemented as an SPM according to the strategies mentioned above. The idea of mechanically detecting magnetic resonance was suggested quite some time ago [42] and demonstrated with a macroscopic setup similar to the one used in the detection of the Einstein-de Haas effect. In 1991, the MRFM was proposed as an instrument being closely related to the SFM [43]. It was suggested that it might be possible to detect a single nuclear spin by a three-way resonant coupling between a precessing spin and a vibrating substrate in the inhomogeneous magnetic field of a magnetic particle mounted to a mechanical oscillator. The first experimental implementation of the basic concept was presented in 1992 [44]. Since that experiment, several groups have started to improve methods based on MRFM. The main driving force is to establish SPM that produce a chemical contrast generally not obtained by the methods mentioned so far. An additional motivation is, of course, that if sufficient sensitivity could be insured, MRFM could allow investigations on samples that are too small for conventional inductive magnetic resonance analysis. Most of the effort, therefore, seems to be directed at single spin detection, which would certainly revolutionize the application of magnetic resonance and probably chemical research in general [45–47]. Single-spin detection, if possible at all, will require reduced pressure, sub-K temperatures, and sub-micron cantilevers with very high Q factors.

This certainly restricts the applicability of the method. In the present context, MRFM should be discussed under less demanding experimental conditions, i.e., on the basis of standard cantilevers and elevated temperatures. It is conceptually important to also include in the present discussion mechanically detected magnetic resonance experiments, where microscopy is not the major goal or where NMR or ESR are the underlying phenomena. On the other hand, FMR and aspects of microscopy are certainly the most important subjects here.

The basic principle of mechanical detection of magnetic resonance is very simple and is shown in Fig. 13.12. Either the sample-on-cantilever or the magnet-on-cantilever configuration is possible. The magnetic dipole moment μ of the sample results from nuclear or electron spins. Diamagnetic contributions are neglected for the moment. Let us first consider the sample-on-cantilever configuration. The magnetic field exerts a force

$$F = \nabla(\mu \cdot B) \tag{13.1}$$

and the torque

$$\tau = \mu \times B. \tag{13.2}$$

The cantilever excitation force is thus proportional to the magnetic field gradient at the site of the sample. In a homogeneous magnetic field there will be no resultant force. Only a torque on the spins will occur, which can also be used for mechanical excitation of the cantilever [48]. The force detection will be performed at maximum sensitivity if the force from Eq. 13.1 is oscillating with a frequency component equal to the mechanical resonant frequency of the cantilever-sample ensemble. While, in principle, both μ and B can be made time-dependent, only a time dependence of the sample magnetization can be utilized to obtain some information on the spins.

An additional rf magnetic excitation causes an adjustable precession of the spins around the B axis. In the original MRFM proposal, the spin larmor precession was supposed to be the source of the fluctuating magnetic moment. However, several methods

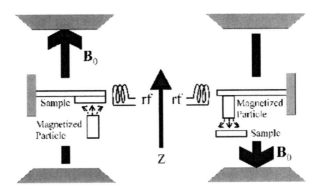

Fig. 13.12. Sample-on-cantilever and magnet-on-cantilever configurations for the mechanical detection of magnetic resonance

to create a time-dependent spin magnetization are available. They can be subdivided into three categories: pulsed rf irradiation, continuous frequency-modulated rf irradiation, and modulation techniques involving the polarizing magnetic field or other parameters. In the present context, there is no principal difference between the two setups shown in Fig. 13.12. The magnet-on-cantilever setup is, of course, much more versatile with respect to imaging procedures and other applications.

An estimation of the obtainable sensitivity provides a very useful insight into the critical parameters of the MRFM [49]. The signal-to-noise ratio SNR is given by

$$SNR \propto \sqrt{\frac{\omega Q}{k_m T}}, \qquad (13.3)$$

where Q is the quality factor of the cantilever, T is the temperature of sample and cantilever, and k_m is a generalized "magnetic spring constant" given by

$$k_m = \frac{k}{\partial B_z/\partial Z}, \qquad (13.4)$$

where k is the mechanical spring constant of the cantilever that oscillates along the z-axis. For a bending bar of length l with a square cross section of width d, and an elastic modulus of Y, the mechanical spring constant is given by

$$k \propto Y \frac{d^4}{l^3}. \qquad (13.5)$$

The above relations show that scaling has a big effect on the mechanical detection scheme. By choosing small, mechanically soft cantilevers and high field gradients, the sensitivity can be considerably increased. The scaling laws in particular show that with smaller and smaller dimensions the sensitivity of the mechanically detected magnetic resonance experiment becomes far superior to that of the inductively detected ones.

13.5.2 Global Experiments

In the very first experimental realization [44], an ESR signal from a DPPH sample of about 30 ng was detected in a magnetic field on the order of 10 mT at room temperature in a sample-on-cantilever setup. The mechanical resonant frequency was 8 kHz at a Q factor of 2000. The field gradient produced by a FeNdB particle amounted to 60 T/m, and the minimum sample volume to $(3\,\mu m)^3$. Later approaches succeeded in obtaining complete spectra utilizing various spin-magnetization modulation techniques according to the aforementioned categories [50,51]. The angular-momentum-based detection set-up [48] is an approach that offers particularly interesting possibilities for the integration of an MRFM setup within the architecture of a normal ESR spectrometer.

At first glance, FMR seems to be especially well-suited to MRFM experiments, because, in this case, a very high degree of electron spin polarization exists even at room temperature. The first mechanical FMR detection was reported in 1996 [52].

A small piece of yttrium-iron garnet – this material has a narrow FMR line – was attached to the cantilever. The experiment was performed under ambient conditions and based on a heterodyne modulation technique. The main problem in FRM experiments on materials like iron or cobalt is that the resonance lines are very broad. This reduces the achievable magnetization modulation amplitudes and creates the necessity to work at relatively high frequency. This, in turn, demands relatively complicated resonator setups. Nevertheless, mechanically detected FMR signals have recently been reported for cobalt samples sputter-deposited on the cantilever [53, 54].

Most of the effort in global mechanical detection of magnetic resonance was focused on NMR. The first successful experiment was already reported in 1994 [55]. Further investigations proved the applicability of mechanical detection in NMR on a variety of samples, involving various magnetization modulation methods [56–59].

13.5.3 Mechanically Detected Magnetic Resonance As an Imaging Technique

The first two-dimensional ESR image based on MRFM was reported in 1993 [60]. For projection-reconstruction methods it is necessary to rotate the gradient with respect to the sample or vice versa. Therefore, it is not straightforward to obtain a three-dimensional image of the sample with the MRFM on the basis of projection-reconstruction methods. Usually, slice-selective excitation of the sensitive volume is performed in order to obtain a modulation of the magnetization in the field direction, which has to be maintained for a sufficiently long time in order to excite a mechanical resonance of the cantilever. For some modulation methods, this leads to signals that correspond to a derivative of the actual image structure. With other excitation methods, the obtained image is similar to an image obtained by an inductive NMR setup.

The simplest form of MRFM imaging is a one-dimensional one in a sample-on-cantilever geometry, in wich the gradient-generating magnet is big compared with the sample. In this case, the sensitive slice of the sample is more or less a plane, and the one-dimensional projection of the NMR signal from the resonant planes onto the gradient direction can be acquired. However, when the sample-layer structure is not parallel to the gradient, the resulting image might be quite blurred [61]. Three-dimensional projection reconstruction imaging, on the other hand, would require one to rotate the gradient of the z-component of the magnetic field with respect to the sample or the sample with respect to the gradient. It is difficult to achieve this in a well-defined way over a wide range of gradient directions in a sample-on-cantilever MRFM, as both the direction of the magnetic field gradient and the cantilever motion detected are predetermined by the direction of the polarizing magnetic field (see Fig. 13.12). For a magnet-on-cantilever approach, it may be possible to rotate a sufficiently small sample. For planar samples, rotation of the samples seems to be impossible, but for such planar samples the depth profile can be determined in other ways.

In a magnet-on-cantilever configuration, it is possible to obtain an image of the sample by stepping the sample through the sensitive resonance volume. The signal detected as a function of the sample position is not directly the spin magnetization distribution. The latter is convolved with a point-spread function [62]. In order to reconstruct the actual image, computer-based deconvolution has to be performed [60].

Very promising results have been obtained on magnetic samples in a magnet-on-cantilever setup. FMR imaging of an yttrium-iron garnet plate was performed at a temperature of 5–10 K under vacuum conditions [50]. A very interesting sample-on-cantilever approach was recently presented by Ruskell et al. [64]. A cantilever with a DPPH sample, as a sensitive detector for mapping magnetic field differences along the sample surface, was used. The results suggest that in the future a spatial resolution on the order of 50 nm at room temperature and a magnetic field resolution of 10 μT could be achieved. A big advantage of this method is that it does not suffer from strong bias fields that would heavily affect almost all other field-sensitive imaging methods.

In conclusion, the MRFM will have some interesting niche applications in the analysis of magnetic nanostructures, especially in terms of approaches based on paramagnetic sensors [64]. The main areas of application will certainly be given for NMR-based MRFM, where the main research directions are on the one hand, to approach single spin detection and, on the other hand, high spatial resolution mapping of sample surfaces.

13.6 Conclusions and Outlook

Scanning probes microscopes have become, in more than 20 years since the invention of the STM, widely distributed, powerful tools applied in many areas of science and technology. The most versatile instrument is the SFM, because it requires minimum sample preparations and is thus particularly suited for technological applications. In contrast, the STM and SNOM are much more sensitive to environmental conditions and sample preparation. Thus, these microscopes have predominant applications in basic research where a considerable preparation effort is justified by the nanoscale information on samples that can be obtained under advantageous conditions.

Among the magnetic imaging methods based on SPM, the MFM is a straight offspring of the SFM and has the same versatility. It is thus not surprising that the MFM became a widely used instrument in fundamental research as well as applied technology. The MOSNOM, the SFM, and the MRFM are microscopes that due to their setup or their operation principle are highly sensitive to environmental conditions and sample preparation. They are clearly based on the general working principle of scanning-probe-based instruments for magnetic imaging. However, in contrast to MFM, they have and will have their main field of application in very dedicated areas of basic research. The considerable effort in setting up the instruments and preparing the samples is justified by the obtained results. These areas are, e.g., the detection of ultrafast dynamic phenomena for MOSNOM or magnetic detection involving a fairly small amount of spins for the SSM and MRFM. These latter applications, which are hardly accessible by most conventional magnetic imaging methods, become at the same time more and more important due to increasing requirements defined by the progress in magneto-electronics in general. Magnetic switching processes enter the GHz regime, and miniaturization leads to high relevance of nanometer-scale characteristics. In this context, MOSNOM, SSM, and MRFM have to be mentioned together with the SPSTM as those instruments that will hardly become widely used tools, but rather have a strategic importance in getting inside magnetic phenomena down to the atomic scale.

References

1. U. Hartmann (ed.): *Magnetic Multilayers and Giant Magnetoresistance*, Springer Ser. in Surf. Sci. Vol. **37** (Springer, Berlin Heidelberg 1999)
2. S. Amelincks, D. van Dyck, J. van Landuyt, and G. van Tendeloo (eds.): *Handbook of Microscopy* (VCH, Weinheim 1997)
3. A. Hubert and R. Schäfer: *Magnetic Domains – The Analysis of Magnetic Micrsostructures* (Springer, Berlin Heidelberg 1998)
4. G. Binnig, H. Rohrer, Ch. Gerber, and E. Weibel: Phys. Rev. Lett. **49**, 57 (1982)
5. R. Wiesendanger and H.-J. Güntherodt (eds.): *Scanning Tunneling Microscopy I–III*, Springer Ser. in Surf. Sci. Vols. **20**, **28**, **29** (Springer, Berlin Heidelberg 1992–95)
6. J. Chen: *Introduction to Scanning Tunneling Microscopy* (Oxford Univ. Press, New York 1993)
7. R. Wiesendanger (ed.): *Scanning Probe Microscopy*, Springer Ser. in NanoSci. Techn. (Springer, Berlin Heidelberg 1998)
8. R. Wiesendanger: *Scanning Probe Microscopy and Spectroscopy* (Cambridge Univ. Press, Cambridge 1994)
9. T. Sakurai and Y. Watanabe (eds.): *Advances in Scanning Probe Microscopy*, Spinger Ser. in Adv. Mat. Res. (Springer, Berlin Heidelberg 2000)
10. G. Binnig, C.F. Quate, and Ch. Gerber: Phys. Rev. Lett. **56**, 930 (1986)
11. D. Pohl, W. Denk, and M. Lanz: Appl. Phys. Lett. **44**, 651 (1984)
12. S.D. Kevan (ed.): *Angle-Resolved Photoemission* (Elsevier, Amsterdam 1992)
13. L. Baumgarten, C.M. Schneider, H. Petersen, F. Schäfer, and J. Kirschner: Phys. Rev. Lett. **23**, 492 (1980)
14. E. Fuchs, Naturwiss. **47**, 392 (1960)
15. J.N. Chapman: J. Phys. D **17**, 623 (1984)
16. A. Tonomura, T. Matsuda, R. Suzuki, A. Fukuhara, N. Osakabe, H. Umezaki, J. Endo, K. Shinigawa, Y. Sugita, and H. Fujiwara: Phys. Rev. Lett. **48**, 1443 (1982)
17. M.S. Altmann, H. Pinkvos, J. Hurst, H. Poppa, G. Marx, and E. Bauer: Mat. Res. Soc. Symp. Proc. **232**, 125 (1991)
18. K. Koike and K. Hayakawa: Jpn. J. Appl. Phys. **23**, L 187 (1984)
19. S.W. Lovesey: *Theory of Neutron Scattering from Condensed Matter* (Clarendon Press, Oxford 1984)
20. R. Wiesendanger, H.-J. Güntherodt, G. Güntherodt, R.J. Gambio, and R. Ruf: Phys. Rev. Lett. **57**, 2442 (1986)
21. Y. Martin and H.K. Wickramasinghe: Appl. Phys. Lett. **50**, 1455 (1987)
22. A. Carey and E.D. Isaac: *Magnetic Domains and Techniques for Their Observation* (Academic Press, New York 1966)
23. E. Betzig, J.K. Trautmann, R. Wolfe, E.M. Gyorgy, and L. Finn: Appl. Phys. Lett. **61**, 142 (1992)
24. A. Oral, S.J. Bending, R.G. Humphreys, and M. Heinini: J. Low Temp. Phys. **105**, 1135 (1996)

25. B.D. Josephson: Phys. Lett. **1**, 251 (1962)
26. W.G. Jenks, I.M. Thomas, and J.P. Wikswo Jr. in: *Encyclopedia of Applied Physics*, G.L. Trigg, E.S. Vera, and W. Greulich (eds.) Vol. 19, p. 457 (VCH, New York 1997)
27. H. Koch, R. Cantor, D. Drung, S.N. Erne, K.P. Matthies, M. Peters, T. Reyhauen, H.J. Scheer, and H.D. Hahlbohm: IEEE Trans Mag. **27**, 2793 (1991)
28. R.H. Koch, C.P. Umbach, G.J. Clarke, P. Chandhari, and R.B. Laibowith: Appl. Phys. Lett. **51**, 200 (1978); H. Nakane, Y. Tarntani, T. Nishino, H. Yamada, and U. Kawabe: Jpn. J. Appl. Phys. **26**, L 1925 (1987)
29. J.G. Bednorz and K.A. Müller: Z. Phys. B **64**, 189 (1986)
30. J.R. Kirtley and J.P. Wikswo Jr.: Annu. Rev. Mater. Sci. **29**, 117 (1999)
31. H. Weinstock (ed.): *SQUID Sensors – Fundamentals, Fabrication and Applications* (Uluwer, Dodrecht 1996)
32. F.P. Rogers, Master's Thesis (MIT, Cambridge 1983)
33. H. Minami, Q. Geng, K. Chiara, J. Yuyama, and E. Goto: Cryogenics **32**, 648 (1992)
34. A. Mathai, D. Song, Y. Gun, and F.C. Wellstood: Appl. Phys. Lett. **61**, 598 (1992)
35. A. Mathai, D. Song, Y. Gun, and F.C. Wellstood: IEEE Trans. Appl. Supercond. **3**, 2609 (1993)
36. L.N. Vu, M.S. Wistrom, and D. van Harlingen: Appl. Phys. Lett. **63**, 1693 (1993)
37. J.R. Kirtley, M.B. Ketchen, C.C. Tsuei, K.G. Stawiasz, J.Z. Sun, W.J. Gallagher, U.S. Yu Jalmes, A. Gupta, and S.J. Wind: IBM J. Res. Develop. **39**, 655 (1995)
38. R.C. Black, F.C. Wellstood, E. Dantsler, A.H. Micklich, J.H. Kingston, D.T. Nemeth, and J. Clarke: Appl. Phys. Lett. **64**, 100 (1994)
39. T.S. Lee, Y.R. Chemla, E. Dantsker, and J. Clarke: IEEE Trans. Appl. Supercond. **7**, 3147 (1997)
40. P. Pitzius, V. Dworak, and U. Hartmann in: *Proc. ISEC'97 Conf.*, H. Koch, S. Knappe (eds.) Vol. 3, p. 359 (PTB, Braunschweig, 1997)
41. U. Hartmann: Annu. Rev. Mat. Sci. **29**, 53 (1999)
42. G. Alzetta, E. Arminando, C. Ascoli, and A. Zozzini: Il Nuovo Cimento B **52**, 392 (1967)
43. J.A. Sidles: Appl. Phys. Lett. **58**, 2854 (1991)
44. D. Rugar, S.C. Yannoni, and J.A. Sidles: Nature **360**, 563 (1992)
45. J.A. Sidles, J.L. Garbini, K.J. Burland, D. Rugar, O. Züger, S. Hoen, and S.C. Yannoni: Reviews of Modern Physics **67**, 249 (1995)
46. C.S. Yannoni, O. Züger, D. Rugar, and J.S. Sidles in *Encyclopedia of Magnetic Resonance*, D.M. Grant and R.K. Harris (eds.), p. 2093 (Wiley, Chichester 1996)
47. J.A. Sidles, J.L. Garbini, and G.L. Drobny: Rev. Sci. Instrum. **63**, 3881 (1992)
48. C. Ascoli, P. Bachieri, C. Frediani. L. Leuci, M. Martinelli, G. Alzetta, R.M. Celli, and L. Pardi: Appl. Phys. Lett. **69**, 3920 (1996)
49. J.A. Sidles and R. Rugar: Phys. Rev. Lett. **70**, 3506 (1993)
50. K. Wago, D. Botkin, S.C. Yannoni, and D. Rugar: Appl. Phys. Lett. **72**, 2757 (1998)
51. K. Wago, O. Züger, J. Wegener, R. Kendrick, S.C. Yannoni, and D. Rugar: Rev. Sci. Instrum. **68**, 1823 (1997)
52. Z. Zhang, P.C. Hammel, and P.E. Wigen: Appl. Phys. Lett. **68**, 2005 (1996)
53. J. Suk, P.C. Hammel, M. Midzor, M.L. Roukes, and J.R. Childress: J. Vac. Sci. Technol. B **16**, 2275 (1998)
54. Z. Zhang, P.C. Hammel, M. Midzor, M.L. Roukes, and J.R. Childress: Appl. Phys. Lett. **73**, 1959 (1998)
55. D. Rugar, O. Züger, S. Hoen, C.S. Yannoni, H.M. Vieth, and R.D. Kendrick: Science **264**, 1560 (1994)
56. K. Wago, O. Züger, R. Kendrick, C.S. Yannoni, and D. Rugar: J. Vac. Sci. Technol. B **14**, 1197 (1996)

57. A. Schaff and W.S. Veeman: J. Magn. Reson. **126**, 200 (1997)
58. T.A. Barrett, C.R. Miers, H.A. Sommer, K. Mochizuki, and J.T. Markert: Appl. Phys. Lett. **83**, 6235 (1998)
59. G. Leskowitz, L.A. Masden, and D.P. Weitekamp: Solid State Nuclear Magnetic Resonance **11**, 73 (1998)
60. O. Züger and D. Rugar: Appl. Phys. Lett. **63**, 2496 (1993)
61. A. Schaff and W.S. Veeman: Appl. Phys. Lett. **70**, 2598 (1997)
62. O. Züger, S.T. Hoen, C.S. Yannoni, and D. Rugar: J. Appl. Phys. **79**, 1881 (1996)
63. O. Züger and D. Rugar: J. Appl. Phys. **63**, 611 (1994)
64. T.G. Ruskell, M. Löhndorf, and J. Moreland: J. Appl. Phys. **86**, 664 (1999)

Index

a-Fe$_2$O$_3$ 34
absorption length   32
AC mode   229
AFM   225
AFM coupling   130
AFM domain structure   30
Aharonov-Bohm effect   94
amplitude mode
   MFM   257
analog operation
   SEMPA   145
angle-resolved photoemission   4
antiferromagnetic coupling   15, 18
antiferromagnetic surfaces
   domain structure   215
Au(111)
   overlayer   199

bar type tip
   MFM   259
barium ferrite   98
biquadratic coupling   129
Bitter colloid technique   288
branching domain structure   191
Brillouin function   35
bulk antiferromagnetic domains   46

cantilever   225, 226
charge microscopy
   MFM   240
chemical contrast   300
chromatic aberrations   37
circular polarization   290
closure domain   191

closure domain pattern   193
Co   34
Co (10 nm)/Au (5 nm)/Ni (10 nm)   102
Co films on W(111)   115
Co on Au(111)   123
Co on W(110)   123
Co/Au(111)   150
Co/Cu(001)   19, 155
Co/Cu(1 1 13)   155
Co$_{80}$Cr$_{16}$Ta$_4$ 170
Co$_{86}$Cr$_{10}$Ta$_4$ 170
Co(0001)   191
Co(0001) surface   115
Co-CoO   173
Co-Pt   95
Co/Au/Co epitaxial sandwiches   129
Co/Cu multilayer film   75
Co/Cu/Co epitaxial sandwiches   129
Co/NiO(001) interface   41
Co/Pt
   multilayer   292
coexistence of states
   Co/Au(111)   154
CoFe/FeMn nanostructures   106
CoFeSiB tip   186
   magnetic configuration   187
   magnetostriction   188
   switching behavior   188
coherent Foucault mode   74
coherent oscillations   63
CoNiO$_x$ 46
constant-height mode   287
constant-interaction mode   287

CoO   32
core hole   32
coupling between Co and NiO   44
Cr(001)   205, 216
CrBr$_3$ 53
CrO$_2$ needles   98
cross-tie wall   156

damped ferromagnetic resonance   63
DC mode   228
decoration technique   160
differential conductivity   206, 216
differential phase contrast   93
differential phase contrast microscopy   76
dipole selection rules   32
domain wall   138
  cross-tie   233
  Néel wall   233, 247
domain wall movement   196
DPC   76
dynamic mode
  MFM   229, 257

eddy current   295
electron column   145
electron emission microscopy   4
electron holography   74, 80, 87
electron spin resonance   285
electron-hole pair creation   116
elemental and chemical contrast   37
epitaxial strain   39
ESR   285
exchange bias   30, 106, 158
exchange bias structures   41
exchange coupled films   157
exchange potential   112

Faraday effect   54
Faraday microscopy   289
Fe   34
Fe double layer on W(100)   114
Fe films on Cu(100)   115
Fe K edge   31
Fe on Cu(100)   123
Fe whiskers   157
Fe/Cr(100)   158
Fe/Cr/Fe   157
Fe/Cu(001)   15
Fe/Mn/Fe   158

Fe/NiO(100)   159
Fe(110) surface   114
Fe-coated tip   206, 207, 221
Fe-Pt   95
Fe/W(110)
  double-layer stripes   211
(FeMoCo)$_{73}$ (BSi)$_{27}$
  soft magnetic tip   296
ferrofluid   297
ferromagnet-antiferromagnet structures   29
ferromagnetic resonance   286
field emission column   146
field emission gun   74
field sensitivity
  SSM   298
field-emission gun
  TEM   69
fine structure   32
flux guide   297
flux noise   293
flux sensitivity   292
flux-closure structure   245
flux-guided scanning SQUID microscope
  296, 298
flux-locked loop   296
FMR   286
Foucault mode   68, 71, 73
frequency imaging
  MFM   230
Fresnel mode   71, 72
Fresnel zone plate   2, 4

Gd(0001)
  exchange-split surface state   204
  spin-split surface state   206
Gd-coated tip   206, 212
GdFe-coated tip   206
giant magnetoresistance   82

Hall bars
  micron-sized   292
Hall probe   285
Hall sensor   243
high coercivity tip
  MFM   278
high-temperature superconductor
  HTSC   293

imaging in magnetic fields
  SEMPA   162

in-plane magnetized films 155
inelastic mean free paths 117
inelastic scattering 116
interface anisotropy
   Co/Au(111) 152
interfacial spin polarization in Co/NiO (001) 46
interfacial spins 30
interlayer exchange coupling 18

Josephson junctions 293

Kerr microscopy 52, 289, 291
Kerr spectroscopy 3

L edges 31
$LaFeO_3$ 32
$LaFeO_3$(001) thin films 37
LEDS-detector 143, 145
LEED spin-polarization analyzer 138
LEED-detector 144
LEEM 5, 111
LEEM intensity oscillations 128
lift mode
   MFM 228
linear polarization 290
local domain wall speed 196
local magnetic susceptibility 195
longitudinal susceptibility 296
Lorentz force 70
Lorentz microscopy 68, 70
low energy diffuse scattering spin analyzer 138
low energy electron microscopy 5, 10
low moment tip
   MFM 233, 245, 278
low-angle electron diffraction mode 69, 71, 74

magnet-on-cantilever 301
magnetic circular dichroism 6
magnetic dissipation image
   MFM 230
magnetic field sensor 289
magnetic force microscope 253, 285
magnetic force microscopy 225
magnetic interferogram 74
magnetic resonance force microscopy 286, 289, 300

magnetic vortex
   core 233, 241
magnetite nanocrystals 90
magnetization dynamics 24
magnetization reversal 72
   spin valve 83
magnetization ripple 72
magnetizing stage
   TEM holder 80
magneto-optic Kerr effect 52
magneto-optic scanning near-field optical microscope 285, 289–291
magneto-optical (MO) recording media 176
magneto-optical Kerr effect 4
magneto-resistive probe 285
magneto-resistors
   micron-sized 292
Magnetospirillum magnetotacticum 99
magnetotactic bacterial cell 90
magnon excitations 116
MFM 226, 253, 285
microspectroscopy 2
Mn/W(110) 205, 220
modulation technique
   SP-STM 185
MOSNOM 285
Mott spin-polarization analyzer 138
Mott-detector 143
MRAM 51
MRFM 286
multiple scattering 113
multiplets 33

Néel temperature 39
nanoparticles 231
nanowires 232
$Nd_2Fe_{14}B$ 96
near-field regime 290
Ni 34
   polycrystalline 190
Ni/Co/Cu(001) 19
Ni/Cu(001) 19
$Ni_{80}Fe_{20}$ 52
$Ni_{81}Fe_{19}$/NiO 53
Ni(100) surface 114
NiFe
   thin film 72
NiFe film 77

NiFe sense layer
  spin valve   73
NiFe/Al$_2$O$_3$/NiFe/MnFe
  tunnel junction   84
NiO   30, 32
NiO(001)   43
NMR   285
noise
  MFM   267
nuclear magnetic resonance   285
nuclear spin   300, 301

off-axis holography   91
optical fiber tip   290
optically pumped
  GaAs   204
orange peel coupling   83
orbital magnetic moment   8
oscillatory exchange coupling   157

partial flux closure domains   193
patterned elements   232
patterned nanostructures   100
PEEM   1
PEEM-2 facility   35
Permalloy   52, 61
perpendicular magnetic anisotropy
  Au/Co/Au(111)   153
perpendicular magnetized films   149
perturbation
  MFM   243
phase diagram
  Co/Au(111)   154
phase imaging
  MFM   230
phase mode
  MFM   257
photoelectron emission microscopy   1, 5, 9
pick-up loop   297
pinning   30
polarization vector analysis   147
polarized X-ray absorption spectroscopy
  31
probing depth   32
pulse counting operation
  SEMPA   145
pump-and-probe   56
pump-probe experiments   24

quantum size effects   115

real-time tuning
  SEMPA   142
resolution
  DPC   76
  MFM   264
  MOSNOM   290, 292
  PEEM   11
  SEMPA   138
  SP-STM   191, 197
  SP-STS   204
  SSM   297
  TEM   67
resonant X-ray absorption   10

S-domain   41
S-state   233
sample-on-cantilever   301
scanning electron microscopy with
    polarization analysis   137
scanning force microscope   285
scanning Hall probe   289
scanning Kerr microscopy   3
scanning probe microscope   285, 287, 289
scanning SQUID microscope   285,
    292–294
scanning transmission electron microscopy
    68, 76
scanning X-ray microscopy   2
scatter plot   147
scattering asymmetry   113
screw dislocation   217
second order surface anisotropy   154
secondary electron emission   139
secondary electron yield   145
selection rules   6
SEMPA   137
sensitivity
  MRFM   302
SFM   285
shear-force sensor   291
Sherman function   139
SHMOKE   53
Si$_3$N$_4$ membrane   70
signal-to-noise ratio
  MRFM   302
single spin detection   300
soft magnetic materials   226

soft X-ray absorption 6
SP-IPE 209
SP-STM 181
SP-STS 203
spatial coherence 74
spectromicroscopy 2
spin larmor precession 301
spin reorientation transition 19, 125, 150, 154
spin scattering length 200
spin-density wave 215
spin-flop canting 31, 46
spin-orbit splitting 32
spin-polarization analyzer 137, 143
spin-polarized scanning tunneling microscopy 181, 185, 204
spin-polarized scanning tunneling spectroscopy 203
spin-resolved density of states 7
spin-resolved inverse photoemission spectroscopy 209
spin-SEM 137, 169
spin-valve 102
spin-valve structure 82
spintronics 29
SPLEED 111
SPLEEM 111
SQUID 289, 292
  pick-up loop 293
SQUID magnetometer
  sensitivity 293
$SrTiO_3(001)$ 34
SSM 285
static mode
  MFM 228, 254
STEM 68
Stoner excitations 116
Stranski-Krastanov mode 124
sum rules 6
summed image differential phase contrast mode 78
superconducting quantum interference device
  SQUID 285
superconductor 292
surface magnetization 114

SXM 2
synchrotron radiation 5

$<xT$ domains 41
TbFeCo recording films 176
TEM 67, 87
thermionic gun 146
tip shape
  MFM 276
tip transfer function
  MFM 259
TiSapphire laser 57
TMR 183
topographic contrast 37
topological antiferromagnetism 216
transition noise 170
transmission electron microscopy 67
transmission X-ray microscope 4
tunnel junction 83, 203
tunneling conductance 182
tunneling magnetoresistance 181, 197

ultrathin films 148
uncompensated spins 46

vacuum barrier 197

washer SQUID 295
workfunction contrast 37

X-PEEM 29
X-ray absorption 31
X-ray absorption cross section 3
X-ray absorption spectroscopy 7, 32
X-ray magnetic circular dichroism 1, 4, 31
X-ray magnetic circular dichroism (XMCD) 34
X-ray magnetic linear dichroism (XMLD) 33
X-ray Photoemission Electron Microscopy 29
XAS 7, 31
XMCD 1, 31, 54
XMLD 31

Zeeman-splitting 204
zigzag bit boundaries 170

Printing: Strauss GmbH, Mörlenbach
Binding: Schäffer, Grünstadt

CPSIA information can be obtained at www.ICGtesting.com
Printed in the USA
LVOW08*1829241114

415382LV00010B/96/P

9 783540 401865